Carpentry and Joinery

NVQ and Technical Certificate Level 2

www.heinemann.co.uk

✓ Free online support
✓ Useful weblinks
✓ 24 hour online ordering

01865 888118

Heinemann

Heinemann is an imprint of Pearson Education Limited, a company incorporated in England and Wales, having its registered office at Edinburgh Gate, Harlow, Essex, CM20 2JE. Registered company number: 872828

www.heinemann.co.uk

Heinemann is a registered trademark of Pearson Education Limited

Text © Carillion Construction Ltd 2008

First published 2006

Second edition published 2008

12 11 10 09 08
10 9 8 7 6 5 4 3 2 1

British Library Cataloguing in Publication Data

A catalogue record for this book is available from the British Library.

ISBN 978 0 435325 72 5

Designed by HL Studios
Typeset by HL Studios
Original illustrations © Pearson Education Limited 2006, 2008
Illustrated by HL Studios
Cover design by GD Associates
Cover photo: © Pearson Education Limited/Ben Nicholson
Printed in the UK by Scotprint

Acknowledgements
Every effort has been made to contact copyright holders of material reproduced in this book. Any omissions will be rectified in subsequent printings if notice is given to the publishers.

Websites
The websites used in this book were correct and up-to-date at the time of publication. There are links to relevant websites in this book. In order to ensure that the links are up-to-date, that the links work, and that the sites are not inadvertently linked to sites that could be considered offensive, we have made the links available on the Heinemann website at www.heinemann.co.uk/hotlinks. When you access the site, use the express code is 5701P.

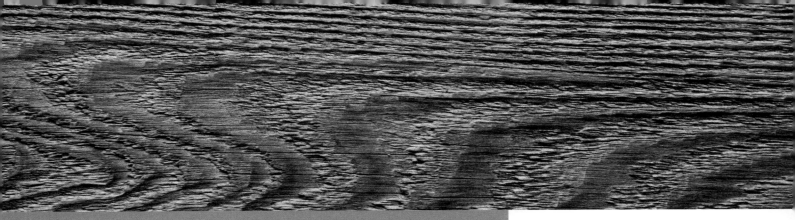

Contents

Acknowledgements

Carillion would like to thank the following people for their contribution to this book: Kevin Jarvis (lead author), Ralph Need, Robert Otterson, Peter Fox and Redvers Johnson.

Pearson Education Limited would like to thank the following for providing technical feedback: Kevan Stephenson and Graeme Fallowfield from Hartlepool College and Brian Bibby and Keith Pickering from Furness College.

Pearson Education Limited would like to thank everyone at the Carillion Construction Training Centre in Sunderland for all their help at the photo shoots.

Special thanks to Chris Ledson at Toolbank for supplying some photos. Visit the Toolbank website at www.toolbank.com.

Photo acknowledgements

The author and publisher would like to thank the following individuals and organisations for permission to reproduce photographs:

Alamy Images/ David R. Frazier Photolibrary, Inc, p264 (bottom right); Alamy Images/ David J. Green, p47 (bottom); Alamy Images/Nic Hamilton, p33; Alamy Images/Roger Hutchings, p233; Alamy Images/Justin Kase, p36; Alamy Images/Niall McDiarmid, p251; Alamy Images/PlainPicture GmbH & Co KG, p46; Alamy Images/Mike Rinnan, p394 (top); Alamy Images/Charles Stirling, p175 (shoulder); Alamy Images/The Photolibrary Wales, p56; Alamy Images/Webstream, p208; Constructionphotography.com, p38, p63; Constructionphotography.com/DIY Photolibrary, p86, p264 (top); Constructionphotography. com/Chris Henderson, p77; Constructionphotography.com/Paul McMullin, p264 (bottom left), p381; Constructinphotography.com/Sally-Ann Norman, p71 (bottom); Constructionphotography.com/Ken Price, p71 (middle); Corbis, p1, p52 (top), p52 (bottom), p59, p62 (top), p111, p131, p161, p329; Corbis/Creasource, p129; Corbis/Holger Scheibe/ Zefa, p301; Digital Vision, p211; Dreamstime.com/Andrzej Tokarski, p167 (top); Getty Images/Photodisc, p8, p9, p11, p13, p28, p31, p39, p42, p61, p91, p277, p419; HNT Gordon Planemakers, Australia/Terry Gordon, p174 (jointer); iStock Photo/Guy Erwood, p62 (bottom); Maria Joannou, p164 (top), p176 (bottom), p180 (bottom), p187 (top), p189 (file), p306 (cast butt), p307 (top left), p307 (top right), p308 (pad bolt), p308 (bottom), p309 (top), p312 (frame fixings), p313 (top), p326 (bottom); Jupiter Images/Photos.com, p327; Pearson Education Ltd/Chris Honeywell, p52 (goggles); Pearson Education Ltd/Jules Selmes, p54, p87, p266 (all), p267 (all), p299, p300 (all), p302 (bottom), p303 (all but bottom), p392; Pearson Education Ltd/Ginny Stroud Lewis, p53 (ear plugs, ear defenders, dust mask); Photographers Direct/Andrew Arthurs, p311 (security); Photographers Direct/ Bjorn Beheydt, p127; Photographers Direct/Martyn F. Chillmaid, p313 (gravity toggles); Photographers Direct/David Griffiths, p125; Photographers Direct/Lasertech, p180 (snail); Photographers Direct/Quayside Graphics, p215; Photographers Direct/Andy Smith, p431; Photographers Direct/Brian Thearle, p 179 (forstner); Rockwool, p263; Science Photo Library/Scott Camazine, p47 (top); Science Photo Library/Garry Watson, p37 (corrosive); Shout, p49 (foam); Shutterstock.com, p3, p195 (top); Shutterstock.com/7505811966, p207 (bottom); Shutterstock.com/Mitch Aunger, p394 (bottom); Shutterstock.com/Dusty Cline, p207 (cartridge); Shutterstock.com/Stephen Coburn, p75; Shutterstock.com/Frances A Miller, p71 (top); Toolbank, p53 (top), p53 (bottom), p163 (top), p164 (marking knife), p165 (mitre square), p168 (top), p168 (bottom), p169 (bottom), p183 (rachet-type), p186 (stair-housing jig), p311 (flange head), p415 (left); Topham Picturepoint, p37 (toxic and explosive); Rachael Williams, p2, p95 (top), p295.

All other photos Pearson Education Ltd/Gareth Boden.

Model on cover: Christopher Barrett.

The contents of this book have been endorsed by Construction Skills.

Introduction

Carpentry and joinery combine many different practical and visual skills with knowledge of specialised materials and techniques. The information contained in these pages covers the carpentry and joinery trade as a whole and in particular the knowledge and skills needed for first and second fixing, structural carcassing and joinery products. This book has been written based on a concept used with Carillion Training Centres for many years. The concept is about providing learners with the necessary information they need to support their studies and at the same time ensuring it is presented in a style which is both manageable and relevant.

This book has been produced to help you build a sound knowledge and understanding of all aspects of the NVQ and Technical Certificate requirements associated with carpentry and joinery. It has also been designed to provide assistance when revising for Technical Certificate end tests and NVQ job knowledge tests.

Each chapter of this book relates closely to a particular unit of the NVQ or Technical Certificate and aims to provide just the right level of information needed to form the required knowledge and understanding of that subject area.

This book provides a basic introduction to the tools, materials and methods of work required to complete work activities effectively and productively. Upon completion of your studies, this book will remain a valuable source of information and support when carrying out your work activities.

For further information on how the content of this student book matches to the unit requirements of the NVQ and Level 2 Diploma, please see the mapping documents found on pages 455–458 of this book.

How this book can help you

You will discover a variety of features throughout this book, each of which has been designed and written to increase and improve your knowledge and understanding. These features are:

- **Photographs** – many photographs that appear in this book are specially taken and will help you to follow a step-by-step procedure or identify a tool or material.

- **Illustrations** – clear and colourful drawings will give you more information about a concept or procedure.

- **Definitions** – new or difficult words are picked out in **bold** in the text and defined in the margin.

- **Remember** – key concepts or facts are highlighted in these margin boxes.

- **Find out** – carry out these short activities and gain further information and understanding of a topic area.

- **Did you know?** – interesting facts about the building trade.

- **Safety tips** – follow the guidance in these margin boxes to help you work safely.

- **FAQs** – frequently asked questions appear in all chapters along with informative answers from the experts.

- **On the job scenarios** – read about a real-life situation and answer the questions at the end. What would you do? (Answers can be found in the Tutor Resource Disc that accompanies this book.)

- **Chapter opener module listings** – the technical certificate modules covered in each the chapter are identified on the first page of the chapter. A detailed table defining each module can be found on page 455 of this book.

- **End of chapter knowledge checks** – test your understanding and recall of a topic by completing these questions.

- **Mapping grids for technical certificate** – a comprehensive list of the technical certificate module outcomes covered in this book can be found on pages 456-458 of this book.

- **Glossary** – at the end of this book you will find a comprehensive glossary that defines all the bold words and phrases found in the text. A great quick reference tool.

- **Links to useful websites** – any websites referred to in this book can be found at www.heinemann.co.uk/hotlinks. Just enter the express code 5701P to access the links.

chapter 1

The construction industry

OVERVIEW

Construction means creating buildings and services. These might be houses, hospitals, schools, offices, roads, bridges, museums, prisons, train stations, airports, monuments – and anything else you can think of that needs designing and building! What about an Olympic stadium? The 2012 London games will bring a wealth of construction opportunity to the UK and so it is an exciting time to be getting involved.

In the UK, 2.2 million people work in the construction industry – more than in any other – and it is constantly expanding and developing. There are more choices and opportunities than ever before and pay and conditions are improving all the time. Your career doesn't have to end in the UK either – what about taking the skills and experience you are developing abroad? Construction is a career you can take with you wherever you go. There's always going to be something that needs building!

This chapter will cover the following topics:

- Understanding the industry
- Communication
- General site paperwork
- Getting involved in the construction industry
- Sources of information and advice.

These topics can be found in the following modules:

CC 1001K	CC 2002K
CC 1001S	CC 2002S

Understanding the industry

The construction industry is made up of countless companies and businesses that all provide different services and materials. An easy way to divide these companies into categories is according to their size.

- A small company is defined as having between 1 and 49 members of staff.
- A medium company consists of between 50 and 249 members of staff.
- A large company has 250 or more people working for it.

A business might only consist of one member of staff (a sole trader).

The different types of construction work

There are four main types of construction work:

1. New work – this refers to a building that is about to be or has just been built.

2. Maintenance work – this is when an existing building is kept up to an acceptable standard by fixing anything that is damaged so that it does not fall into disrepair.

3. Refurbishment/renovation work – this generally refers to an existing building that has fallen into a state of disrepair and is then brought up to standard by repair. It also refers to an existing building that is to be used for a different purpose, for example changing an old bank into a pub.

4. Restoration work – this refers to an existing building that has fallen into a state of disrepair and is then brought back to its original condition or use.

These four types of work can fall into one of two categories depending upon who is paying for the work:

1. Public – the government pays for the work, as is the case with most schools and hospitals etc.

2. Private – work is paid for by a private client and can range from extensions on existing houses to new houses or buildings.

Job and careers

Jobs and careers in the construction industry fall mainly into one of four categories:

- building
- civil engineering
- electrical engineering
- mechanical engineering.

Find out

Think of an example of a small, medium and large construction company. Do you know of any construction companies that have only one member of staff?

New work is just one type of construction area

Building involves the physical construction (making) of a structure. It also involves the maintenance, restoration and refurbishment of structures.

Civil engineering involves the construction and maintenance of work such as roads, railways, bridges etc.

Electrical engineering involves the installation and maintenance of electrical systems and devices such as lights, power sockets and electrical appliances etc.

Mechanical engineering involves the installation and maintenance of things such as heating, ventilation and lifts.

The category that is the most relevant to your course is building.

Job types

The construction industry employs people in four specific areas:

1. professionals

2. technicians

3. building craft workers

4. building operatives.

Professionals

Professionals are generally of graduate level (i.e. people who have a degree from a university) and may have one of the following types of job in the construction industry:

- architect – someone who designs and draws the building or structure
- structural engineer – someone who oversees the strength and structure of the building
- land surveyor – someone who checks the land for suitability to build on
- building surveyor – someone who provides advice on construction projects
- service engineer – someone who plans the services needed within the building, for example gas, electricity and water supplies.

Technicians

Technicians link professional workers with craft workers and are made up of the following people:

- architectural technician – someone who looks at the architect's information and makes drawings that can be used by the builder
- building technician – someone who is responsible for estimating the cost of the work and materials and general site management

- quantity surveyor – someone who calculates ongoing costs and payment for work done.

Building craft workers

Building craft workers are the skilled people who work with materials to physically construct the building. The following jobs fall into this category:

- carpenter or joiner – someone who works with wood but also other construction materials such as plastic and iron. A carpenter primarily works on site while a joiner usually works off site, producing components such as windows, stairs, doors, kitchens, and **trusses**, which the carpenter then fits into the building

- bricklayer – someone who works with bricks, blocks and cement to build the structure of the building

- plasterer – someone who adds finish to the internal walls and ceilings by applying a **plaster skim**. They also make and fix plaster **covings** and plaster decorations

- painter and decorator – someone who uses paint and paper to decorate the internal plaster and timberwork such as walls, ceilings, windows and doors, as well as **architraves** and **skirting**

- electrician – someone who fits all electrical systems and fittings within a building, including power supplies, lights and power sockets

- plumber – someone who fits all water services within a building, including sinks, boilers, water tanks, radiators, toilets and baths. The plumber also deals with lead work and rainwater fittings such as guttering

- slater and tiler – someone who fits tiles on to the roof of a building, ensuring that the building is watertight

- woodworking machinist – someone who works in a machine shop, converting timber into joinery components such as window sections, spindles for stairs, architraves and skirting boards, amongst other things. They use a variety of machines such as lathes, bench saws, planers and sanders.

Building operatives

There are two different building operatives working on a construction site.

1. Specialist building operative – someone who carries out specialist operations such as dry wall lining, asphalting, scaffolding, floor and wall tiling and glazing.

2. General building operative – someone who carries out non-specialist operations such as kerb laying, concreting, path laying and drainage. These operatives also support other craft workers and do general labouring. They use a variety of hand tools and power tools as well as **plant**, such as dumper trucks and JCBs.

Definition

Trusses – prefabricated components of a roof which spread the load of a roof over the outer walls and form its shape

Plaster skim – a thin layer of plaster that is put on to walls to give a smooth and even finish

Coving – a decorative moulding that is fitted to the top of a wall where it meets the ceiling

Architrave – a decorative moulding, usually made from timber, that is fitted around door and window frames to hide the gap between the frame and the wall

Skirting – a decorative moulding that is fitted at the bottom of a wall to hide the gap between the wall and the floor

Definition

Plant – industrial machinery

The building team

Constructing a building or structure is a huge task that needs to be done by a team of people who all need to work together towards the same goal. The team of people is often known as the building team and is made up of the following people.

Client

The client is the person who requires the building or refurbishment. This person is the most important person in the building team because they finance the project fully and without the client there is no work. The client can be a single person or a large organisation.

Architect

The architect works closely with the client, interpreting their requirements to produce contract documents that enable the client's wishes to be realised.

Clerk of works

Selected by the architect or client to oversee the actual building process, the clerk of works ensures that construction sticks to agreed deadlines. They also monitor the quality of workmanship.

Local Authority

The Local Authority is responsible for ensuring that construction projects meet relevant planning and building legislation. Planning and building control officers approve and inspect building work.

Quantity surveyor

The quantity surveyor works closely with the architect and client, acting as an accountant for the job. They are responsible for the ongoing evaluation of cost and interim payments from the client, establishing whether or not the contract is on budget. The quantity surveyor will prepare and sign off final accounts when the contract is complete.

Specialist engineers

Specialist engineers assist the architect in specialist areas, such as civil engineering, structural engineering and service engineering.

Health and safety inspectors

Employed by the Health and Safety Executive (HSE), health and safety inspectors ensure that the building contractor fully implements and complies with government health and safety legislation. For more information on health and safety in the construction industry, see Chapter 2 (page 31).

Building contractors

The building contractors agree to carry out building work for the client. Contractors will employ the required workforce based on the size of the contract.

Estimator

The estimator works with the contractor on the cost of carrying out the building contract, listing each item in the bill of quantities (e.g. materials, labour and plant). They calculate the overall cost for the contractor to complete the contract, including further costs as overheads, such as site offices, management administration and pay, not forgetting profit.

Site agent

The site agent works for the building contractor and is responsible for the day-to-day running of the site such as organising deliveries etc.

Suppliers

The suppliers work with the contractor and estimator to arrange the materials that are needed on site and ensure that they are delivered on time and in good condition.

General foreman

The general foreman works for the site manager and is responsible for co-ordinating the work of the ganger (see below), craft foreman and subcontractors. They may also be responsible for the hiring and firing of site operatives. The general foreman also liaises with the clerk of works.

Craft foreman

The craft foreman works for the general foreman organising and supervising the work of particular crafts. For example, the carpentry craft foreman will be responsible for all carpenters on site.

Ganger

The ganger supervises general building operatives.

Chargehand

The chargehand is normally employed only on large building projects, being responsible for various craftsmen and working with joiners, bricklayers, and plasterers.

Operatives

Operatives are the workers who carry out the building work, and are divided into three subsections:

1. Craft operatives are skilled tradesman such as joiners, plasterers, bricklayers.

2. Building operatives include general building operatives who are responsible for drain laying, mixing concrete, unloading materials and keeping the site clean.

3. Specialist operatives include tilers, pavers, glaziers, scaffolders and plant operators.

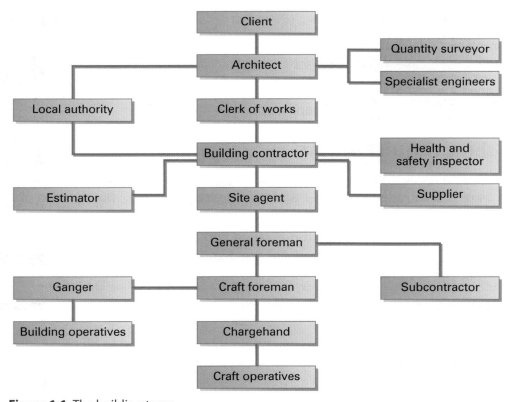

Figure 1.1 The building team

A low rise residential building

The different types of building

There are of course lots of very different types of building, but the main types are:

- residential – houses and flats etc.
- commercial – shops and supermarkets etc.
- industrial – warehouses and factories etc.

These types of building can be further broken down by the height or number of storeys that they have (one storey being the level from floor to ceiling):

- low rise – a building with one to three storeys
- medium rise – a building with four to seven storeys
- high rise – a building with seven storeys or more.

Buildings can also be categorised according to the number of other buildings they are attached to:

- detached – a building that stands alone and is not connected to any other building
- semi-detached – a building that is joined to one other building and shares a dividing wall, called a party wall
- terraced – a row of three or more buildings that are joined together, of which the inner buildings share two party walls.

Building requirements

Every building must meet the minimum requirements of the *Building Regulations*, which were first introduced in 1961 and then updated in 1985. The purpose of building regulations is to ensure that safe and healthy buildings are constructed for the public and that **conservation** is taken into account when they are being constructed. Building regulations enforce a minimum standard of building work and ensure that the materials used are of a good standard and fit for purpose.

What makes a good building?

When a building is designed, there are certain things that need to be taken into consideration, such as:

- security
- safety
- privacy
- warmth
- light
- ventilation.

A well-designed building will meet the minimum standards for all of the considerations above and will also be built in line with building regulations.

The properties and principles of building work will be covered in greater detail in Chapter 4.

Definition

Conservation
– preservation of the environment and wildlife

Communication

Communication, in the simplest of terms, is a way or means of passing on information from one person to another. Communication is very important in all areas of life and we often do it without even thinking about it. You will need to communicate well when you are at work, no matter what job you do. What would happen if someone couldn't understand something you had written or said? If we don't communicate well, how will other people know what we want or need and how will we know what other people want?

Companies that do not establish good methods of communicating with their workforce or with other companies, will not function properly and will end up with bad working relationships. Good working relationships can *only* be achieved with co-operation and good communication.

Methods of communication

There are many different ways of communicating with others and they all generally fit into one of these four categories:

1. speaking (verbal communication), for example talking face to face or over the telephone

2. writing, for example sending a letter or taking a message

3. body language, for example the way we stand or our facial expressions

4. electronic, for example email, fax and text messages.

Each method of communicating has good points (advantages) and bad points (disadvantages).

Verbal communication is probably the method you will use most

Verbal communication

Verbal communication is the most common method we use to communicate with each other. If two people don't speak the same language or if someone speaks very quietly or not very clearly, verbal communication cannot be effective. Working in the construction industry you may communicate verbally with other people face to face, over the telephone or by radio/walkie-talkie.

Advantages	Disadvantages
Verbal communication is instant, easy and can be repeated or rephrased until the message is understood.	Verbal communication can be easily forgotten as there is no physical evidence of the message. Because of this it can be easily changed if passed to other people. Someone's accent or use of slang language can sometimes make it difficult to understand what they are saying.

Written communication

Written communication can take the form of letters, faxes, messages, notes, instruction leaflets, text messages, drawings and emails, amongst others.

Advantages	Disadvantages
There is physical evidence of the communication and the message can be passed on to another person without it being changed. It can also be read again if it is not understood.	Written communication takes longer to arrive and understand than verbal communication and body language. It can also be misunderstood or lost. If it is handwritten, the reader may not be able to read the writing if it is messy.

Messages

To Andy Rodgers ....

Date Tues 10 Nov   Time 11.10 am   ..

Message: Mark from Stokes called with a query about the recent order. Please phone asap (tel 01234 567 890)   .

....

Message taken by: Lee Barber
....

Figure 1.2 A message is a form of written communication

Body language

It is said that, when we are talking to someone face to face, only 10 per cent of the communication is verbal. The rest of the communication is body language and facial expression. This form of communication can be as simple as the shaking of a head from left to right to mean 'no' or as complex as the way someone's face changes when they are happy or sad or the signs given in body language when someone is lying.

We often use hand gestures as well as words to get across what we are saying, to emphasise a point or give a direction. Some people communicate entirely through a form of body language called sign language.

Advantages	Disadvantages
If you are aware of your own body language and know how to use it effectively, you can add extra meaning to what you say. For example, say you are talking to a client or a work colleague. Even if the words you are using are friendly and polite, if your body language is negative or unfriendly, the message that you are giving out could be misunderstood. By simply maintaining eye contact, smiling and not folding your arms, you have made sure that the person you are communicating with has not got a mixed or confusing message.

Body language is quick and effective. A wave from a distance can pass on a greeting without being close, and using hand signals to direct a lorry or a load from a crane is instant and doesn't require any equipment such as radios. | Some gestures can be misunderstood, especially if they are given from very far away, and gestures that have one meaning in one country or culture can have a completely different meaning in another. |

Try to be aware of your body language

Electronic communication

Electronic communication is becoming more and more common with the advances in technology allowing us to communicate more easily. Electronic communication can take many forms, such as email and fax. It is now even possible to send and receive emails via a mobile phone, which allows important information to be sent or received from almost anywhere in the world.

Advantages	Disadvantages
Electronic communication takes the best parts from verbal and written communication in as much as it is instant, easy and there is a record of the communication being sent. Electronic communication goes even further as it can tell the sender if the message has been received and even read. Emails in particular can be used to send a vast amount of information and can even give links to websites or other information. Attachments to emails allow anything from instructions to drawings to be sent with the message.	There are few disadvantages to electronic communication, the obvious ones being no signal or a flat battery on a mobile phone and servers being down which prevent emails etc. Not every one is up to speed on the latest technology and some people are not comfortable using electronic communication. You need to make sure that the person receiving your message is able to understand how to access the information. Computer viruses can also be a problem as can security where hackers can tap into your computer and read your emails and other private information. A good security set-up and anti-virus software are essential.

Which type of communication should I use?

Of the many different types of communication, the type you should use will depend upon the situation. If someone needs to be told something formally, then written communication is generally the best way. If the message is informal, then verbal communication is usually acceptable.

The way that you communicate will also be affected by who it is you are communicating with. You should of course always communicate in a polite and respectful manner with anyone you have contact with, but you must also be aware of the need to sometimes alter the style of your communication. For example, when talking to a friend, it may be fine to talk in a very informal way and use slang language, but in a work situation with a client or a colleague, it is best to alter your communication to a more formal style in order to show professionalism. In the same way, it may be fine to leave a message or send a text to a friend that says 'C U @ 8 4 work', but if you wrote this down for a work colleague or a client to read, it would not look very professional and they may not understand it.

Communicating with other trades

Communicating with other trades is vital because they need to know what you are doing and when, and you need to know the same information from them. Poor communication can lead to delays and mistakes, which can both be costly. It is quite possible for poor communication to result in work having to be stopped or redone. Say you are decorating a room in a new building. You are just about to finish when you find out that the electrician, plumber and carpenter have to finish off some work in the room. This information didn't reach you and now the decorating will have to be done again once the other work has been finished. What a waste of time and money. A situation like this can be avoided with good communication between the trades.

You will work with people from other trades

Common methods of communicating in the construction industry

A career in construction means that you will often have to use written documents such as drawings, specifications and schedules. These documents can be very large and seem very complicated but, if you understand what they are used for and how they work, using such documents will soon become second nature.

General site paperwork

No building site could function properly without a certain amount of paperwork. Here is a brief, but not exhaustive, description of some of the other documents you may encounter. Some companies will have their own forms to cover such things as scaffolding checks.

Timesheet

Timesheets record hours worked, and are completed by every employee individually. Some timesheets are basic, asking just for a brief description of the work done each hour, but some can be complicated. In some cases timesheets may be used to work out how many hours the client will be charged for.

P. Gresford Building Contractors

Timesheet _____

Employee _____ **Project/site** _____

Date	Job no.	Start time	Finish time	Total time	Travel time	Expenses
M						
Tu						
W						
Th						
F						
Sa						
Su						
Totals						

Employee's signature _____

Supervisor's signature _____

Date _____

Figure 1.3 Timesheet

Day worksheets

Day worksheets are often confused with timesheets, but are different as they are used when there is no price or estimate for the work, to enable the contractor to charge for the work. Day worksheets record work done, hours worked and sometimes materials used.

P. Gresford Building Contractors

Day worksheet

Customer _Chris MacFarlane_   Date _____

Description of work being carried out _____
Hang internal door in kitchen.

Labour	Craft	Hours	Gross rate	TOTALS
Materials	Quantity	Rate	% addition	
Plant	Hours	Rate	% addition	

Comments

Signed _____   Date _____

Site manager/foreman signature _____

Figure 1.4 Day worksheet

P. Gresford Building Contractors

Job sheet

Customer   Chris MacFarlane

Address   1 High Street
         Any Town
         Any County

Work to be carried out
Hang internal door in kitchen

Special conditions/instructions
Fit with door closer
3 × 75mm butt hinges

Figure 1.5 Job sheet

Job sheet

A job sheet is similar to a day worksheet – it records work done – but is used when the work has already been priced. Job sheets enable the worker to see what needs to be done and the site agent or working foreman to see what has been completed.

Variation order

This sheet is used by the architect to make any changes to the original plans, including omissions, alterations and extra work.

VARIATION TO PROPOSED WORKS AT 123 A STREET

REFERENCE NO:

DATE _____

FROM _____

TO _____

POSSIBLE VARIATIONS TO WORK AT 123 A STREET

ADDITIONS

OMISSIONS

SIGNED ---------------------------------------

Figure 1.6 Variation order

Confirmation notice

This is a sheet given to the contractor to confirm any changes made in the variation order, so that the contractor can go ahead and carry out the work.

CONFIRMATION FOR VARIATION TO PROPOSED WORKS AT 123 A STREET

REFERENCE NO:

DATE _____

FROM _____

TO _____

I CONFIRM THAT I HAVE RECEIVED WRITTEN INSTRUCTIONS FROM _____
POSITION _____
TO CARRY OUT THE FOLLOWING POSSIBLE VARIATIONS TO THE ABOVE NAMED CONTRACT

ADDITIONS

OMISSIONS

SIGNED ---------------------------------------

Figure 1.7 Confirmation notice

Orders/requisitions

A requisition form or order is used to order materials or components from a supplier.

P. Gresford Building Contractors

Requisition form

Supplier _____   Order no. _____

_____   Serial no. _____

Tel no. _____   Contact _____

Fax no. _____   Our ref _____

Contract/Delivery address/Invoice address   **Statements/applications**

_____   **for payments to be sent to**

_____   _____

Tel no. _____   _____

Fax no. _____   _____

Item no.	Quantity	Unit	Description	Unit price	Amount

Total £ _____

Payment terms _____   Date

_____

Originated by

_____

Authorised by

_____

Figure 1.8 Requisition form

Bailey & Sons Ltd

Building materials supplier

Tel: 01234 567890

Your ref: AB00671

Our ref: CT020

Order no: 67440387

Date: 17 Jul 2006

Invoice address:
Carillion Training Centre,
Deptford Terrace, Sunderland

Delivery address:
Same as invoice

Description of goods	Quantity	Catalogue no.
OPC 25kg	10	OPC1.1

Comments:

Date and time of receiving goods:
Name of recipient (caps):
Signature:

Figure 1.9 Delivery note

Remember

Invoices may need paying by a certain date – fines for late payment can sometimes be incurred – so it is important that they are passed on to the finance office or financial controller promptly

Delivery notes

Delivery notes are given to the contractor by the supplier, and list all the materials and components being delivered. Each delivery note should be checked for accuracy against the order (to ensure what is being delivered is what was asked for) and against the delivery itself (to make sure that the delivery matches the delivery note). If there are any discrepancies or if the delivery is of a poor quality or damaged, you must write on the delivery note what is wrong *before* signing it and ensure the site agent is informed so that he/she can rectify the problem.

Invoices

Invoices come from a variety of sources such as suppliers or subcontractors, and state what has been provided and how much the contractor will be charged for it.

INVOICE			**JARVIS BUILDING SUPPLIES** *3rd AVENUE THOMASTOWN*	

L Weeks Builders
4th Grove
Thomastown

Quantity	Description	Unit price	Vat rate	Total
30	Galvanised joint hangers	£1.32	17.5%	£46.53
				TOTAL £46.53

To be paid within 30 days from receipt of this invoice

Please direct any queries to 01234 56789

Figure 1.10 Invoice

Delivery records

Delivery records list all deliveries over a certain period (usually a month), and are sent to the contractor's Head Office so that payment can be made.

JARVIS BUILDING SUPPLIES
3rd AVENUE
THOMASTOWN

Customer ref_____

Customer order date_____

Delivery date_____

Item no	Qty Supplied	Qty to follow	Description	Unit price
1	30	0	Galvanised joinst hangers	£1.32

Delivered to: L Weeks builders
4th Grove
Thomastown
Customer signature -----------------------------

Figure 1.11 Delivery record

DAILY REPORT/SITE DIARY

PROJECT_____
DATE_____

Identify any of the following factors, which are affecting or may affect the daily work activities and give a brief description in the box provided

WEATHER ( )   ACCESS ( )   ACCIDENTS ( )   SERVICES ( )
DELIVERIES ( )   SUPPLIES ( )   LABOUR ( )   OTHER ( )

SIGNED -----------------------------
POSITION -----------------------------

Figure 1.12 Daily report or site diary

Daily report/ site diary

This is used to pass general information (deliveries, attendance etc.) on to a company's Head Office.

Remember

You should always check a delivery note against the order and the delivery itself, then write any discrepancies or problems on the delivery note *before* signing it

Accident and near miss reports

It is a legal requirement that a company has an accident book, in which reports of all accidents must be made. Reports must also be made when an accident nearly happened, but did not in the end occur – known as a 'near miss'. It is everyone's responsibility to complete the accident book. If you are also in a supervisory position you will have the responsibility to ensure all requirements for accident reporting are met.

Safety tip

If you are involved in or witness an accident or near miss, make sure it is entered in the book – for your own safety and that of others on the site. If you don't report it, it's more likely to happen again

Report of an Accident, Dangerous Occurrence or Near Miss

Date of incident _____   Time of incident _____

Location of incident _____

Details of person involved in accident

Name _____   Date of birth _____

Address _____

_____

_____   Occupation _____

Date off work (if applicable) _____   **Date returning to work** _____

Nature of injury _____

Management of injury   ☐ First Aid only   ☐ Advised to see doctor

                                 ☐ Sent to casualty   ☐ Admitted to hospital

Account of accident, dangerous occurrence or near miss
(Continued on separate sheet if necessary)

[ ]

Witnesses to the incident
(Names, addresses and occupations)

[ ]

Was the injured person wearing PPE? If yes, what PPE? _____

_____

Signature of person completing form _____

Occupation _____   **Date** _____

Figure 1.13 Accident/ near miss report

Drawings

Drawings are done by the architect and are used to pass on the client's wishes to the building contractor. Drawings are usually done to scale because it would be impossible to draw a full-sized version of the project. A common scale is 1:10, which means that a line 10 mm long on the drawing represents 100 mm in real life. Drawings often contain symbols instead of written words to get the maximum amount of information across without cluttering the page.

Specifications

Specifications accompany a drawing and give you the sizes that are not available on the drawing, as well as telling you the type of material to be used and the quality that the work has to be finished to.

Schedules

A schedule is a list of repeated design information used on big building sites when there are several types of similar room or house. For example, a schedule will tell you what type of door must be used and where. Another form of schedule used on building sites contains a detailed list of dates by which work must be carried out and materials delivered etc.

See Chapter 6 *Drawings* for more information.

Work programme

A work programme is a method of showing very easily what work is being carried out on a building and when. The most common form of work programme is a bar chart. Used by many site agents or supervisors, a bar chart lists the tasks that need to be done down the left side and shows a timeline across the top. A work programme is used to make sure that the relevant trade is on site at the correct time and that materials are delivered when needed. A site agent or supervisor can quickly tell from looking at the chart if work is keeping to schedule or falling behind.

Bar charts

The bar or Gantt chart is the most popular work programme as it is simple to construct and easy to understand. Bar charts have tasks listed in a vertical column on the left and a horizontal timescale running along the top.

Did you know?

The Gantt chart is named after the first man to publish it. This was Henry Gantt, an American engineer, in 1910

Time in days										
Activity	1	2	3	4	5	6	7	8	9	10
Dig for foundation and service routes										
Lay foundations										
Run cabling, piping etc. to meet existing services										
Build up to DPC										
Lay concrete floor										

Figure 1.14 Basic bar chart

Each task is given a proposed time, which is shaded in along the horizontal timescale. Timescales often overlap as one task often overlaps another.

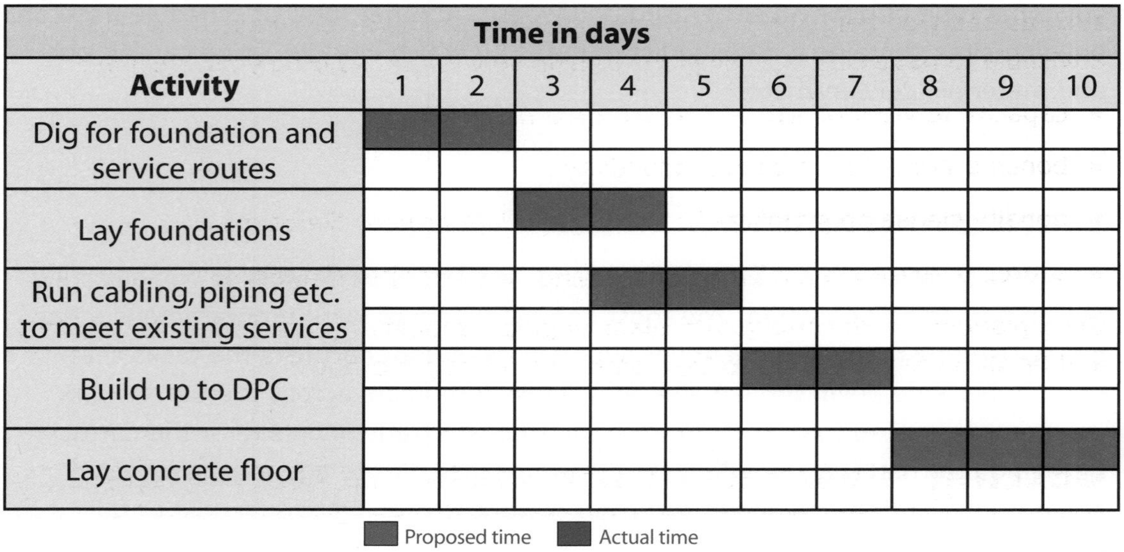

Time in days										
Activity	1	2	3	4	5	6	7	8	9	10
Dig for foundation and service routes	▓	▓								
Lay foundations			▓	▓						
Run cabling, piping etc. to meet existing services				▓	▓					
Build up to DPC						▓	▓			
Lay concrete floor								▓	▓	▓

▓ Proposed time ▓ Actual time

Figure 1.15 Bar chart showing proposed time for a contract

The bar chart can then be used to check progress. Often the actual time taken for a task is shaded in underneath the proposed time (in a different way or colour to avoid confusion). This shows how what *has* been done matches up to what *should have* been done.

As you can see, a bar chart can help you plan when to order materials or plant, see what trade is due in and when, and so on. A bar chart can also tell you if you are behind on a job; if you have a penalty clause written into your contract, this information is vital.

	Time in days									
Activity	1	2	3	4	5	6	7	8	9	10
Dig for foundation and service routes	███	███								
Lay foundations			███	███						
Run cabling, piping etc. to meet existing services				███	███	███				
Build up to DPC						███	███	███		
Lay concrete floor									███	███

███ Proposed time ███ Actual time

Figure 1.16 Bar chart showing actual time half way through a contract

When creating a bar chart, you should build in some extra time to allow for things such as bad weather, labour shortages, delivery problems or illness. It is also advisable to have contingency plans to help solve or avoid problems, such as:

- capacity to work overtime to catch up time
- bonus scheme to increase productivity
- penalty clause on suppliers to try to avoid late or poor deliveries
- source of extra labour (for example from another site) if needed.

Good planning, with contingency plans in place, should allow a job to run smoothly and finish on time, leading to the contractor making a profit.

Getting involved in the construction industry

There are many ways of entering the construction industry, but the most common way is as an apprentice.

Apprenticeships

You can become an apprentice by:

1. being employed directly by a construction company who will send you to college
2. being employed by a training provider, such as Carillion, which combines construction training with practical work experience.

Did you know?

Bad weather is the main external factor responsible for delays on building sites in the UK. A Met Office survey showed that the average UK construction company experiences problems caused by the weather 26 times a year

On 1 August 2002, the construction industry introduced a mandatory induction programme for all apprentices joining the industry. The programme has four distinct areas:

1. apprenticeship framework requirements

2. the construction industry

3. employment

4. health and safety.

Apprenticeship frameworks are based on a number of components designed to prepare people for work in a particular construction occupation.

Construction frameworks are made up of the following mandatory components:

- NVQs

- technical certificates (construction awards)

- key skills.

However, certain trades require additional components. Bricklaying, for example, requires abrasive wheels certification.

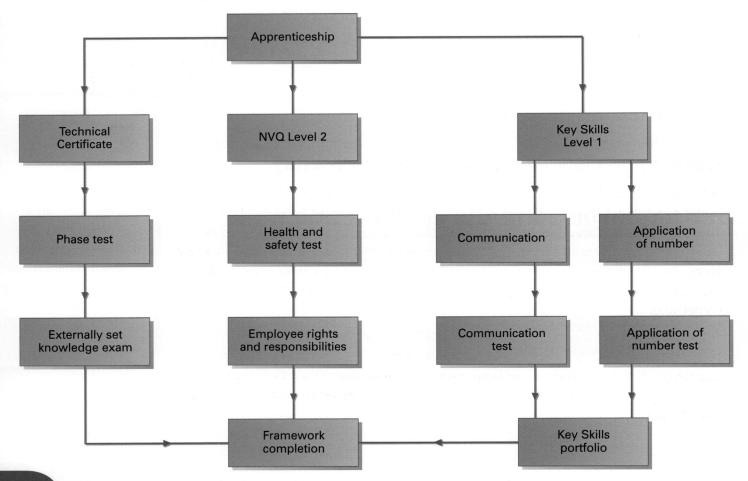

Figure 1.17 Apprenticeship framework

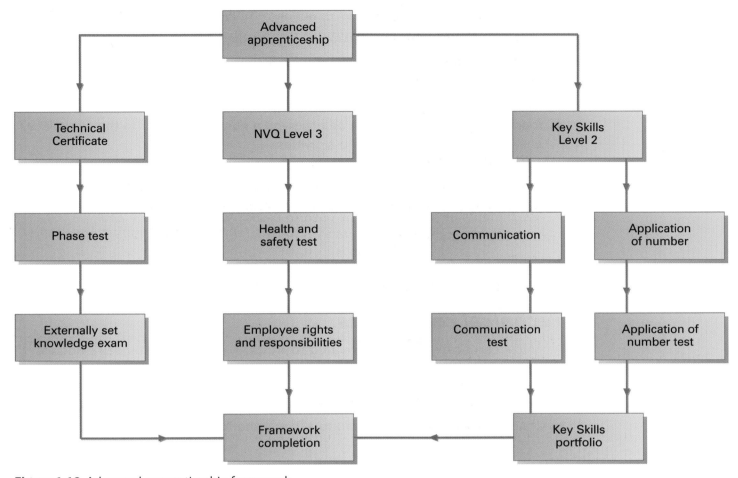

Figure 1.18 Advanced apprenticeship framework

National Vocational Qualifications (NVQs)

NVQs are available to anyone, with no restrictions on age, length or type of training, although learners below a certain age can only perform certain tasks. There are different levels of NVQ (for example 1, 2, 3), which in turn are broken down into units of competence. NVQs are not like traditional examinations in which someone sits an exam paper. An NVQ is a 'doing' qualification, which means it lets the industry know that you have the knowledge, skills and ability to actually 'do' something.

The Construction Industry Training Board (CITB) is the national training organisation for construction in the UK and is responsible for setting training standards. NVQs are made up of both mandatory and optional units and the number of units that you need to complete for an NVQ depends on the level and the occupation.

NVQs are assessed in the workplace, and several types of evidence are used:

- Witness testimony consists of evidence provided by various individuals who have first-hand knowledge of your work and performance relating to the NVQ. Work colleagues, supervisors and even customers can provide evidence of your performance.

- Your natural performance can be observed a number of times in the workplace while carrying out work-related activities.

- The use of historical evidence means that you can use evidence from past achievements or experience, if it is directly related to the NVQ.

- Assignments or projects can be used to assess your knowledge and understanding of a subject.

- Photographic evidence showing you performing various tasks in the workplace can be used, providing it is authenticated by your supervisor.

Technical certificates

Technical certificates are often related to NVQs. A certificate provides evidence that you have the underpinning knowledge and understanding required to complete a particular task. An off-the-job training programme, either in a college or with a training provider, may deliver technical certificates. You generally have to sit an end-of-programme exam to achieve the full certificate.

Key skills

Some students have key skills development needs, so learners and apprentices must achieve key skills at Level 1 or 2 in both Communications and Application of number. Key skills are signposted in each level of the NVQ and are assessed independently, so you will need to be released from your training to attend a key skills test.

Employment

Conditions of employment are controlled by legislation and regulations. The Department of Trade and Industry (DTI) publishes most of this legislation. To find out more about your working rights, visit the DTI website. A quick link has been made available at www.heinemann.co.uk/hotlinks – just enter the express code 5701P.

The main pieces of legislation that will apply to you are:

- The Employment Act 2002 which gives extra rights to working parents and gives new guidance on resolving disputes, amongst other things.

- The Employment Relations Act 1999 covers areas such as trade union membership and disciplinary and grievance proceedings.

- The Employment Rights Act 1996 details the rights an employee has by law, including the right to have time off work and the right to be given notice if being dismissed.

- The Sex Discrimination Acts of 1975 and 1986 state that it is illegal for an employee to be treated less favourably because of their sex, for example paying a man more than a woman or offering a woman more holiday than a man, even though they do the same job.

- The Race Relations Act 1976 states that it is against the law for someone to be treated less favourably because of their skin colour, race, nationality or ethnic origin.

- The Disability Discrimination Act 2005 makes it illegal for someone to be treated less favourably just because they have a physical or mental disability.

- The National Minimum Wage Act 1998 makes sure that everyone in the UK is paid a minimum amount. How much you must be paid depends on how old you are and whether or not you are on an Apprenticeship Scheme. The national minimum wage is periodically assessed and increased so it is a good idea to make sure you know what it is. At the time of writing, under 18s and those on Apprenticeship Schemes do not qualify for the minimum wage. For those aged 18–21, the minimum wage is £4.60 per hour and for adult workers aged 22 or over, the minimum wage is £5.52 per hour.

Men and women must be treated equally at work

Contract of employment

Within two months of starting a new job, your employer must give you a contract of employment. This will tell you the terms of your employment and should include the following information:

- job title
- place of work
- hours of work
- rates of pay
- holiday pay
- overtime rates
- statutory sick pay
- pension scheme
- discipline procedure

Find out

What is the national minimum wage at the moment? You can find out from lots of different places, including the DTI website. You can find a link to the site at www.heinemann.co.uk/hotlinks – just enter the express code 5701P

- termination of employment
- dispute procedure.

If you have any questions about information contained within your contract of employment, you should talk to your supervisor before you sign it.

When you start a new job, you should also receive a copy of the safety policy and an employee handbook containing details of the general policy, procedures and disciplinary rules.

Discrimination in the workplace

Discrimination means treating someone unjustly, and in the workplace it can range from bullying, intimidation or harassment to paying someone less money or not giving them a job. Discriminating against people within the working environment is against the law. This includes discrimination on the grounds of:

- sex, gender or sexual orientation
- race, colour, nationality or ethnic origin
- religious beliefs
- disability.

The law states that employment, training and promotion should be open to all employees regardless of any of the above. Pay should be equal for men and women if they are required to do the same job.

The Race Relations Act protects people of all skin colours, races and nationalities

Sources of information and advice

There are many places you can go to get information and advice about a career in the construction industry. If you are already studying, you can speak to your tutor, your school or college careers adviser or you can get in touch with Connexions for careers advice especially for young people. Visit www.heinemann. co.uk/hotlinks and enter the express code 5701P for a link to Connexions' website. You can also find their telephone number in your local phone book.

Organisations such as those listed below are very good sources of careers advice specific to the construction industry.

- CITB (Construction Industry Training Board) – the industry's national training organisation
- City and Guilds – a provider of recognised vocational qualifications
- The Chartered Institute of Building Services Engineers
- The Institute of Civil Engineers

- Trade unions such as GMB (Britain's General Union), UCATT (Union of Construction, Allied Trades and Technicians), UNISON (the public services union), Amicus (the manufacturing union, previously MSF).

Links to all these organisations' websites can be found by visiting www.heinemann.co.uk/hotlinks and entering the express code 5701P.

FAQ

Why do I need to learn about different trades?

It is very important that you have some basic knowledge of what other trades do. This is because you will often work with people from other trades and their work will affect yours and vice versa.

What options do I have once I have gained my NVQ Level 2 qualification?

Once you are qualified, there is a wide range of career opportunities available to you. For example, you could progress from a tradesman to a foreman and then to a site agent. There may also be the opportunity to become a clerk of works, an architect or a college lecturer. Some tradesmen are happy to continue as tradesmen and some start up their own businesses.

Knowledge check

1. How many members of staff are there in a small company, a medium company and a large company?

2. Give an example of a public construction project. Who pays for public work?

3. Name a job in each of the four construction employment areas: professional; technician; building craft worker; building operative.

4. Why is the client the most important member of the building team?

5. Explain the meaning of the following building types: residential; low rise; semi-detached.

6. What are the four different methods of communication?

7. What information might a schedule give you?

8. What does NVQ stand for?

9. What information must be in your contract of employment?

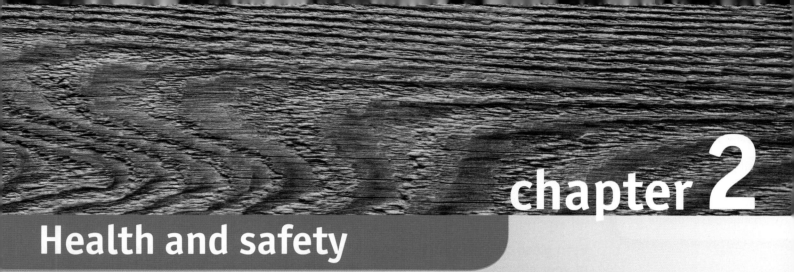

chapter 2

Health and safety

OVERVIEW

Every year in the construction industry over 100 people are killed and thousands more are seriously injured as a result of the work that they do. There are thousands more who suffer from health problems, such as dermatitis, asbestosis, industrial asthma, vibration white finger and deafness. You can therefore see why learning as much as you can about health and safety is very important.

This chapter will cover the following topics:

- Health and safety legislation
- Health and welfare in the construction industry
- Manual handling
- Hazards
- Fire and fire-fighting equipment
- Safety signs
- Personal protective equipment (PPE)
- Emergencies
- First aid
- Reporting accidents
- Risk assessments.

These topics can be found in the following modules:

CC 1001K   CC 1001S

Health and safety legislation

While at work, whatever your location or type of work, you need to be aware that there is important legislation you must comply with. Health and safety legislation is there not just to protect you, but also states what you must and must not do to ensure that no workers are placed in a situation **hazardous** to themselves or others.

Each piece of legislation covers your own responsibilities as an employee and those of your employer – it is vital that you are aware of both. As a Level 2 candidate, you not only have to think of your responsibilities for you own actions, but must also consider your supervisory responsibilities for others. These may involve ensuring that others are aware of legislation and considering such legislation when you are overseeing others' work.

What is legislation?

Legislation means a law or set of laws passed by Parliament, often called an Act. There are hundreds of Acts covering all manner of work from hairdressing to construction. Each Act states the duties of the **employer** and **employee**. If an employer or employee does something they shouldn't – or doesn't do something they should – they can end up in court and be fined or even imprisoned.

Approved code of practice, guidance notes and safety policies

As well as Acts, there are two sorts of codes of practice and guidance notes: those produced by the **Health and Safety Executive (HSE)**, and those created by companies themselves. Most large construction companies – and many smaller ones – have their own guidance notes, which go further than health and safety law. For example, the law states that that everyone must wear safety boots in a hazardous area, but a company's code may state that everyone must wear safety boots at all times. This is called taking a **proactive** approach, rather than a **reactive** one.

Most companies have some form of safety policy outlining the company's commitment and stating what they plan to do to ensure that all work is carried out as safely as possible. As an employee, you should make sure you understand the company's safety policy as well as their codes of practice. If you act against company policy you may not be prosecuted in court, but you could still be disciplined by the company or even fired.

Health and safety legislation you need to be aware of

There are some 20 pieces of legislation you will need to be aware of, each of which sets out requirements for employers and often employees. One phrase often comes up here – '*so far as is reasonably practicable*'. This means that health and safety must be adhered to at all times, but must take a common sense, practical approach.

For example, the Health and Safety at Work Act 1974 states that an employer must *so far as is reasonably practicable* ensure that a safe place of work is provided. Yet employers are not expected to do everything they can to protect their staff from lightning strikes, as there is only a 1 in 800,000 chance of this occurring – this would not be reasonable!

We will now look at the regulations that will affect you most.

The Health and Safety at Work Act 1974 (HASAW)

HASAW applies to all types and places of work and to employers, employees, the self-employed, subcontractors and even suppliers. The act is there to protect not only the people at work but also the general public, who may be affected in some way by the work that has been or will be carried out.

The main objectives of the health and safety at work act are to:

- ensure the health, safety and welfare of all persons at work
- protect the general public from all work activities
- control the use, handling, storage and transportation of explosives and highly flammable substances
- control the release of noxious or offensive substances into the atmosphere.

Legislation is there to protect employees and the public alike

To ensure that these objectives are met there are duties for all employers, employees and suppliers.

Employer's duties

Employers must:

- provide safe **access** and **egress** to and within the work area
- provide a safe place to work
- provide and maintain plant and machinery that is safe and without risks to health

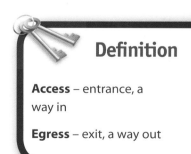

Definition

Access – entrance, a way in

Egress – exit, a way out

- provide information, instruction, training and supervision to ensure the health and safety at work of all employees

- ensure safety and the absence of risks to health in connection with the handling, storage and transportation of articles and substances

- have a written safety policy that must be revised and updated regularly, and ensure all employees are aware of it

- involve trade union safety representatives, where appointed, in all matters relating to health and safety

- provide and not charge for **personal protective equipment (PPE)**.

Employee's duties

The employee must:

- take reasonable care for his/her own health and safety

- take reasonable care for the health and safety of anyone who may be affected by his/her acts or **omissions**

- co-operate with his/her employer or any other person to ensure legal **obligations** are met

- not misuse or interfere with anything provided for their health and safety

- use any equipment and safeguards provided by his/her employer.

Employees cannot be charged for anything that has been done or provided for them to ensure that legal requirements on health and safety are met. The self-employed and subcontractors have the same duties as employees – and if they have employees of their own, they must obey the duties set down for employers.

Supplier's duties

Persons designing, manufacturing, importing or supplying articles or substances for use at work must ensure that:

- articles are designed and constructed so that they will be safe and without risk to health at all times while they are being used or constructed

- substances will be safe and without risk to health at all times when being used, handled, transported and stored

- tests on articles and substances are carried out as necessary

- adequate information is provided about the use, handling, transporting and storing of articles or substances.

Definition

PPE – personal protective equipment, such as gloves, a safety harness or goggles

Omission – something that has not been done or has been missed out

Obligation – something you have a duty or a responsibility to do

---

HASAW, like most of the other Acts mentioned, is enforced by the Health and Safety Executive (HSE). HSE inspectors visit sites and have the power to:

- enter any premises at any reasonable time
- take a police constable with them
- examine and investigate anything on the premises
- take samples
- take possession of any dangerous article or substance
- issue improvement notices giving a company a certain amount of time to sort out a health and safety problem
- issue a prohibition notice stopping all work until the site is deemed safe
- **prosecute** people who break the law including employers, employees, self-employed, manufacturers and suppliers.

Provision and Use of Work Equipment Regulations 1998 (PUWER)

These regulations cover all new or existing work equipment – leased, hired or second-hand. They apply in most working environments where the HASAW applies, including all industrial, offshore and service operations.

PUWER covers starting, stopping, regular use, transport, repair, modification, servicing and cleaning.

'Work equipment' includes any machinery, appliance, apparatus or tool, and any assembly of components that are used in non-domestic premises. Dumper trucks, circular saws, ladders, overhead projectors and chisels would all be included, but substances, private cars and structural items all fall outside this definition.

The general duties of the Act require equipment to be:

- suitable for its intended purpose and only to be used in suitable conditions
- maintained in an efficient state and maintenance records kept
- used, repaired and maintained only by a suitably trained person, when that equipment poses a particular risk
- able to be isolated from all its sources of energy
- constructed or adapted to ensure that maintenance can be carried out without risks to health and safety
- fitted with warnings or warning devices as appropriate.

In addition, the Act requires:

- all those who use, supervise or manage work equipment to be suitably trained

Definition

Prosecute – to accuse someone of committing a crime, which usually results in the accused being taken to court and, if found guilty, being punished

- access to any dangerous parts of the machinery to be prevented or controlled
- injury to be prevented from any work equipment that may have a very high or low temperature
- suitable controls to be provided for starting and stopping the work equipment
- suitable emergency stopping systems and braking systems to be fitted to ensure the work equipment is brought to a safe condition as soon as reasonably practicable
- suitable and sufficient lighting to be provided for operating the work equipment.

Control of Substances Hazardous to Health Regulations 2002 (COSHH)

These regulations state how employees and employers should work with, handle, store, transport and dispose of potentially hazardous substances (substances that might negatively affect your health) including:

- substances used directly in work activities (e.g. adhesives or paints)
- substances generated during work activities (e.g. dust from sanding wood)
- naturally occurring substances (e.g. sand dust)
- biological agents (e.g. bacteria).

These substances can be found in nearly all work environments. All are covered by COSHH regulations except asbestos and lead paint, which have their own regulations.

To comply with COSHH regulations, eight steps must be followed:

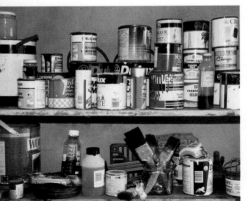

Hazardous substances

Step 1 Assess the risks to health from hazardous substances used or created by your activities.

Step 2 Decide what precautions are needed.

Step 3 Prevent employees from being exposed to any hazardous substances. If prevention is impossible, the risk must be adequately controlled.

Step 4 Ensure control methods are used and maintained properly.

Step 5 Monitor the exposure of employees to hazardous substances.

Step 6 Carry out health surveillance to ascertain if any health problems are occurring.

Step 7 Prepare plans and procedures to deal with accidents such as spillages.

Step 8 Ensure all employees are properly informed, trained and supervised.

Identifying a substance that may fall under the COSHH regulations is not always easy, but you can ask the supplier or manufacturer for a COSHH data sheet, outlining the risks involved with a substance. Most substance containers carry a warning sign stating whether the contents are corrosive, harmful, toxic or bad for the environment.

Toxic hazard

Risk of explosion

Common safety signs for corrosive, toxic and explosive materials

The Personal Protective Equipment at Work Regulations 1992 (PPER)

These regulations cover all types of PPE, from gloves to breathing apparatus. After doing a risk assessment and once the potential hazards are known, suitable types of PPE can be selected. PPE should be checked prior to issue by a trained and competent person and in line with the manufacturer's instructions. Where required, the employer must provide PPE free of charge along with a suitable and secure place to store it.

The employer must ensure that the employee knows:

- the risks the PPE will avoid or reduce
- its purpose and use
- how to maintain and look after it
- its limitations.

The employee must:

- ensure that they are trained in the use of the PPE prior to use
- use it in line with the employer's instructions
- return it to storage after use
- take care of it, and report any loss or defect to their employer.

Remember

PPE must only be used as a last line of defence

Noise at work

The Control of Noise at Work Regulations 2005

At some point in your career in construction, you are likely to work in a noisy working environment. These regulations help protect you against the consequences of being exposed to high levels of noise, which can lead to permanent hearing damage.

Damage to hearing has a range of causes, from ear infections to loud noises, but the regulations deal mainly with the latter. Hearing loss can result from one very loud noise lasting only a few seconds, or from relatively loud noise lasting for hours, such as a drill.

The regulations state that the employer must:

- assess the risks to the employee from noise at work
- take action to reduce the noise exposure that produces these risks
- provide employees with hearing protection or, if this is impossible, reduce the risk by other methods
- make sure the legal limits on noise exposure are not exceeded
- provide employees with information, instruction and training
- carry out health surveillance where there is a risk to health.

The Work at Height Regulations 2005

Construction workers often work high off the ground, on scaffolding, ladders or roofs. These regulations make sure that employers do all that they can to reduce the risk of injury or death from working at height.

The employer has a duty to:

- avoid work at height where possible
- use any equipment or safeguards that will prevent falls
- use equipment and any other methods that will minimise the distance and consequences of a fall.

As an employee, you must follow any training given to you, report any hazards to your supervisor and use any safety equipment made available to you.

The Electricity at Work Regulations 1989

These regulations cover any work involving the use of electricity or electrical equipment. An employer has the duty to ensure that the electrical systems their employees come into contact with are safe and regularly maintained. They must also have done everything the law states to reduce the risk of their employees coming into contact with live electrical currents.

Did you know?

Noise is measured in decibels (dB). The average person may notice a rise of 3dB, but with every 3dB rise, the noise is doubled. What may seem like a small rise is actually very significant

Remember

Hearing loss affects the young as well as the old

The Manual Handling Operations Regulations 1992

These regulations cover all work activities in which a person does the lifting rather than a machine. They state that, wherever possible, manual handling should be avoided, but where this is unavoidable, a risk assessment should be done.

In a risk assessment, there are four considerations:

- *Load* – is it heavy, sharp-edged, difficult to hold?
- *Individual* – is the individual small, pregnant, in need of training?
- *Task* – does the task require holding goods away from the body, or repetitive twisting?
- *Environment* – is the floor uneven, are there stairs, is it raining?

After the assessment, the situation must be monitored constantly and updated or changed if necessary.

The Reporting of Injuries, Diseases and Dangerous Occurrences Regulations 1995 (RIDDOR)

Under RIDDOR, employers have a duty to report accidents, diseases or dangerous occurrences. The HSE use this information to identify where and how risk arises and to investigate serious accidents.

Other Acts to be aware of

You should also be aware of the following pieces of legislation:

- The Fire Precautions (Workplace) Regulations 1997
- The Fire Precautions Act 1991
- The Highly Flammable Liquids and Liquid Petroleum Gases Regulations 1972
- The Lifting Operations and Lifting Equipment Regulations 1998
- The Construction (Health, Safety and Welfare) Regulations 1996
- The Environmental Protection Act 1990
- The Confined Spaces Regulations 1997
- The Working Time Regulations 1998
- The Health and Safety (First Aid) Regulations 1981
- The Construction (Design and Management) Regulations 1994.

You can find out more at the library or online.

Find out

Look into the other regulations listed here via the Government website www.hse.gov.uk

Health and welfare in the construction industry

Jobs in the construction industry have one of the highest injury and accident rates and as a worker you will be at constant risk unless you adopt a good health and safety attitude. By following the rules and regulations set out to protect you and by taking reasonable care of yourself and others, you will become a safe worker and thus reduce the chance of any injuries or accidents.

The most common risks to a construction worker

What do you think these might be? Think about the construction industry you are working in and the hazards and risks that exist.

The most common health and safety risks a construction worker faces are:

- accidents
- ill health.

Accidents

We often hear the saying 'accidents will happen', but when working in the construction industry, we should not accept that accidents just happen sometimes. When we think of an accident, we quite often think about it as being no one's fault and something that could not have been avoided. The truth is that most accidents are caused by human error, which means someone has done something they shouldn't have done or, just as importantly, not done something they should have done.

Accidents often happen when someone is hurrying, not paying enough attention to what they are doing or they have not received the correct training.

If an accident happens, you or the person it happened to may be lucky and will not be injured. More often, an accident will result in an injury which may be minor (e.g. a cut or a bruise) or possibly major (e.g. loss of a limb). Accidents can also be fatal. The most common causes of fatal accidents in the construction industry are:

- falling from scaffolding
- being hit by falling objects and materials
- falling through fragile roofs
- being hit by forklifts or lorries
- cuts
- infections
- burns
- electrocution.

Remember

Health and safety laws are there to protect you and other people. If you take shortcuts or ignore the rules, you are placing yourself and others at serious risk

Accidents can happen if your work area is untidy

Ill health

While working in the construction industry, you will be exposed to substances or situations that may be harmful to your health. Some of these health risks may not be noticeable straight away and it may take years for **symptoms** to be noticed and recognised.

Ill health can result from:

- exposure to dust (such as asbestos), which can cause breathing problems and cancer

- exposure to solvents or chemicals, which can cause **dermatitis** and other skin problems

- lifting heavy or difficult loads, which can cause back injury and pulled muscles

- exposure to loud noise, which can cause hearing problems and deafness

- exposure to sunlight, which can cause skin cancer

- using vibrating tools, which can cause **vibration white finger** and other problems with the hands.

Everyone has a responsibility for health and safety in the construction industry but accidents and health problems still happen too often. Make sure you do what you can to prevent them.

Substance abuse

Substance abuse is a general term and mainly covers things such as drinking alcohol and taking drugs.

Taking drugs or inhaling solvents at work is not only illegal, but is also highly dangerous to you and everyone around you as reduced concentration problems can lead to accidents. Drinking alcohol is also dangerous at work; going to the pub for lunch and having just one drink can lead to slower reflexes and reduced concentration.

Although not a form of abuse as such, drugs prescribed by your doctor as well as over the counter painkillers can be dangerous. Many of these medicines carry warnings such as "may cause drowsiness" or "do not operate heavy machinery". It is better to be safe than sorry, so always ensure you follow any instructions on prescriptions and, if you feel drowsy or unsteady, then stop work immediately.

Definition

Symptom – a sign of illness or disease (e.g. difficulty breathing, a sore hand or a lump under the skin)

Dermatitis – a skin condition where the affected area is red, itchy and sore

Vibration white finger – a condition that can be caused by using vibrating machinery (usually for very long periods of time). The blood supply to the fingers is reduced which causes pain, tingling and sometimes spasms (shaking)

Always wash your hands to prevent ingesting hazardous substances

Staying healthy

As well as keeping an eye out for hazards, you must also make sure that you look after yourself and stay healthy. One of the easiest ways to do this is to wash your hands on a regular basis. By washing your hands you are preventing hazardous substances from entering your body through ingestion (swallowing). You should always wash your hands after going to the toilet and before eating or drinking.

Other precautions that you can take are ensuring that you wear **barrier cream**, the correct PPE and only drink water that is labelled as drinking water. Ensure that you are protected from the sun with a good sunscreen, and ensure your back, arms and legs are covered by suitable clothing. Remember that some health problems do not show symptoms straight away and what you do now can affect you much later in life.

Welfare facilities

Welfare facilities are things such as toilets, which must be provided by your employer to ensure a safe and healthy workplace. There are several things that your employer must provide to meet welfare standards and these are:

- Toilets – the number of toilets provided depends upon the amount of people who are intended to use them. Males and females can use the same toilets providing there is a lock on the inside of the door. Toilets should be flushable with water or, if this is not possible, with chemicals.

- Washing facilities – employers must provide a basin large enough to allow people to wash their hands, face and forearms. Washing facilities must have hot and cold running water as well as soap and a means of drying your hands. Showers may be needed if the work is very dirty or if workers are exposed to **corrosive** and **toxic** substances.

- Drinking water – there should be a supply of clean drinking water available, either from a tap connected to the mains or from bottled water. Taps connected to the mains need to be clearly labelled as drinking water and bottled drinking water must be stored in a separate area to prevent **contamination**.

- Storage or dry room – every building site must have an area where workers can store the clothes that they do not wear on site, such as coats and motorcycle helmets. If this area is to be used as a drying room then adequate heating must also be provided in order to allow clothes to dry.

Definition

Barrier cream – a cream used to protect the skin from damage or infection

Definition

Corrosive – a substance that can damage things it comes into contact with (e.g. material, skin)

Toxic – poisonous

Contamination – when harmful chemicals or substances pollute something (for example water)

- Lunch area – every site must have facilities that can be used for taking breaks and lunch well away from the work area. These facilities must provide shelter from the wind and rain and be heated as required. There should be access to tables and chairs, a kettle or urn for boiling water and a means of heating food, such as a microwave.

When working in an occupied house, you can make arrangements with the client to use the facilities in their house.

Safety tip

When placing clothes in a drying room, do not place them directly on to heaters as this can lead to fire

On the job: Manual handling

Glynn and Frankie are unloading bags of plaster from a wheelbarrow. While handling a bag of plaster, Glynn gets a sharp pain in his back and drops the bag. Frankie goes and tells their supervisor, who comes over to where Glynn is sitting in a great deal of pain. What do you think should happen next? Do you think this incident could have been prevented?

Manual handling

Manual handling means lifting and moving a piece of equipment or material from one place to another without using machinery. Lifting and moving loads by hand is one of the most common causes of injury at work. Most injuries caused by manual handling result from years of lifting items that are too heavy, are awkward shapes or sizes, or from using the wrong technique. However, it is also possible to cause a lifetime of back pain with just one single lift.

Poor manual handling can cause injuries such as muscle strain, pulled ligaments and hernias. The most common injury by far is spinal injury. Spinal injuries are very serious because there is very little that doctors can do to correct them and, in extreme cases, workers have been left paralysed.

What you can do to avoid injury

The first and most important thing you can do to avoid injury from lifting is to receive proper manual handling training. Kinetic lifting is a way of lifting objects that reduces the chance of injury and is covered in more detail on the next page.

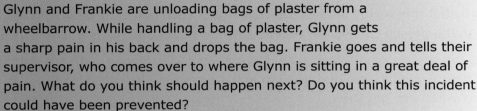

Poor manual handling techniques can lead to serious permanent injury

Before you lift anything you should ask yourself some simple questions:

- Does the object need to be moved?

- Can I use something to help me lift the object? A mechanical aid such as a forklift or crane or a manual aid such as a wheelbarrow may be more appropriate than a person.

- Can I reduce the weight by breaking down the load? Breaking down a load into smaller and more manageable weights may mean that more journeys are needed, but it will also reduce the risk of injury.

- Do I need help? Asking for help to lift a load is not a sign of weakness and team lifting will greatly reduce the risk of injury.

- How much can I lift safely? The recommended maximum weight a person can lift is 25 kg but this is only an average weight and each person is different. The amount that a person can lift will depend on their physique, age and experience.

- Where is the object going? Make sure that any obstacles in your path are out of the way before you lift. You also need to make sure there is somewhere to put the object when you get there.

- Am I trained to lift? The quickest way to receive a manual handling injury is to use the wrong lifting technique.

Lifting correctly (kinetic lifting)

When lifting any load it is important to keep the correct posture and to use the correct technique.

The correct posture before lifting:

- feet shoulder width apart with one foot slightly in front of the other

- knees should be bent

- back must be straight

- arms should be as close to the body as possible

- grip must be firm using the whole hand and not just the finger tips.

The correct technique when lifting:

- approach the load squarely facing the direction of travel

- adopt the correct posture (as above)

- place hands under the load and pull the load close to your body

- lift the load using your legs and not your back.

When lowering a load you must also adopt the correct posture and technique:

- bend at the knees, not the back
- adjust the load to avoid trapping fingers
- release the load.

Think before lifting

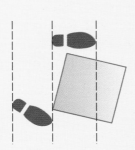

Adopt the correct posture before lifting

Get a good grip on the load

Adopt the correct posture when lifting

Move smoothly with the load

Adopt the correct posture and technique when lowering

Hazards

The building industry can be a very dangerous place to work and there are certain hazards that all workers need to be aware of.

The main types of hazards that you will face are:

- falling from height
- tripping
- chemical spills
- burns
- electrical
- fires.

Falling from height will be covered in detail in Chapter 3 *Working at height* so here we will look at the remaining hazards.

An untidy work site can present many trip hazards

Tripping

The main cause of tripping is poor housekeeping. Whether working on scaffolding or on ground level, an untidy workplace is an accident waiting to happen. All workplaces should be kept tidy and free of debris. All offcuts should be put either in a wheelbarrow (if you are not near a skip) or straight into the skip. Not only will this prevent trip hazards, but it will also prevent costly clean-up operations at the end of the job and will promote a good professional image.

Chemical spills

Chemical spillages can range from minor inconvenience to major disaster. Most spillages are small and create minimal or no risk. If the material involved is not hazardous, it simply can be cleaned up by normal operations such as brushing or mopping up the spill. However, on some occasions the spill may be on a larger scale and may involve a hazardous material. It is important to know what to do before the spillage happens so that remedial action can be prompt and harmful effects minimised. Of course, when a hazardous substance is being used a COSHH or risk assessment will have been made, and it should include a plan for dealing with a spillage. This in turn should mean that the materials required for dealing with the spillage should be readily available.

Burns

Burns can occur not only from the obvious source of fire and heat but also from materials containing chemicals such as cement or painter's solvents. Even electricity can cause burns. It is vital when working with materials that you are aware of the hazards it may present and take the necessary precautions.

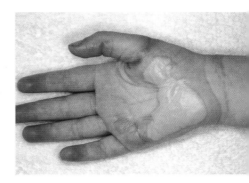

Fire, heat, chemicals and electricity can cause burns

Electricity

Electricity is a killer. Around 30 workers a year die from electricity related accidents, with over 1000 more being seriously injured (source: HSE). One of the main problems with electricity is that it is invisible. You don't even have to be working with an electric tool to be electrocuted. Working too close to live overhead cables, plastering a wall with electric sockets, carrying out maintenance work on a floor, or drilling into a wall can all lead to an electric shock.

Electric shocks may not always be fatal; electricity can also cause burns, muscular problems and cardiac (heart) problems. The level of voltage is not a direct guide to the level of injury or danger of death, despite this common misconception; a small shock from static electricity may contain thousands of volts but has very little current behind it. Generally, the lower the voltage, the less chance of death occurring.

There are two main types of voltage in use in the UK. These are 230 V and 110 V. The standard UK power supply Is 230 V and this is what all the sockets in your house are. Contained within the wiring there should be three wires: the live and neutral, which carry the alternating current, and the earth wire, which acts as a safety device. The three wires are colour-coded as follows:

> Live – Brown
>
> Neutral – Blue
>
> Earth – Yellow and green

These have been changed recently to comply with European colours. Some older properties will have the following colours:

> Live – Red
>
> Neutral – Black
>
> Earth – Yellow and green

230 V has been deemed as unsafe on construction sites so 110 V must be used here. 110 V, identified by a yellow cable and different style plug, works from a transformer which converts the 230 V to 110 V.

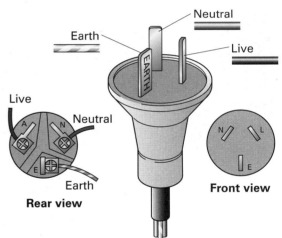

Figure 2.1 Colour coding of the wires in a 230 V plug

A 110 V plug

Dealing with electric shocks

In helping a victim of an electric shock, the first thing you must do is disconnect the power supply, if it is safe to do and will not take long to find; touching the power source may put you in danger. If the victim is in contact with something portable, such as a drill, attempt to move it away using a non-conductive object such as a wooden broom. Time is precious and separating the victim from the source can prove an effective way to speed the process. Do not attempt to touch the affected person until they are free and clear of the supplied power. Don't even touch the victim until you are sure the power supply is turned off. Be especially careful in wet areas, such as bathrooms, since most water will conduct electricity and electrocuting yourself is also possible.

People 'hung up' in a live current flow may think they are calling out for help but most likely no sound will be heard from them. When the muscles contract under household current (most electrocutions happen from house current at home), the person affected will appear in locked-up state, unable to move or react to you. Using a wooden object, swiftly and strongly knock the person free, trying not to injure them, and land them clear of the source. The source may also be lifted or removed, if possible, with the same wooden item. This is not recommended on voltages that exceed 500 V. Do not attempt any of this without rubber or some form of insulated sole shoes; bare or socked feet will allow the current to flow to ground through your body as well.

First aid procedures for an electric shock victim

- Check if you are alone. If there are other people around, instruct them to call an ambulance right away.

- Check for a response and breathing.

- If the area is safe for you to be in, and you have removed the object or have cut off its power supply, yell to the person to see if they are conscious. At this stage, do not touch the victim.

- Check once again to see if the area is safe. If you are satisfied that it is safe, start resuscitating the victim. If you have no first aid knowledge, call emergency services for an ambulance.

Fire and fire-fighting equipment

Fires can start almost anywhere and at any time but a fire needs three things to burn. These are:

1. fuel

2. heat

3. oxygen.

Figure 2.2 The triangle of fire

This can be shown in what is known as 'the triangle of fire'. If any of the sides of the triangle are removed, the fire cannot burn and it will go out.

Remember:

- Remove the fuel and there is nothing to burn so the fire will go out.
- Remove the heat and the fire will go out.
- Remove the oxygen and the fire will go out as fire needs oxygen to survive.

Fires can be classified according to the type of material that is involved:

- Class A – wood, paper, textiles etc.
- Class B – flammable liquids, petrol, oil etc.
- Class C – flammable gases, liquefied petroleum gas (**LPG**), propane etc.
- Class D – metal, metal powder etc.
- Class E – electrical equipment.

Fire-fighting equipment

There are several types of fire-fighting equipment, such as fire blankets and fire extinguishers. Each type is designed to be the most effective at putting out a particular class of fire and some types should never be used in certain types of fire.

Fire extinguishers

A fire extinguisher is a metal canister containing a substance that can put out a fire. There are several different types and it is important that you learn which type should be used on specific classes of fires. This is because if you use the wrong type, you may make the fire worse or risk severely injuring yourself.

Fire extinguishers are now all one colour (red) but they have a band of colour which shows what substance is inside.

Water

The coloured band is red and this type of extinguisher can be used on Class A fires. Water extinguishers can also be used on Class C fires in order to cool the area down.

A water fire extinguisher should *never* be used to put out an electrical or burning fat/oil fire. This is because electrical current can carry along the jet of water back to the person holding the extinguisher, electrocuting them. Putting water on to burning fat or oil will make the fire worse as the fire will 'explode', potentially causing serious injury.

Find out

What fire risks are there in the construction industry? Think about some of the materials (fuel) and heat sources that could make up two of the sides of 'the triangle of fire'

Water fire extinguisher

Foam fire extinguisher

Carbon dioxide (CO$_2$) extinguisher

Foam

The coloured band is cream and this type of extinguisher can also be used on Class A fires. A foam extinguisher can also be used on a Class B fire if the liquid is not flowing and on a Class C fire if the gas is in liquid form.

Carbon dioxide (CO$_2$)

The coloured band is black and the extinguisher can be used on Class A, B, C and E fires.

Dry powder

The coloured band is blue and this type of extinguisher can be used on all classes of fire. The powder puts out the fire by smothering the flames.

Fire blankets

Fire blankets are normally found in kitchens or canteens as they are good at putting out cooking fires. They are made of a fireproof material and work by smothering the fire and stopping any more oxygen from getting to it, thus putting it out. A fire blanket can also be used if a person is on fire.

It is important to remember that when you put out a fire with a fire blanket, you need to take extra care as you will have to get quite close to the fire.

What to do in the event of a fire

During **induction** to any workplace, you will be made aware of the fire procedure as well as where the fire assembly points (also known as **muster points**) are and what the alarm sounds like. On hearing the alarm you must stop what you are doing and make your way to the nearest muster point. This is so that everyone can be accounted for. If you do not go to the muster point or if you leave before someone has taken your name, someone may risk their life to go back into the fire to get you.

When you hear the alarm, you should not stop to gather any belongings and you must not run. If you discover a fire, you must only try to fight the fire if it is blocking your exit or if it is small. Only when you have been given the all-clear can you re-enter the site or building.

Dry powder extinguisher

A fire blanket

Safety signs

Safety signs can be found in many areas of the workplace and they are put up in order to:

- warn of any **hazards**
- prevent accidents
- inform where things are
- tell you what to do in certain areas.

Types of safety sign

There are many different safety signs but each will usually fit into one of four categories:

1. Prohibition signs – these tell you that something MUST NOT be done. They always have a white background and a red circle with a red line through it.

2. Mandatory signs – these tell you that something MUST be done. They are also circular but have a white symbol on a blue background.

3. Warning signs – these signs are there to alert you to a specific hazard. They are triangular and have a yellow background and a black border.

4. Information signs – these give you useful information like the location of things (e.g. a first aid point). They can be square or rectangular and are green with a white symbol.

Most signs only have symbols that let you know what they are saying. Others have some words as well, for example a no smoking sign might have a cigarette in a red circle, with a red line crossing through the cigarette and the words 'No smoking' underneath.

Definition

Hazard – a danger or risk

Figure 2.3 A prohibition sign

Figure 2.4 A mandatory sign

Figure 2.5 A warning sign

Figure 2.6 An information sign

Figure 2.7 A safety sign with both symbol and words

Remember

Make sure you take notice of safety signs in the workplace – they have been put up for a reason!

Personal protective equipment (PPE)

Personal protective equipment (PPE) is a form of defence against accidents or injury and comes in the form of articles of clothing. This is not to say that PPE is the only way of preventing accidents or injury. It should be used together with all the other methods of staying healthy and safe in the workplace (i.e. equipment, training, regulations and laws etc.).

PPE must be supplied by your employer free of charge and you have responsibility as an employee to look after it and use it whenever it is required.

Remember

PPE only works properly if it is being used and used correctly!

Types of PPE

There are certain parts of the body that require protection from hazards during work and each piece of PPE must be suitable for the job and used properly.

Head protection

A safety helmet

There are several different types of head protection but the one most commonly used in construction is the safety helmet (or hard hat). This is used to protect the head from falling objects and knocks and has an adjustable strap to ensure a snug fit. Some safety helmets come with attachments for ear defenders or eye protection. Safety helmets are meant to be worn directly on the head and must not be worn over any other type of hat.

Eye protection

Safety goggles

Eye protection is used to protect the eyes from dust and flying debris. The three main types are:

1. Safety goggles – made of a durable plastic and used when there is a danger of dust getting into the eyes or a chance of impact injury.

2. Safety spectacles – these are also made from a durable plastic but give less protection than goggles. This is because they don't fully enclose the eyes and so only protect from flying debris.

3. Facemasks – again made of durable plastic, facemasks protect the entire face from flying debris. They do not, however, protect the eyes from dust.

Safety spectacles

Foot protection

Safety boots or shoes are used to protect the feet from falling objects and to prevent sharp objects such as nails from injuring the foot. Safety boots should have a steel toe-cap and steel mid-sole.

Safety boots

Hearing protection

Hearing protection is used to prevent damage to the ears caused by very loud noise. There are several types of hearing protection available but the two most common types are ear-plugs and ear defenders.

1. Ear-plugs – these are small fibre plugs that are inserted into the ear and used when the noise is not too severe. When using ear-plugs, make sure that you have clean hands before inserting them and never use plugs that have been used by somebody else.

2. Ear defenders – these are worn to cover the entire ear and are connected to a band that fits over the top of the head. They are used when there is excessive noise and must be cleaned regularly.

Ear-plugs

Respiratory protection

Respiratory protection is used to prevent the worker from breathing in any dust or fumes that may be hazardous. The main type of respiratory protection is the dust mask.

Dust masks are used when working in a dusty environment and are lightweight, comfortable and easy to fit. They should be worn by only one person and must be disposed of at the end of the working day. Dust masks will only offer protection from non-toxic dust, so if the worker is to be exposed to toxic dust or fumes, a full respiratory system should be used.

Ear defenders

Hand protection

There are several types of hand protection and each type must be used for the correct task. For example, wearing lightweight rubber gloves to move glass will not offer much protection so leather gauntlets must be used. Plastic–coated gloves will protect you from certain chemicals and Kevlar® gloves offer cut resistance. To make sure you are wearing the most suitable type of glove for the task, you need to look first at what is going to be done and then match the type of glove to that task.

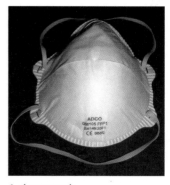

A dust mask

Safety gloves

Emergencies

We have so far covered most of the emergencies that occur on site, such as accidents and fires, but there are other emergencies that you need to be aware of, such as security alerts and bomb scares.

At your site induction, it should be made perfectly clear to you what you should do in the event of an emergency. You also should be made aware of any sirens or warning noises that accompany each and every type of emergency such as bomb scares or fire alarms. Some sites may have different variations on sirens or emergency procedures, so it is vital that you pay attention and listen to all instructions. If you are unsure always ask.

First Aid

In the unfortunate event of an accident on site, first aid may have to be administered. If there are more than five people on a site, then a qualified first aider must be present at all times. On large building sites there must be several first aiders. During your site induction you will be made aware of who the first aiders are and where the first aid points are situated. A first aid point must have the relevant first aid equipment to deal with the types of injuries that are likely to occur. However, first aid is only the first step and, in the case of major injuries, the emergency services should be called.

A good first aid box should have plasters, bandages, antiseptic wipes, latex gloves, eye patches, slings, wound dressings and safety pins. Other equipment, such as eye wash stations, must also be available if the work being carried out requires it.

Remember

Health and safety is everyone's duty. If you receive first aid treatment and notice that there are only two plasters left, you should report it to your line manager.

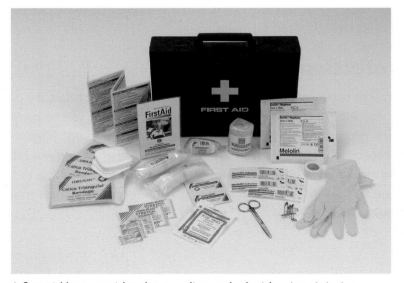

A first aid box provides the supplies to deal with minor injuries

Reporting accidents

When an accident occurs, there are certain things that must be done. All accidents need to be reported and recorded in the accident book and the injured person must report to a trained first aider in order to receive treatment. Serious accidents must be reported under the Reporting of Injuries, Diseases and Dangerous Occurrences Regulations 1995 (RIDDOR). Under RIDDOR your employer must report to the HSE any accident that results in:

- death

- major injury

- an injury that means the injured person is not at work for more than three consecutive days.

The accident book

The accident book is completed by the person who had the accident or, if this is not possible, someone who is representing the injured person.

The accident book will ask for some basic details about the accident, including:

- who was involved

- what happened

- where it happened

- the day and time of the accident

- any witnesses to the accident

- the address of the injured person

- what PPE was being worn

- what first aid treatment was given.

As well as reporting accidents, 'near misses' must also be reported. This is because near misses are often the accidents of the future. Reporting near misses might identify a problem and can prevent accidents from happening in the future. This allows a company to be proactive rather than reactive.

Report of an Accident, Dangerous Occurrence or Near Miss

Date of incident _____ Time of incident _____

Location of incident _____

Details of person involved in accident

Name _____ Date of birth _____ Sex _____

Address _____

_____ Occupation _____

Date off work (if applicable) _____ Date returning to work _____

Nature of injury _____

Management of injury ☐ First Aid only ☐ Advised to see doctor

☐ Sent to casualty ☐ Admitted to hospital

Account of accident, dangerous occurrence or near miss
(Continued on separate sheet if necessary)

Witnesses to the incident
(Names, addresses and occupations)

Was the injured person wearing PPE? If yes, what PPE? _____

Signature of person completing form _____

Occupation _____ Date _____

Figure 2.8 A typical accident book page

Risk assessments

You will have noticed that most of the legislation we have looked at requires risk assessments to be carried out. The Management of Health and Safety at Work Regulations 1999 require every employer to make suitable and sufficient assessment of:

- the risks to the health and safety of his/her employees to which they are exposed while at work
- the risks to the health and safety of persons not in his/her employment arising out of or in connection with his/her work activities.

As a Level 2 candidate, it is vital that you know how to **carry out a risk assessment**. Often you may be in a position where you are given direct responsibility for this, and the care and attention you take over it may have a direct impact on the safety of others. You must be aware of the dangers or hazards of any task, and know what can be done to prevent or reduce the risk.

There are five steps in a risk assessment – here we use cutting the grass as an example:

Step 1 Identify the hazards

When cutting the grass the main hazards are from the blades or cutting the wire, electrocution and any stones that may be thrown up.

Step 2 Identify who will be at risk

The main person at risk is the user but passers-by may be struck by flying debris.

Step 3 Calculate the risk from the hazard against the likelihood of an accident taking place

The risks from the hazard are quite high: the blade or wire can remove a finger, electrocution can kill and the flying debris can blind or even kill. The likelihood of an accident happening is medium: you are unlikely to cut yourself on the blades, but the chance of cutting through the cable is medium, and the chance of hitting a stone high.

Step 4 Introduce measures to reduce the risk

Training can reduce the risks of cutting yourself; training and the use of an **RCD** can reduce the risk of electrocution; and raking the lawn first can reduce the risk of sending up stones.

Step 5 Monitor the risk

Constantly changing factors mean any risk assessment may have to be modified or even changed completely. In our example, one such factor could be rain.

Definition

Carry out a risk assessment – measure the dangers of an activity against the likelihood of accidents taking place.

Even an everyday task like cutting the grass has its own dangers

Definition

RCD – residual current device, a device that will shut the power down on a piece of electrical equipment if it detects a change in the current, thus preventing electrocution

On the job: Scaffold safety

Ralph and Vijay are working on the second level of some scaffolding clearing debris. Ralph suggests that, to speed up the task, they should throw the debris over the edge of the scaffolding into a skip below. The building Ralph and Vijay are working on is on a main road and the skip is not in a closed off area. What do you think of Ralph's idea? What are your reasons for this answer?

FAQ

How do I find out what safety legislation is relevant to my job?

Ask your employer or manager, or contact the HSE at www.hse.gov.uk.

When do I need to do a risk assessment?

A risk assessment should be carried out if there is any chance of an accident happening as a direct result of the work being done. To be on the safe side, you should make a risk assessment before starting each task.

Do I need to read and understand every regulation?

No. It is part of your employer's duty to ensure that you are aware of what you need to know.

Knowledge check

1. Name five pieces of health and safety legislation that affect the construction industry.

2. What does HSE stand for? What does it do?

3. What does COSHH stand for?

4. What does RIDDOR stand for?

5. What might happen to you or your employer if a health and safety law is broken?

6. What are the two most common risks to construction workers?

7. State two things that you can do to avoid injury when lifting loads using manual handling techniques.

8. What are the two main types of voltage in use in the UK?

9. What three elements cause a fire and keep it burning?

10. What class(es) of fire can be put out with a carbon dioxide (CO_2) extinguisher?

11. What does a prohibition sign mean?

12. Describe how you would identify a warning sign.

13. Name the six different types of PPE.

14. Name five items that should be included in a first aid kit.

15. Who fills in an accident report form?

16. Why is it important to report 'near misses'?

17. Briefly explain what a risk assessment is.

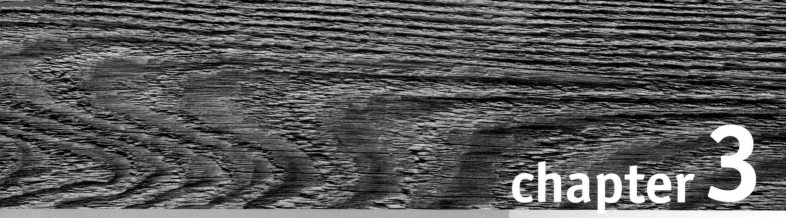

chapter 3

Working at height

OVERVIEW

Most construction trades require frequent use of some type of working platform or access equipment. Working off the ground can be dangerous and the greater the height the more serious the risk of injury. This chapter will give you a summary of some of the most common types of access equipment and provide information on how they should be used, maintained and checked to ensure that the risks to you and others are minimal.

This chapter will cover the following topics:

- General safety considerations
- Stepladders and ladders
- Roof work
- Trestle platforms
- Hop-ups
- Scaffolding
- Fall protection.

These topics can be found in the following modules:

CC 1001K   CC 1001S

General safety considerations

You will need to be able to identify potential hazards associated with working at height, as well as hazards associated with equipment. It is essential that access equipment is well maintained and checked regularly for any deterioration or faults, which could compromise the safety of someone using the equipment and anyone else in the work area. Although obviously not as important as people, equipment can also be damaged by the use of faulty access equipment. When maintenance checks are carried out they should be properly recorded. This provides very important information that helps to prevent accidents.

Risk assessment

Before any work is carried out at height, a thorough risk assessment needs to be completed. Your supervisor or someone else more experienced will do this while you are still training, but it is important that you understand what is involved so that you are able to carry out an assessment in the future.

For a working at height risk assessment to be valid and effective a number of questions must be answered:

1. How is access and egress to the work area to be achieved?
2. What type of work is to be carried out?
3. How long is the work likely to last?
4. How many people will be carrying out the task?
5. How often will this work be carried out?
6. What is the condition of the existing structure (if any) and the surroundings?
7. Is adverse weather likely to affect the work and workers?
8. How competent are the workforce and their supervisors?
9. Is there a risk to the public and work colleagues?

Duties

Your employer has a duty to provide and maintain safe plant and equipment, which includes scaffold access equipment and systems of work.

You have a duty:

- to comply with safety rules and procedures relating to access equipment
- to take positive steps to understand the hazards in the workplace and report things you consider likely to lead to danger, for example a missing handrail on a working platform
- not to tamper with or modify equipment.

Stepladders and ladders

Stepladders

A stepladder has a prop, which when folded out allows the ladder to be used without having to lean it against something. Stepladders are one of the most frequently used pieces of access equipment in the construction industry and are often used every day. This means that they are not always treated with the respect they demand. Stepladders are often misused – they should only be used for work that will take a few minutes to complete. When work is likely to take longer than this, a sturdier alternative should be found.

When stepladders are used, the following safety points should be observed:

- Ensure the ground on which the stepladder is to be placed is firm and level. If the ladder rocks or sinks into the ground it should not be used for the work.

- Always open the steps fully.

- Never work off the top tread of the stepladder.

- Always keep your knees below the top tread.

- Never use stepladders to gain additional height on another working platform.

- Always look for the kitemark, which shows that the ladder has been made to British Standards.

A number of other safety points need to be observed depending on the type of stepladder being used.

Wooden stepladder

Before using a wooden stepladder:

- Check for loose screws, nuts, bolts and hinges.

- Check that the tie ropes between the two sets of **stiles** are in good condition and not frayed.

- Check for splits or cracks in the stiles.

- Check that the treads are not loose or split.

Never paint any part of a wooden stepladder as this can hide defects, which may cause the ladder to fail during use, causing injury.

Figure 3.1 British Standards Institution kitemark

Definition

Stiles – the side pieces of a stepladder into which the steps are set

Wooden stepladder

Safety tip

If any faults are revealed when checking a stepladder, it should be taken out of use, reported to the person in charge and a warning notice attached to it to stop anyone using it

Find out

What are the advantages and disadvantages of each type of stepladder?

Did you know?

Stepladders should be stored under cover to protect from damage such as rust or rotting

Safety tip

Ladders must NEVER be repaired once damaged and must be disposed of

Aluminium stepladder

Before using an aluminium stepladder:

- check for damage to stiles and treads to see whether they are twisted, badly dented or loose
- avoid working close to live electricity supplies as aluminium will conduct electricity.

Fibreglass stepladder

Before using a fibreglass stepladder, check for damage to stiles and treads. Once damaged, fibreglass stepladders cannot be repaired and must be disposed of.

Aluminium stepladder

Ladders

A ladder, unlike a stepladder, does not have a prop and so has to be leant against something in order for it to be used. Together with stepladders, ladders are one of the most common pieces of equipment used to carry out work at heights and gain access to the work area.

As with stepladders, ladders are also available in timber, aluminium and fibreglass and require similar checks before use.

Ladder types

Pole ladder

These are single ladders and are available in a range of lengths. They are most commonly used for access to scaffolding platforms. Pole ladders are made from timber and must be stored under cover and flat, supported evenly along their length to prevent them sagging and twisting. They should be checked for damage or defects every time before being used.

Extension ladder

Extension ladders have two or more interlocking lengths, which can be slid together for convenient storage or slid apart to the desired length when in use.

Pole ladder

Extension ladders are available in timber, aluminium and fibreglass. Aluminium types are the most favoured as they are lightweight yet strong and available in double and triple extension types. Although also very strong, fibreglass versions are heavy, making them difficult to manoeuvre.

Erecting and using a ladder

The following points should be noted when considering the use of a ladder:

- As with stepladders, ladders are not designed for work of long duration. Alternative working platforms should be considered if the work will take longer than a few minutes.

- The work should not require the use of both hands. One hand should be free to hold the ladder.

- You should be able to do the work without stretching.

- You should make sure that the ladder can be adequately secured to prevent it slipping on the surface it is leaning against.

Pre-use checks

Before using a ladder check its general condition. Make sure that:

- no rungs are damaged or missing
- the stiles are not damaged
- no **tie-rods** are missing
- no repairs have been made to the ladder.

In addition, for wooden ladders ensure that:

- they have not been painted, which may hide defects or damage
- there is no decay or rot
- the ladder is not twisted or warped.

Erecting a ladder

Observe the following guidelines when erecting a ladder:

- Ensure you have a solid, level base.
- Do not pack anything under either (or both) of the stiles to level it.
- If the ladder is too heavy to put it in position on your own, get someone to help.
- Ensure that there is at least a four-rung overlap on each extension section.

Aluminium extension ladder

Definition

Tie-rods – metal rods underneath the rungs of a ladder that give extra support to the rungs

Did you know?

On average in the UK, 14 people a year die at work falling from ladders; nearly 1,200 suffer major injuries (source: Health and Safety Executive)

- Never rest the ladder on plastic guttering as it may break, causing the ladder to slip and the user to fall.

- Where the base of the ladder is in an exposed position, ensure it is adequately guarded so that no one knocks it or walks into it.

- The ladder should be secured at both the top and bottom. The bottom of the ladder can be secured by a second person, however this person must not leave the base of the ladder whilst it is in use.

- The angle of the ladder should be a ratio of 1:4 (or 75°). This means that the bottom of the ladder is 1 m away from the wall for every 4 m in height (see Figure 3.2).

- The top of the ladder must extend at least 1 m, or 5 rungs, above its landing point.

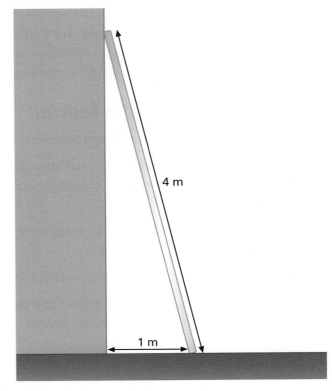

Figure 3.2 Correct angle for a ladder

Roof work

When carrying out any work on a roof, a roof ladder or **crawling board** must be used. Roof work also requires the use of edge protection or, where this is not possible, a safety harness.

The roof ladder is rolled up the surface of the roof and over the ridge tiles, just enough to allow the ladder to be turned over and the ladder hook allowed to bear on the tiles on the other side of the roof. This hook prevents the roof ladder sliding down the roof once it is accessed.

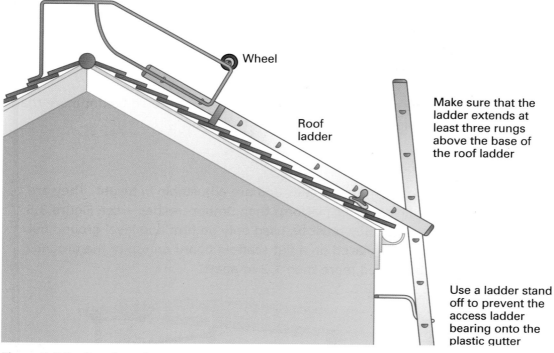

Figure 3.3 Roof work equipment

Wheel

Roof
ladder

Make sure that the
ladder extends at
least three rungs
above the base of
the roof ladder

Use a ladder stand
off to prevent the
access ladder
bearing onto the
plastic gutter

Trestle platforms

A trestle is a frame upon which a platform or
other type of surface (e.g. a table top) can
be placed. A trestle should be used rather
than a ladder for work that will take longer
than a few minutes to complete. Trestle
platforms are composed of the frame and
the platform (sometimes called a stage).

Frames

A-frames

These are most commonly
used by carpenters and painters. As the
name suggests, the frame is in the shape
of a capital A and can be made from timber,
aluminium or fibreglass. Two are used
together to support a platform (a scaffold or
staging board). See Figure 3.4.

Figure 3.4 A-frame trestles
with scaffold board

When using A-frames:

- they should always be opened fully and, in the same way as stepladders, must be placed on firm, level ground
- the platform width should be no less than 450 mm
- the overhang of the board at each end of the platform should be not more than four times its thickness.

Steel trestles

These are sturdier than A-frame trestles and are adjustable in height. They are also capable of providing a wider platform than timber trestles – see Figure 3.5. As with the A-frame type, they must be used only on firm and level ground but the trestle itself should be placed on a flat scaffold board on top of the ground. Trestles should not be placed more than 1.2 m apart.

Figure 3.5 Steel trestle with staging board

Platforms

Scaffold boards

To ensure that scaffold boards provide a safe working platform, before using them check that they:

- are not split
- are not twisted or warped
- have no large knots, which cause weakness.

Safety tip

A-frame trestles should never be used as a stepladder as they are not designed for this purpose

Staging boards

These are designed to span a greater distance than scaffold boards and can offer a 600 mm wide working platform. They are ideal for use with trestles.

Hop-ups

Also known as step-ups, these are ideal for reaching low-level work that can be carried out in a relatively short period of time. A hop-up needs to be of sturdy construction and have a base of not less than 600 mm by 500 mm. Hop-ups have the disadvantage that they are heavy and awkward to move around.

Scaffolding

Tubular scaffold is the most commonly used type of scaffolding within the construction industry. There are two types of tubular scaffold:

1. Independent scaffold – free-standing scaffold that does not rely on any part of the building to support it (although it must be tied to the building to provide additional stability).

2. Putlog scaffold – scaffolding that is attached to the building via the entry of some of the poles into holes left in the brickwork by the bricklayer. The poles stay in position until the construction is complete and give the scaffold extra support.

No one other than a qualified **carded scaffolder** is allowed to erect or alter scaffolding. Although you are not allowed to erect or alter this type of scaffold, you must be sure it is safe before you work on it. You should ask yourself a number of questions to assess the condition and suitability of the scaffold before you use it:

- Are there any signs attached to the scaffold which state that it is incomplete or unsafe?
- Is the scaffold overloaded with materials such as bricks?
- Are the platforms cluttered with waste materials?
- Are there adequate guardrails and scaffold boards in place?
- Does the scaffold actually *look* safe?
- Is there the correct access to and from the scaffold?
- Are the various scaffold components in the correct place (see Figure 3.6)?
- Have the correct types of fittings been used (see Figure 3.7)?

Safety tip

Do not use items as hop-ups that are not designed for the purpose (e.g. milk crates, stools or chairs). They are usually not very sturdy and can't take the weight of someone standing on them, which may result in falls and injury

Definition

Carded scaffolder – someone who holds a recognised certificate showing competence in scaffold erection

Did you know?

It took 14 years of experimentation to finally settle on 48 mm as the diameter of most tubular scaffolding poles

Remember

If you have any doubts about the safety of scaffolding, report them. You could very well prevent serious injury or even someone's death

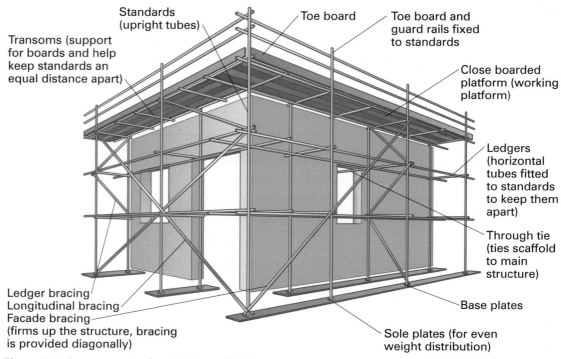

Standards (upright tubes)

Transoms (support for boards and help keep standards an equal distance apart)

Toe board

Toe board and guard rails fixed to standards

Close boarded platform (working platform)

Ledgers (horizontal tubes fitted to standards to keep them apart)

Through tie (ties scaffold to main structure)

Base plates

Ledger bracing
Longitudinal bracing
Facade bracing
(firms up the structure, bracing is provided diagonally)

Sole plates (for even weight distribution)

Figure 3.6 Components of a tubular scaffolding structure

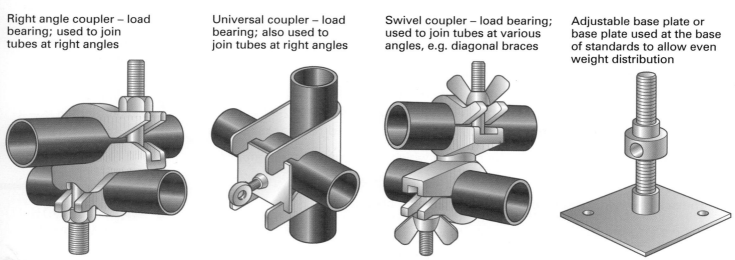

Right angle coupler – load bearing; used to join tubes at right angles

Universal coupler – load bearing; also used to join tubes at right angles

Swivel coupler – load bearing; used to join tubes at various angles, e.g. diagonal braces

Adjustable base plate or base plate used at the base of standards to allow even weight distribution

Figure 3.7 Types of scaffold fittings

Mobile tower scaffolds

Mobile tower scaffolds are so called because they can be moved around without being dismantled. Lockable wheels make this possible and they are used extensively throughout the construction industry by many different trades. A tower can be made from either traditional steel tubes and fittings or aluminium, which is lightweight and easy to move. The aluminium type of tower is normally specially designed and is referred to as a 'proprietary tower'.

Low towers

These are a smaller version of the standard mobile tower scaffold and are designed specifically for use by one person. They have a recommended working height of no more than 2.5 m and a safe working load of 150 kg. They are lightweight and easily transported and stored.

These towers require no assembly other than the locking into place of the platform and handrails. However, you still require training before you use one and you must ensure that the manufacturer's instructions are followed when setting up and working from this type of platform.

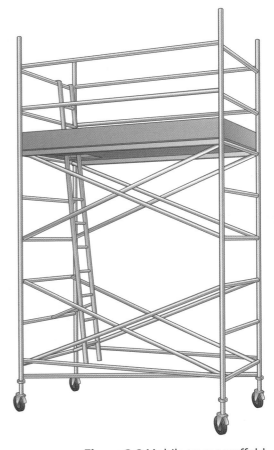

Figure 3.8 Mobile tower scaffold

Figure 3.9 Low tower scaffold

Erecting a tower scaffold

It is essential that tower scaffolds are situated on a firm and level base. The stability of any tower depends on the height in relation the size of the base:

- For use inside a building, the height should be no more than three-and-a-half times the smallest base length.

- For outside use, the height should be no more than three times the smallest base length.

The height of a tower can be increased provided the area of the base is increased **proportionately**. The base area can be increased by fitting outriggers to each corner of the tower.

For mobile towers, the wheels must be in the locked position whilst they are in use and unlocked only when they are being repositioned.

There are several important points you should observe when working from a scaffold tower:

- Any working platform above 2 m high must be fitted with guardrails and toe boards. Guard rails may also be required at heights of less than 2 m if there is a risk of falling onto potential hazards below, for example reinforcing rods. Guardrails must be fitted at a minimum height of 950 mm.

- If guardrails and toe boards are needed, they must be positioned on all four sides of the platform.

- Any tower higher than 9 m must be secured to the structure.

- Towers must not exceed 12 m in height unless they have been specifically designed for that purpose.

- The working platform of any tower must be fully boarded and be at least 600 mm wide.

- If the working platform is to be used for materials then the minimum width must be 800 mm.

- All towers must have their own access and this should be by an internal ladder.

Fall protection

With any task that involves working at height, the main danger to workers is falling. Although scaffolding, etc. should have edge protection to prevent falls, there are certain tasks where edge protection or scaffolding simply cannot be used.

In these instances some form of fall protection must be in place to prevent the worker falling, keep the fall distance to a minimum or ensure the landing point is cushioned.

There are various different types of fall protection available but the most common used are:

- harness and lanyards
- safety netting
- air bags.

Harness and lanyards

Harness and lanyards are a type of fall-arrest system, which means that, in the event of a slip or fall, the worker will only fall a few feet at most.

The system works with a harness that is attached to the worker and a lanyard attached to a secure beam/eyebolt. If the worker slips, then they will only fall as far as the length of cord/lanyard and will be left hanging, rather than falling to the ground.

A harness and lanyard can prevent a worker from falling to the ground

Safety netting

Safety netting is also a type of fall-arrest system but is used mainly on the top floor where there is no higher point to attach a lanyard.

Primarily used when decking roofing, the nets are attached to the joists/beams and are used to catch any worker who may slip or fall. Safety netting is also used on completed buildings where there is a fragile roof.

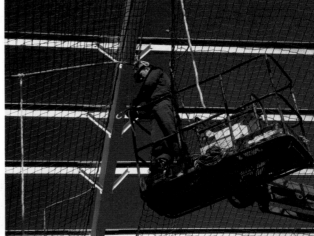

Safety netting is used when working at the highest point

Safety netting can be used under fragile roofs

Air bags

An airbag safety system is a form of soft fall-arrest and is comprised of interlinked modular air mattresses. The modules are connected by push connectors and/or flexible couplings and are inflated by a pump-driven fan, which can be electric, petrol, or butane gas powered. As the individual airbags fill with low-pressure air, they expand together to form a continuous protective safety surface, giving a cushioned soft fall and preventing serious injury.

The system must be kept inflated and, if it is run on petrol or gas, should be checked regularly to ensure that it is still functioning. This system is ideal for short fall jobs, but should not be used where a large fall could occur.

FAQ

Am I protected from electrocution if I am working on a wooden stepladder?

No. If you are working near a live current on a wooden stepladder, if any metal parts of the ladder, such as the tie rods, come into contact with the current, they will conduct the flow of electricity and you may be electrocuted. Take every precaution possible in order to avoid the risk of electrocution – the simplest precaution is turning off the electricity supply.

What determines the type of scaffolding used on a job?

As you will have read in this chapter, only a carded scaffolder is allowed to erect or alter scaffolding. They will select the scaffolding to be used according to the ground condition at the site, whether or not people will be working on the scaffolding, the types of materials and equipment that will be used on the scaffolding and the height to which access will be needed.

On the job: Attending to fascia boards

Pete has been asked by a client to take a look at all the fascia boards on a two-storey building. Depending on the condition of the fascia boards, they will need either repairing or replacing. The job will probably take Pete between two and six hours, depending on what he has to do.

What types of scaffolding do you think might be suitable for Pete's job? Can you think of anything Pete will have to consider while he prepares for and carries out this task? Think about things like egress and exit points, whether or not the area is closed off to the public and how long Pete will be working at height etc.

Knowledge check

1. Name four different methods of gaining height while working.

2. What must be done before any work at height is carried out?

3. What are your three health and safety duties when working at height?

4. As a rule, what is the maximum time you should work from a ladder or stepladder?

5. How should a wooden stepladder be checked before use?

6. When storing a wooden pole ladder, why does it need to be evenly supported along its length?

7. Explain the 1:4 (or 75°) ratio rule which should be observed when erecting a ladder.

8. When should a trestle platform be used?

9. What two types of board can be used as a platform with a trestle frame?

10. Why should you only use a specially designed hop-up?

11. There are two types of tubular scaffolding – what are they and how do they differ?

12. What are the eight questions you should ask yourself before using scaffolding?

13. In order to increase the height of a tower scaffold, what else has to be increased and by how much?

14. How high should scaffold guardrails be?

15. What is the only way you should access scaffolding?

16. When can safety netting be used?

17. How does an airbag safety system work?

chapter 4

Principles of building

OVERVIEW

Whatever type of building is being constructed there are certain principles/elements that must be included, for example a block of flats and a warehouse will all have foundations, a roof, etc.

In Chapter 1, you learned about different types of building. In this chapter, you will have a more in-depth look at the elements behind the main principles of building work.

This chapter will cover the following topics:

- Structural loading
- Substructure
- Superstructure
- Primary elements
- Secondary elements
- Finishing elements
- Services.

These topics can be found in the following modules:

CC 2003K   CC 2003S

This chapter will only look briefly at the components contained within buildings. For more detailed information on carpentry components, check the relevant chapter in this book For all other components, check the relevant book from Heinemann's Carillion Construction series.

Structural loading

The main parts of a building that are in place to carry a load are said to be in a constant state of **stress**.

There are three main types of stress:

Definition

stress - a body that has a constant force or system of forces exerted upon it resulting in strain or deformation

- Tension pulls or stretches a material and can have a lengthening effect.

- Compression squeezes the material and can have a shortening effect.

- Shear occurs when one part of a component slips or slides over another causing a slicing effect.

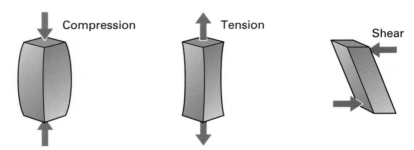

Figure 4.1 The three types of stress

To cause one of these types of stress a component or member must be under the strain of a load. Within construction there are two main types of loading:

- Dead load – the weight of the building itself and the materials used to construct the building, covering components such as floors and roofs.

- Imposed loads – any moveable loads like furniture as well as natural forces such as wind, rain and snow.

To cope with the loads that a building must withstand there are load-bearing structural members strategically placed throughout the building.

There are three main types of load bearing members:

- Horizontal members – One of the most common type of horizontal members is a floor joist, which carries the load and transfers it back to its point of support. The horizontal member, when under loading, can bend and be in all three types of stress, with the top in compression, the bottom in tension and the ends in shear.

Remember

Where you are in the country will determine what materials you use for constructing. For example, some places with a lot of snowfall will require stronger structures to deal with the extra load from the snow

The bending can be contained by using correctly stress-graded materials or by adding a load-bearing wall to support the floor.

- Vertical members – Any walls or columns that are in place to transfer the loads from above (including horizontal members) down to the substructure and foundations have vertical members. Vertical members are usually in a compression state.

- Bracing members – Bracing members are usually fitted diagonally to form a triangle which stiffens the structure. Bracing members can be found in roofs and even on scaffolding. Bracing is usually in compression or tension.

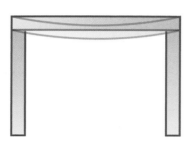

Figure 4.2 Horizontal structural members

Figure 4.3 Vertical structural members

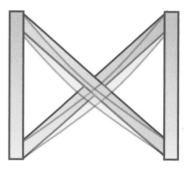

Figure 4.4 Bracing structural members

Substructure

All buildings will start with the substructure – that is, all of the structure below ground and up to and including the damp proof course (DPC). The purpose of the substructure is to receive the loads from the main building (superstructure) and transfer them safely down to a suitable load-bearing layer of ground.

The main part of the substructure is the foundations. When a building is at the planning stage, the entire area – including the soil – will be surveyed to check what depth, width and size of foundation will be required. This is vital: the wrong foundation could lead to the building subsiding or even collapsing.

All buildings have a substructure

Did you know?

During the surveying of the soil, the density and strength of the soil are tested and laboratory tests check for harmful chemicals contained within the soil

The main type of foundation is a strip foundation. Depending on the survey reports and the type of building, one of four types of foundation will usually be used.

- Narrow strip foundation – the most common foundation used for most domestic dwellings and low-rise structures.

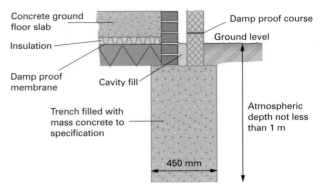

Figure 4.5 Narrow strip foundation

- Wide strip foundation – used for heavier structures or where weak soil is found.

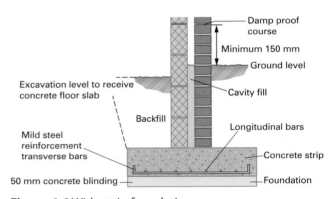

Figure 4.6 Wide strip foundation

- Raft foundation – used where very poor soil is found. This is basically a slab of concrete that is thicker around the edges.

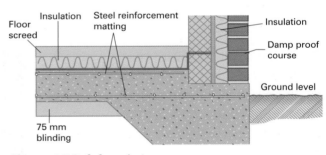

Figure 4.7 Raft foundation

- Pad foundation – where pads are placed at strategic points, with concrete beams placed across the pad to spread the load.

Once the substructure is in place, the building is then built on top of it.

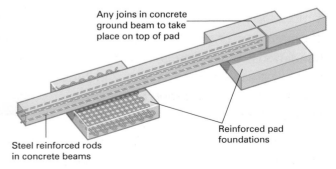

Figure 4.8 Pad foundation

Superstructure

The superstructure covers everything above the substructure, from walls to floors to roofing. The purpose of the superstructure is to enclose and divide space, as well as spread loads safely into the substructure.

Within the superstructure, you will find the primary, secondary and finishing elements, as well as the services.

Primary elements

The primary elements are the main supporting, enclosing and protecting elements of the superstructure. They divide space and provide floor-to-floor access.

The main primary elements are:

- walls
- floors
- roofs
- stairs.

Walls

There are two main types of wall within a building: external and internal.

External walls

External walls come in a variety of styles, but the most common is cavity walling. Cavity walling is simply two brick walls built parallel to each other, with a gap between acting as the cavity. The cavity wall acts as a barrier to weather, with the outer leaf preventing rain and wind penetrating the inner leaf. The cavity is usually filled with insulation to prevent heat loss.

Timber kit houses are becoming more and more common as they can be erected to a wind and watertight stage within a few days. The principle is similar to a cavity wall: the inner skin is a timber frame clad in timber sheet material, covered in a breathable membrane to prevent water and moisture penetrating the timber. The outer skin is usually face brickwork.

There are also other types of exterior walling, such as solid stone or log cabin style. Industrial buildings may have steel walls clad in sheet metal.

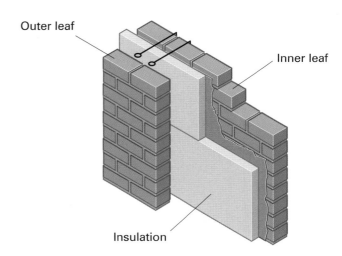

Figure 4.9 A cavity wall

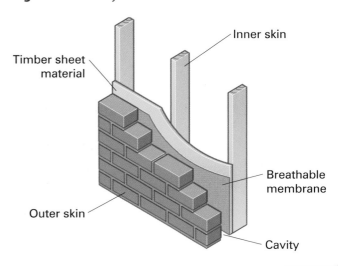

Figure 4.10 A timber and cavity wall

Internal walls

Internal walls are either load bearing – meaning they support any upper floors or roof – or are in place to divide rooms into shapes and sizes.

Internal walls also come in a variety of styles. Here is a list of the most common types.

- Solid block walls – simple block work, either covered with plasterboard or plastered over to give a smooth finish, to which wallpaper or paint is applied. Solid block walls offer low thermal and sound insulation qualities but advances in technology and materials means that blocks such as thermalite blocks can give better sound and heat insulation.

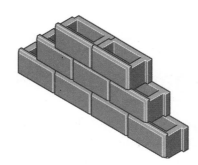

Figure 4.11 Solid block wall

- Solid brick walling – usually made with face brickwork as a decorative finish. It is unusual for all walls within a house to be made from brickwork.

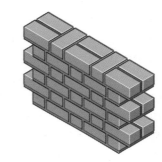

Figure 4.12 Solid brick wall

- Timber stud walling – more common in timber kit houses and newer buildings. Timber stud walling is also preferred when dividing an existing room, as it is quicker to erect. Clad in plasterboard and plastered to a smooth finish, timber stud partitions can be made more fire resistant and sound/thermal qualities can be improved with the addition of insulation or different types of plasterboard. Another benefit of timber stud walling is that timber noggins can be placed within the stud to give additional fixings for components such as radiators or wall units. Timber stud walling can also be load bearing, in which case thicker timbers are used.

Figure 4.13 Timber stud wall

- Metal stud walling – similar to timber stud, except metal studs are used and the plasterboard is screwed to the studding.

Figure 4.14 Metal stud wall

- Grounds lats – timber battens that are fixed to a concrete or stone wall to provide a flat surface, to which plasterboard is attached and a plaster finish applied.

Floors

There are two main types of floor: ground and upper.

Ground floors

There are a few main types of ground floor. These are the ones you will most often come across.

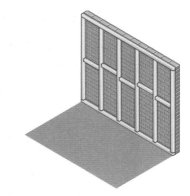

Figure 4.15 Ground lats

- Suspended timber floor – a floor where timber joists are used to span the floor. The size of floor span determines the depth and thickness of the timbers used. The joists are either built into the inner skin of brickwork, sat upon small walls (dwarf/sleeper wall), or some form of joist hanger is used. The joists should span the shortest distance and sometimes dwarf/sleeper walls are built in the middle of the span to give extra support or to go underneath load-bearing walls. The top of the floor is decked with a suitable material (usually chipboard or solid pine tongue and groove boards). As the floor is suspended, usually with crawl spaces underneath, it is vital to have air bricks fitted, allowing air to flow under the floor, preventing high moisture content and timber rot.

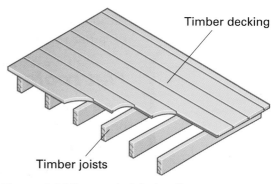

Figure 4.16 Suspended timber floor

- Solid concrete floor – concrete floors are more durable and are constructed on a sub-base incorporating hardcore, damp proof membranes and insulation. The depth of the hardcore and concrete will depend on the building and will be set by the *Building Regulations* and the local authority. Underfloor heating can be incorporated into a solid concrete floor. Great care must be taken when finishing the floor to ensure it is even and level.

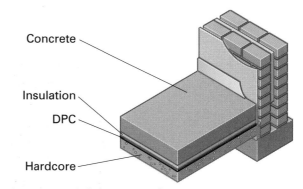

Figure 4.17 Solid concrete floor

- Floating floor – basic timber floor constructions that are laid on a solid concrete floor. The timbers are laid in a similar way to joists, though they are usually 50 mm thick maximum as there is no need for support. The timbers are laid on the floor at predetermined centres, and are not fixed to the concrete base (hence floating floor); the decking is then fixed on the timbers. Insulation or underfloor heating can be placed between the timbers to enhance the thermal and acoustic properties.

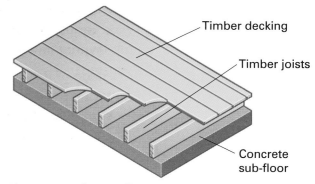

Figure 4.18 Floating floor

Upper floors

Again, solid concrete slabs can be used in larger buildings, but the most common type of upper floor is the suspended timber floor. As before, the joists are either built into the inner skin of brickwork or supported on some form of joist hanger. Spanning the shortest distance, with load-bearing walls acting as supports, it is vital that **regularised joists** are used as a level floor and ceiling are required. The tops of the joists are again decked out, with the underside being clad in plasterboard and insulation placed between the joists to help with thermal and sound properties.

Roofs

Although there are several different types of roofing, all roofs will either technically be a flat roof or a pitched roof.

Flat roofs

A flat roof is a roof with a pitch of 10° or less. The pitch is usually achieved through laying the joists at a pitch, or by using **firring pieces**.

The main construction method for a flat roof is similar to that for a suspended timber floor, with the edges of the joists being supported either via a hanger or built into the brickwork, or even a combination of both. Once the joists are laid and firring pieces are fitted (if required), insulation and a vapour barrier are put in place. The roof is then decked on top and usually plasterboarded on the underside. The decking on a flat roof must be waterproof, and can be made from a wide variety of materials, including fibreglass or bitumen-covered boarding with felt layered on it.

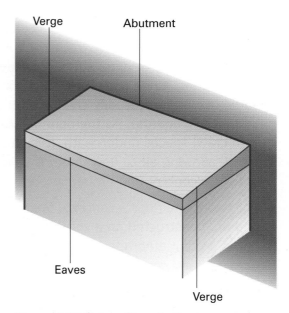

Figure 4.19 Flat roof terminology

Drainage of flat roofs is vital. The edge where the fall leads to must have suitable guttering to allow rainwater to run away, and not down the face of the wall.

Pitched roofs

There are several types of pitch roof, from the basic gable roof to more complex roofs such as mansard roofs. Whichever type of roof is being fitted to a building, it will most likely be constructed in one of the following ways.

● Prefabricated truss roof – as the name implies, this is a roof that has prefabricated members called trusses. Trusses are used to spread the load of the roof and to give it the required shape. Trusses are factory-made, delivered to site and lifted into place, usually by a crane. They are also easy and quick to fit: either they are nailed to a wall plate or held in place by truss clips. Once fitted, bracing is attached to keep the trusses level and secure from wind. Felt is then fixed to the trusses and tiles or slate are used to keep the roof and dwelling waterproof.

Figure 4.20 Duo pitch roof with gable ends

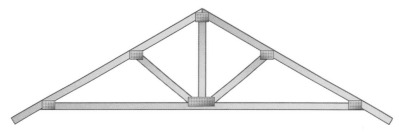

Figure 4.21 Prefabricated wooden roof truss

● Traditional/cut roof – an alternative to trusses, the cut roof uses loose timbers that are cut in-situ to give the roof its shape and spread the relevant load. More time-consuming and difficult to fit than trusses, the cut roof uses rafters that are individually cut and fixed in place, with two rafters forming a sort of truss. Once the rafters are all fixed, the roof is finished with felt and tiles or slate.

Metal trusses can also be used for industrial or more complex buildings.

To finish a roof where it meets the exterior wall (eaves), you must fix a vertical timber board (fascia) and a

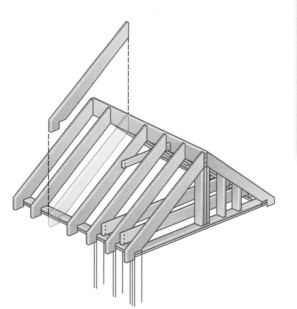

Figure 4.22 Individually cut rafters

horizontal board (soffit) to the foot of the rafters/trusses. The fascia and soffit are used to close off the roof space from insects and birds.

Did you know?

Due to the fact that heat rises, the majority of heat loss that occurs is through a building's roof. Insulation such as mineral wool or polystyrene must be fitted to roof spaces and ideally any intermediate floors

Ventilators are attached to the soffits to allow air into the roof space, preventing rot, and guttering is attached to the fascia board to channel the rainwater into a drain.

Stairs

Stairs are used to provide access between different floors of a dwelling or to gain access to a higher/lower area. Stairs are made up of a number of steps, and each continuous set of steps running in the same direction is known as a flight. Steps are made of vertical boards called risers and horizontal boards called treads.

There are various types of stair, ranging from spiral staircases(often fitted where there is a lack of space) to multi-flight staircases, such as dog-leg or half-turn stairs.

Stairs are strictly governed by the *Building Regulations*, and there are numerous requirements that must be adhered to when constructing and installing them.

Stairs are generally made from three types of material:

- timber – the most common type of stair, used widely in almost all buildings

- in-situ-cast concrete – a wooden frame is constructed around the stairwell and concrete is poured into the frame, forming the staircase

- pre-cast concrete – concrete cast in large moulds to form the staircase, usually found in flat stairwells and other areas of heavy use

- steel – usually found on the exterior of buildings in the form of fire escapes, etc.

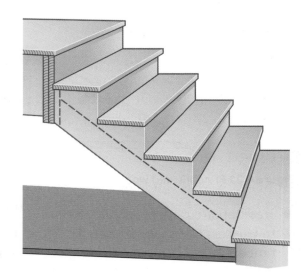

Figure 4.23 A simple staircase

Secondary elements

The secondary elements are not essential to the building's strength or structure, but provide a particular function, such as completing openings in walls, etc.

The main secondary elements are:

- frames and linings
- doors
- windows
- architrave and skirting.

Frames and linings

Frames and linings are fitted around openings, and are used to allow components such as windows and doors to be fitted. The frame or lining is fitted to the wall and usually finished flush with the walls; the joint between the frame or lining and the wall is covered by the architrave.

Doors

The main purpose of a door is to provide access from one room to another, and to allow a space to be closed off for security/thermal/sound reasons.

Doors come in many varieties, shapes and sizes; the type you need will be determined by where the door is being fitted and what for. Exterior doors are generally thicker and are fitted with more ironmongery such as letter plates and locks. Some interior doors will have locks fitted as well, such as bathrooms or doors that need to be secure.

Doors can be solid timber or have glass in them and may be graded for fire resistance.

Windows

Windows are fitted to allow natural light to enter the building with minimal loss of heat. Again, windows come in a variety of shapes and styles. Glass that is fitted in the window can be decorative and heat-loss resistant.

Architraves and skirting

Architraves are decorative mouldings used to hide the gap between frames and the wall finish. Skirting is moulding that covers the gap between the floor and base of a wall. These mouldings come in a variety of **profiles** such as torus and ogee.

Other mouldings can also be used, such as picture and dado rails.

Definition

Profile – the shape of the moulding when you cut through it

Finishing elements

Finishing elements are the final surfaces of an element, which can be functional or decorative.

The main finishing elements are:

- plaster
- render
- paint
- wallpaper.

Plaster

Plaster can be used on a variety of wall surfaces to give a smooth and even finish. The plaster comes in powder form, usually bagged, and is mixed with water until it reaches a consistency that allows it to be applied to the surface and trowelled smooth. Ready-mix plaster is also available, but is more expensive, especially when a lot of surfaces have to be plastered.

These are the main surfaces to which plaster is applied.

- Brick/block work – prior to application, a bonding agent must first be applied to the wall (usually a coat of watered down PVA), to help the plaster adhere to the surface. Usually a first coat of bonding plaster is applied to the wall to give it a level and flat surface; when this is dry, a second, finish coat is applied. As the finish coat is drying, the plasterer will work on the wall, smoothing it out until it is as smooth as glass.

- Plasterboard – as plasterboard is a flat surface to begin with, a bonding coat is rarely used. Generally, the plasterboard is fixed with the back face (the face with the writing) exposed to give better adhesion. Whether it is a wall or a ceiling, the plasterer will again work the finish coat to a very smooth surface.

- Lath and plaster – this is usually found in old properties. The laths are thin strips of wood, which are fixed to the wall with small gaps between to give the plaster a **key**. Once the laths are fixed the plasterer will apply bonding and finish coats the same as before.

Plasterboard with a tapered edge can also be fixed to the walls. In this case, instead of plastering the entire wall, the plasterer will simply fill the nail/screw holes, fit tape where the plasterboard joins are, and fill only the joints. Pre-mixed plaster is usually used for this; when it is dry, a light sanding is required to give a smooth finish. This method is preferred in newer buildings, especially timber kit houses.

Lath and plaster

Not all walls are plastered smooth, as some clients may require a rough or patterned finish. Although not technically a plaster, Artex™ is often used to give decorative finishes, especially on ceilings.

Render

Render is similar to plaster in that it is trowelled on to brick or block work to give a finish. Applied to external walls, the render must be waterproof to prevent damage to the walls. Different finishes are available, from stippling to patterned.

Plaster being applied and trowelled

Paint

Paint is applied to various surfaces and is available in many different types to suit the job they are required for. Paint is applied for a variety of reasons, the most common being to:

- protect – steel can be prevented from corroding due to rust, and wood can be prevented from rotting due to moisture and insect attack

- decorate – the appearance of a surface can be improved or given a special effect (for example marbling, wood graining)

- sanitise – a surface can be made more hygienic with the application of a surface coating, preventing penetration and accumulation of germs and dirt, and allowing easier cleaning.

Paint is either water-based or solvent-based. When a paint is water-based, it means that the main liquid part of the paint is water; with a solvent-based paint, a chemical has been used instead of water to dissolve the other components of the paint.

Water-based paint is generally used on walls and ceilings, while solvent-based paint is used on timber mouldings, doors, metals, etc.

There are other surface finishes besides paint such as varnish (used on wood), masonry paint (used on exterior walls) and preservatives, which are used to protect wood from weather and insect attack.

Wallpaper

Wallpapers are used to decorate walls; thicker wallpapers can also hide minor defects.

Basic wallpapers are made from either wood pulp or vinyl.

Wood-pulp papers can be used as preparatory papers or finish papers. Preparatory papers are usually painted with emulsion to provide a finish, or they can be used as a base underneath finish papers. Types of wood-pulp paper include plain, coloured and reinforced lining paper as well as wood chip.

Vinyl wallpaper is a hard-wearing wallpaper made from a PVC layer attached to a pulp backing paper. Types of vinyl paper include patterned, sculptured or blown vinyl.

Wallpaper is hung on a wall using a paste. Not all pastes have the same strength, so make sure you choose the correct paste for the type of paper you are using.

Services

The services are specialist components within a building ranging from running water to electricity.

The main services in a standard house are:

- Electrical – covers all electrical components within the building from lights to sockets. Electrical installation and maintenance work must be undertaken by a fully trained specialist as electricity can kill.

- Mechanical – covers things such as lifts. As with electrical services, work on mechanical services should only be undertaken by a specialist.

- Plumbing – can cover running water as well as gas, but only if the plumber has been recognised and qualified as a gas installation expert.

Remember

All service work must be carried out by a fully trained and competent person

FAQ

How do I know if the materials I am using are strong enough to carry the load?

On the specification you will find details of the sizes and type of materials to be used.

Do I have to fix battens to a wall before I plasterboard it?

No – the method called dot and dab can be used where plaster is dabbed onto the back of the plasterboard and then pushed onto the wall.

On the job: Identifying load-bearing walls

Jay and Ella are out pricing up a job to place a doorway into a solid wall. Jay says that it will only take a few hours to knock through the brickwork and put a frame in. Ella is not so sure: she thinks the wall may be load bearing. How can they check the wall is load bearing? And what should be done if it is?

Knowledge check

1. State the four main principles of building.

2. List the three main types of stress.

3. What is the main purpose of the substructure?

4. List three different types of foundation.

5. What are the four main primary elements?

6. Give a brief description of external walling.

7. What is the difference between a truss roof and a cut roof?

8. What are the four main secondary elements? Why are they secondary?

9. List four main finishing elements.

10. Give a brief description of the process involved with lath and plaster.

11. What is vinyl wallpaper made from?

12. Give three reasons why paint is used.

13. What are the three main services?

chapter 5

Handling and storage of materials

OVERVIEW

Work in any building trade involves handling and storing materials, tools and equipment, sometimes under difficult conditions. The cost of injuries from poor or careless handling practice is enormous and careless storage risks damage, loss and theft of materials and equipment. This can cause delays and unnecessary cost to contractors. These risks can be minimised by following a few simple guidelines and applying a level of common sense when moving and storing materials and equipment.

This chapter will cover the following topics:

- Safe handling

- Wood and sheet materials

- Joinery components (doors, frames, units)

- Ironmongery

- Adhesives

- Bricks, blocks and other bricklaying materials

- Aggregates and bagged materials (sand, cement, plaster)

- Paint and decorating equipment

- Chemicals

- Highly flammable liquids

- Glass.

These topics can be found in the following modules:

CC 1001K   CC 1001S

Safe handling

Chapter 2 *Health and safety* explains safe **manual handling** methods in more detail – see pages 43–45. When handling any materials or equipment, always think about the health and safety issues involved and remember the manual handling practices explained to you during your induction.

You are not expected to remember everything but basic common sense will help you to work safely.

- Always wear your safety helmet and boots at work.
- Wear gloves and ear defenders when necessary.
- Keep your work areas free from debris and materials, tools and equipment not being used.
- Wash your hands before eating.
- Use barrier cream before starting work.
- Always use correct lifting techniques.

Ensure you follow instructions given to you at all times when moving any materials or equipment. The main points to remember are:

- always try to avoid manual handling (or use mechanical means to aid the process)
- always assess the situation first to establish the best method of handling the load
- always reduce any risks as much as possible (for example, split a very heavy load, move obstacles from your path before lifting)
- tell others around you what you are doing
- if you need help with a load, get it. Do not try to lift something beyond what you can manage.

Wood and sheet materials

There are various types of wood and sheet materials available, but the most common are as follows.

Carcassing timber

Carcassing timber is wood used for non-load-bearing jobs such as ceiling and floorboard supports, stud wall partitions and other types of framework. It should normally be stored outside under a covered framework. It should be placed on timber bearers clear of the ground. The ground should be free of vegetation and ideally covered over with concrete. This reduces the risk of absorption of ground moisture, which can damage the timber and cause wet rot. Piling sticks or cross-bearers should be placed between each layer of timber, about 600 mm apart, to provide support and allow air circulation. Tarpaulins or plastic covers can be used to protect the timber from the elements, however care must be taken to allow air to flow freely through the stack. See Figure 5.1.

Remember

The storage racks used to store wood must take account of the weight of the load. Access to the materials being stored is another important consideration

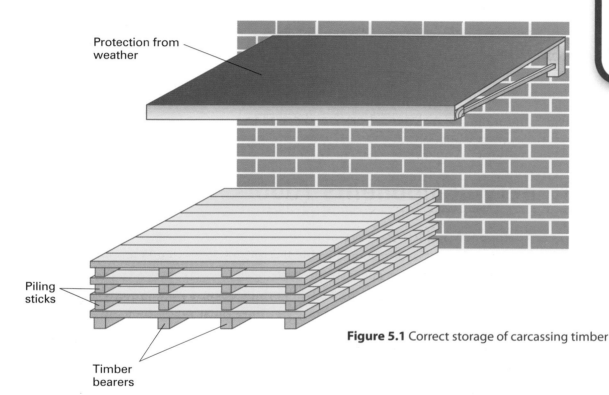

Protection from weather

Piling sticks

Timber bearers

Figure 5.1 Correct storage of carcassing timber

Joinery grade and hardwoods

These timbers should be stored under full cover wherever possible, preferably in a storage shed. Good ventilation is needed to avoid build-up of moisture through absorption. This type of timber should also be stored on bearers on a well-prepared base.

Plywood and other sheet materials

All sheet materials should be stored in a dry, well-ventilated environment. Specialised covers are readily available to give added protection for most sheet materials, helping to prevent condensation caused when non-specialised types of sheeting are used.

Find out

What are the PPE requirements when moving sheet materials?

Storage of sheet materials

Sheet materials should be stacked flat on timber cross-bearers, spaced close enough together to prevent sagging. Alternatively, where space is limited, sheet materials can be stored on edge in purpose-made racks which allow the sheets to rest against the backboard. There should be sufficient space around the plywood for easy loading and removal. The rack should be designed to allow sheets to be removed from either the front or the ends.

Leaning sheets against walls is not recommended, as this makes them bow, which is difficult to correct.

For sheet materials with faces or decorative sides, the face sides should be placed against each other to minimise the risk of damage due to friction of the sheets when they are moved. Different sizes, grades and qualities of sheet materials should be kept separate with off-cuts stacked separately from the main stack.

Sheet materials are awkward, heavy and prone to damage so extra care is essential when transporting them. Always ensure that the correct PPE is worn.

Joinery components (doors, frames, units)

Joinery components, such as doors or kitchen units, must be stored safely and securely to prevent damage. Doors, windows, frames, etc. should ideally be stored flat on timber bearers under cover, to protect them from exposure to the weather. Where space is limited, they can be stored upright using a rack system (similar to the way sheet materials are stored); however, they must never be leant against a wall, as this will bow the door/frame and make it very hard to fit.

Wall and floor units – whether they be kitchen, bedroom or bathroom units – must be stacked on a flat surface, and no more than two units high. As units can be made from porous materials such as chipboard, it is vital they are stored inside, preferably in the room where they are to be fitted to avoid double handling. Protective sheeting should be used to cover units to prevent damage or staining from paints, etc.

Ideally all components and timber products such as architrave should be stored in the room where they are to be fitted: this will allow them to acclimatise to the room and prevent shrinkage or distortion after being fitted. This process is known as 'second seasoning'.

Safety tip

Due to the size, shape and weight of sheet materials, always get help to lift and carry them. If possible, use a purpose-made plywood trolley which will transport the load for you

Did you know?

Wood and wood-based materials are susceptible to rot if the moisture content is too high, to insect attack and to many other defects such as bowing or warping. Proper seasoning and chemical sprays can prevent any defects from occurring

Figure 5.2 Doors should be stacked on a flat surface

Ironmongery

Ironmongery includes not just fittings and fixtures made of iron, but also hardware made from other materials including brass, chrome, porcelain and glass. Items you might come across include door handles, hooks, locks, hinges, window fittings, screws and bolts.

Door furniture

Door furniture such as locks, bolts, letter boxes, knockers and handles etc. are 'desirable' items, which means that they are very likely to be stolen unless stored securely. A store person is usually responsible for the storage and distribution of such items and keeps a check on how many are given out and to whom.

In addition to being stored securely, door furniture should also be kept in separate compartments of a racking storage system or at least on shelving. Where possible all door furniture should be retained in the manufacturer's packaging until needed. This prevents damage and loss of components such as screws and keys. Large heavy items should be stored on lower shelves to avoid unnecessary lifting.

Examples of door furniture

Fixings

Each type of fixing is designed for a specific purpose and includes items such as nails, screws, pins, bolts, washers, rivets and plugs. As with door furniture, fixings tend to disappear if their storage is not supervised and controlled by a store person.

Fixings must be stored appropriately to keep them in good condition and to make them easy to find. Where possible, they should be kept in bags or boxes clearly marked with their size and type. Storing them in separate compartments is the most convenient method of storage. This enables easy and fast selection when required and prevents the wrong fixing being used. Time taken to sort out different types and sizes of fixings that have become mixed up is a waste of your time and your employer's time.

Different types of fixings should be stored separately

Safety tip

Carelessly discarded fixings, such as nails, can be costly, but can also create health and safety hazards

Remember

Broken fixings should be disposed of carefully

Adhesives

Adhesives are substances used to bond (stick) surfaces together. Because of their chemical nature, there are a number of potentially serious risks connected with adhesives if they are not stored, used and handled correctly.

All adhesives should be stored and used in line with the manufacturer's instructions. This usually involves storing them on shelving, with labels facing outward, in a safe, secure area (preferably a lockable store room). It is important to keep the labels facing outwards so that the correct adhesive can be selected.

The level of risk associated with adhesive use is dependent on the type of adhesive. Some of the risks include:

- explosion
- poisoning
- skin irritation
- disease.

As explained in Chapter 2 *Health and safety*, these types of material are closely controlled by COSHH, which aims to minimise the risks involved with their storage and use.

All adhesives have a recommended **shelf life**. This must be taken into account when storing to ensure the oldest stock is stored at the front and used first. Remember to refer to the manufacturer's guidelines as to how long the adhesive will remain fit for purpose once opened. Adhesives can be negatively affected by poor storage, including loss in adhesive strength and extended setting time.

Adhesives should be stored according to the manufacturer's instructions

Definition

Shelf life – how long something will remain fit for its purpose while being stored

Bricks, blocks and other bricklaying materials

Storage of bricks

Safety tip

Take care and stand well clear of a crane used for offloading bricks on delivery

Most bricks delivered to sites are now prepacked and banded using either plastic or metal bands to stop the bricks from separating until ready for use. The edges are also protected by plastic strips to help stop damage during moving, usually by forklift or crane. They are then usually covered in shrink-wrapped plastic to protect them from the elements.

On arrival to site they should be stored on level ground and stacked no more than two packs high, to prevent overreaching or collapse, which could result in injury to workers. They should be stored close to where they are required so further movement is kept to a minimum. On large sites they may be stored further away and moved by telescopic lifting vehicles to the position required for use.

Great care should be taken when using the bricks from the packs as, once the packaging is cut, the bricks can collapse causing injury and damage to the bricks, especially on uneven ground. Bricks should be taken from a minimum of three packs and mixed to stop changes in colour, as the position of the bricks during the kiln process can cause slight colour differences: the nearer the centre of the kiln, the lighter the colour; the nearer the edge of the kiln, the darker the colour as the heat is stronger. If the bricks are not mixed, you could get sections of brickwork in slightly different shades; this is called banding and in most cases is visible to the most inexperienced eye.

If bricks are unloaded by hand they should be stacked on edge in rows, on firm, level and well-drained ground. The ends of the stacks should be bonded and no higher than 1.8 m. To protect the bricks from rain and frost, all stacks should be covered with a tarpaulin or polythene sheets.

Blocks

Blocks are made from concrete, which may be dense or lightweight. Lightweight blocks could be made from a fine aggregate that contains lots of air bubbles. The storage of blocks is the same as for bricks.

Paving slabs

Paving slabs are made from concrete or stone and are available in a variety of sizes, shapes and colours. They are used for pavements and patios, with some slabs given a textured top to improve appearance.

Storage of paving slabs

Paving slabs are normally delivered to sites by lorry and crane offloaded, some in wooden crates covered with shrink-wrapped plastic, or banded and covered on pallets. They should not be stacked more than two packs high for safety reasons and to prevent damage to the slabs due to weight pressure.

Paving slabs unloaded by hand are stored outside and stacked on edge to prevent the lower ones, if stored flat, from being damaged by the weight of the stack. The stack is started by laying about 10 to 12 slabs flat with the others leaning against

Paving slabs stacked flat

Paving slabs on pallet

Did you know?

Efflorescence caused by soluble salts within the bricks can leave a stain. Brushing off the stain and coating with a suitable sealer can resolve the problem

Did you know?

The temperature must be taken into account when laying bricks and blocks as, if it is too cold, this will prevent the cement going off properly

Safety tip

When working with blocks, make sure you always wear appropriate PPE (personal protective equipment), that is boots, safety hat, gloves, goggles and face mask these. If only a small number of slabs are to be stored, they can be stored flat (since the weight will be less).

Slabs should be stored on firm, level ground with timber bearers below to prevent the edges from getting damaged if placed on a solid surface. To provide protection from rain and frost, it is advisable to keep the slabs under cover, by placing a tarpaulin or polythene sheet over the top.

Kerbs

Kerbs are concrete units laid at the edges of roads and footpaths to give straight lines or curves and retain the finished surfaces. The size of a common kerb is 100 mm wide, 300 mm high and 600 mm long. Path edgings are 50 mm wide, 150 mm high and 600 mm long.

Kerbs should be stacked on timber bearers or with overhanging ends, which provides a space for hands or lifting slings if machine lifting is to be used. When they are stacked on top of each other, the stack must not be more than three kerbs high. To protect the kerbs from rain and frost it is advisable to cover them with a tarpaulin or sheet.

Stacked kerbs

Stacked lintels

Pre-cast concrete lintels

Lintels are components placed above openings in brick and block walls to bridge the opening and support the brick or block work above. Lintels made from concrete have a steel reinforcement placed near the bottom for strength, which is why pre-cast concrete lintels will have a 'T' or 'Top' etched into the top surface. Pre-cast concrete lintels come in a variety of sizes to suit the opening size. The stacking and storage methods are the same as for kerbs.

Safety tip

It is good practice to put an intermediate flat stack in long rows to prevent rows from toppling

---

Drainage pipes

Drainage pipes are made from **vitrified** clay or plastic. Pipes made from clay may have a socket and spigot end or be plain and joined by plastic couplings. Plastic pipes are plain ended and joined by couplings, using a lubricant to help jointing.

Storage of drainage pipes and fittings

Pipes should be stored on a firm, level base, and prevented from rolling by placing wedges or stakes on either side of the stack. Do not stack pipes any higher than 1.5 m, and taper the stack towards the top.

Clay pipes with socket and spigot ends should be stored by alternating the ends on each row. They should also be stacked on shaped timber cross-bearers to prevent them from rolling.

Fittings and special shaped pipes, like bends, should be stored separately and, if possible, in a wooden crate until needed.

Stacked drainage pipes

Definition

Vitrified – a material that has been converted into a glass-like substance via exposure to high temperatures

Remember

Clay pipes are easily broken if misused, so care must be taken when handling these items

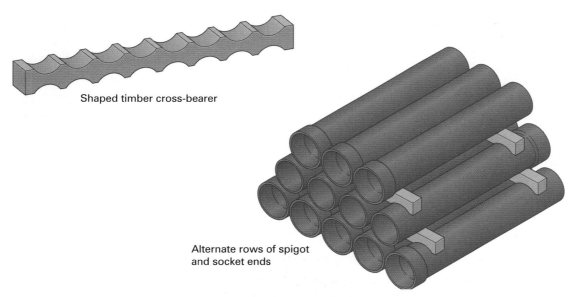

Shaped timber cross-bearer

Alternate rows of spigot and socket ends

Figure 5.3 Timber cross-bearer and pipes stacked on cross-bearer

Half round ridge tile

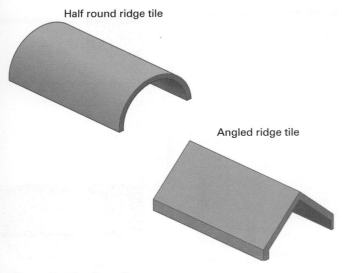

Angled ridge tile

Figure 5.4 Roofing tiles

Roofing tiles

Roofing tiles are made from either clay or concrete. They may be machine-made or handmade, and are available in a variety of shapes and colours. Many roofing tiles are able to interlock to prevent rain from entering the building. Ridge tiles are usually half round but sometimes they may be angled.

Storage of roofing tiles

Roofing tiles are stacked on edge to protect their 'nibs', and in rows on level, firm, well-drained ground. See Figure 5.4. They should not be stacked any higher than six rows high. The stack should be tapered to prevent them from toppling. The tiles at the end of the rows should be stacked flat to provide support for the rows.

Ridge tiles may be stacked on top of each other, but not any higher than ten tiles.

To protect roofing tiles from rain and frost before use, they should be covered with a tarpaulin or plastic sheeting.

Damp proof course (DPC)

A damp proof course (or DPC) and damp proof membranes are used to prevent damp from penetrating into a building. Flexible DPC may be made from polythene, bitumen or lead and is supplied in rolls of various widths for different uses.

Slate can also be used as a damp proof course – older houses often have slate but modern houses normally have polythene.

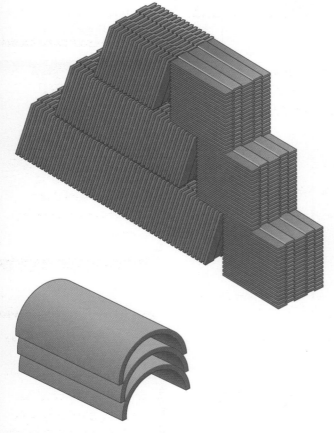

Figure 5.5 Stacks of roofing tiles

Damp proof course (DPC)

Damp proof membrane is used as a waterproof barrier over larger areas, such as under the concrete on floors etc. It is normally made of 1000-gauge polythene, and comes in large rolls, normally black or blue in colour.

Storage of rolled materials

Rolled materials, for example damp proof course or roofing felt, should be stored in a shed on a level, dry surface. Narrower rolls may be best stored on shelves but in all cases they should be stacked on end to prevent them from rolling and to reduce the possibility of them being damaged by compression. See Figure 5.5. In the case of bitumen, the layers can melt together under pressure.

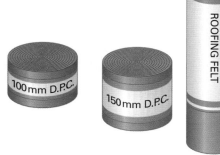

Figure 5.6 Rolled materials stored on end

Aggregates and bagged materials (sand, cement, plaster)

Aggregates

Aggregates are granules or particles that are mixed with cement and water to make mortar and concrete. Aggregates should be hard, durable and should not contain any form of plant life, or anything that can be dissolved in water.

Aggregates are classed in two groups:

- Fine aggregates are granules that pass through a 5 mm sieve.
- Coarse aggregates are particles that are retained by a 5 mm sieve.

The most commonly used fine aggregate is sand. Sand may be dug from pits and riverbeds, or dredged from the sea.

Did you know?

Sea sand contains salts, which may affect the quality of a mix. It should not be used unless it has been washed and supplied by a reputable company

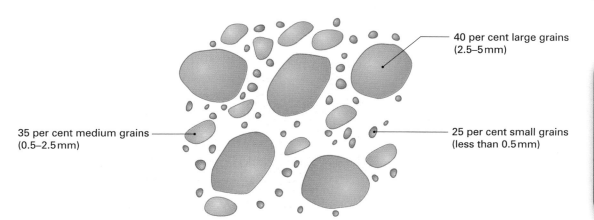

40 per cent large grains (2.5–5 mm)

35 per cent medium grains (0.5–2.5 mm)

25 per cent small grains (less than 0.5 mm)

Figure 5.7 Sand particles

Remember

Poorly graded sands, with single size particles, contain a greater volume of air and require cement to fill the spaces

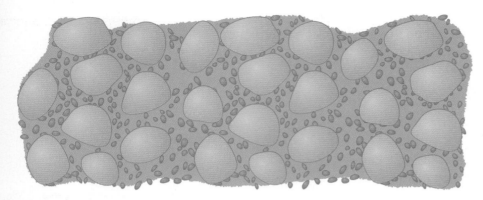

Figure 5.8 Mortar particles

Good mortar should be mixed using 'soft' or 'building' sand. It should be well graded, which means having an equal quantity of fine, medium and large grains.

Concrete should be made using 'sharp' sand, which is more angular and has a coarser feel than soft sand, which has more rounded grains.

When concreting, we also need 'coarse aggregate'. The most common coarse aggregate is usually limestone chippings, which are quarried and crushed to graded sizes, 10 mm, 20 mm or even larger.

Remember

The different sizes of aggregates should be stored separately to prevent the different aggregates getting mixed together

Figure 5.9 Concrete particles

Storage of aggregates

Aggregates are normally delivered in tipper lorries, although nowadays one-tonne bags are available and may be crane handled. The aggregates should be stored on a concrete base, with a fall to allow for any water to drain away.

In order to protect aggregates from becoming contaminated with leaves and litter it is a good idea to situate stores away from trees and cover aggregates with a tarpaulin or plastic sheets.

SAND
10mm
20mm

Figure 5.10 Bays for aggregates

Base laid to a fall for drainage of the aggregates

Bagged materials

Aggregates can be supplied as bagged materials, as can cement and plaster.

Plaster

Plaster is made from gypsum, water and cement or lime. Aggregates can also be added depending on the finish desired. Plaster provides a joint-less, smooth, easily decorated surface for internal walls and ceilings.

Gypsum plaster

This is for internal use and contains different grades of gypsum. Plasters are available depending on the background finish. Browning is usually used as an undercoat on brickwork but in most cases, a one-coat plaster is used, and on plasterboard, board finish is used.

Cement-sand plaster

This is used for external rendering, internal undercoats and waterproofing finishing coats.

Lime-sand plaster

This is mostly used as an undercoat, but may sometimes be used as a finishing coat.

Storage of cement and plaster

Both cement and plaster are usually available in 25 kg bags. The bags are made from multi-wall layers of paper with a polythene liner. Care must be taken not to puncture the bags before use. Each bag, if offloaded manually, should be stored in a ventilated, waterproof shed, on a dry floor on pallets. If offloaded by crane they should be transferred to the shed and the same storage method used.

The bags should be kept clear of the walls, and piled no higher than five bags. It is most important that the bags are used in the same order as they were delivered. This minimises the length of time that the bags are in storage, preventing the contents from setting in the bags, which would require extra materials and cause added cost to the company.

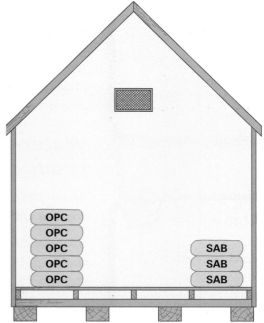

- Dry, ventilated shed

- Stock must be rotated so that old stock is used before new

- Not more than five bags high

- Clear of walls

- Off floor

Figure 5.11 Storage of cement and plaster bags in a shed

Paint and decorating equipment

Liquid material

Oil-based products

Oil-based products such as gloss and varnish should be stored on clearly marked shelves and with their labels turned to the front. They should always be used in date order, which means that new stock should be stored at the back with old stock at the front.

Oil-based products should be **inverted** at regular intervals to stop settlement and separation of the ingredients. They must also be kept in tightly sealed containers to stop the product **skinning**. Storage at a constant temperature will ensure the product retains its desired consistency.

Correct storage of paints

Water-based products

Water-based products, such as emulsions and acrylics, should also be stored on shelves with labels to the front and in date order.

Some water-based products have a very limited shelf life and must be used before their use-by date. As with oil-based products, water-based products keep best if stored at a constant temperature. It is also important to protect them from frost to prevent the water component of the product from freezing.

Powdered materials

Powdered materials a decorator might use include Artex®, fillers, paste and sugar soap.

Large items such as heavy bags should be stored at ground or platform level. Smaller items can be stored on shelves while sealed containers, such as a bin, are ideal for loose materials.

Did you know?

Storage instructions can usually be found on the product label or in a manufacturer's information leaflet

Definition

Inverted – tipped and turned upside down

Skinning – the formation of a skin which occurs when the top layer dries out

Remember

Heavy materials should be stored at low levels to aid manual handling and should never be stacked more than two levels high

Note

It is not a good idea to leave emulsions in a garage or shed over the winter

Powdered materials can have a limited shelf life and can set in high humidity conditions. They must also be protected from frost and exposure to any moisture, including condensation. These types of materials must not be stored in the open air.

Substances hazardous to health

Some substances the decorator will work with are potentially hazardous to health, with **volatile** and highly flammable characteristics. The Control of Substances Hazardous to Health (COSHH) Regulations apply to such materials and detail how they must be stored and handled (see Chapter 2 *Health and safety*, pages 36–37, for general information about COSHH).

Decorating materials that might be hazardous to health include spirits (i.e. methylated and white), turpentine (turps), paint thinners and varnish removers. These should be stored out of the way on shelves, preferably in a suitable locker or similar room that meets the requirements of COSHH. The temperatures must be kept below 15°C as a warmer environment may cause storage containers to expand and blow up.

The storage of liquefied petroleum gas and other highly flammable liquids is covered later in this chapter on pages 107–108.

Paper material

Wallpaper

Wallpaper rolls must be stored in racks with their ends protected from damage. They should also be stored in batches with their identifying number clearly marked. Rolls should be kept wrapped and protected from dust and they should never be in direct sunlight, which can result in fading of colours.

Correct storage of wallpaper

Some special wall coverings such as **lincrusta** have a shelf life that should be taken into consideration, ensuring that the oldest stock is used first.

Safety tip

Some larger bags of powdered materials are heavier than they first appear. Make sure you use the correct manual handling techniques (see Chapter 2)

Definition

Volatile – quick to evaporate (turn to a gas)

Definition

Lincrusta – a wall-hanging with a relief pattern used to imitate wood panelling

Abrasive papers

Abrasive papers should be stored in packets that clearly identify them with respect to grade (most abrasive papers are labelled on the back with the grade of grit and thickness of paper). They should be stored on shelves ensuring that sheets are kept flat and rolls are stored upright or on reels.

Abrasive papers must be kept away from excessive heat as this makes them brittle. Dampness and condensation must also be avoided as it weakens glass papers and some garnet papers by softening the glue that binds the grit to the paper.

Decorating tools and equipment

Dust sheets

Dust sheets should be folded neatly and stored on shelving. It is best to store clean sheets and dirty sheets separately in order to avoid contamination. Sheets must always be stored dry in order to prevent **mildew** attack and fabric fatigue (when fabric disintegrates in the hand like damp paper).

Brushes

Used brushes should be cleaned thoroughly and either hung up to dry or laid flat on well-ventilated shelving. New brushes should be kept wrapped until they are needed. Mildew can grow on wet brushes that are not left to dry thoroughly, which can destroy the filling. Brushes that are stored for long periods without use should be treated with some form of insecticide to protect them from moth attack. Moth balls are ideal for this protection.

Rollers

After cleaning to remove all traces of paint and cleaner, roller sleeves should be hung up to dry in well-ventilated areas. Mohair and lambswool sleeves will need extra consideration and you should always refer to the manufacturer's instructions.

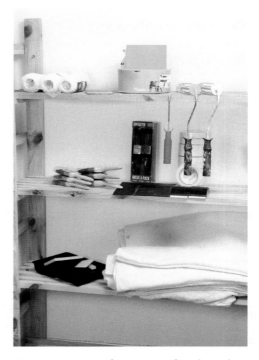

Correct storage of a variety of tools and equipment

Steel tools

Before storing, steel tools must be thoroughly cleaned and lightly oiled. A rustproof, oiled paper wrapping may be appropriate for long-term storage.

Chemicals

Chemicals such as brick cleaner or certain types of adhesive can be classified as dangerous chemicals. All chemicals should be stored in a locked area to prevent abuse or cross-contamination. You should refer to COSHH and/or the manufacturer to check storage details for any chemicals that you come across.

Highly flammable liquids

Liquefied petroleum gas (LPG), petrol, cellulose thinners, methylated spirits, chlorinated rubber paint and white spirit are all highly flammable liquids. These materials require special storage in order to ensure they do not risk injury to workers.

- Containers should only be kept in a special storeroom which has a floor made of concrete that slopes away from the storage area. This is to prevent leaked materials from collecting under the containers.
- The actual storeroom should be built of concrete, brick or some other fireproof material.
- The roof should be made from an easily shattered material in order to minimise the effect of any explosion.
- Doors should be at least 50 mm thick and open outwards.
- Any glass used in the structure should be wired and not less than 6 mm thick.
- The standing area should have a sill surrounding it that is deep enough to contain the contents of the largest container stored.
- Containers should always be stored upright.
- The area should not be heated.
- Electric lights should be intrinsically safe.
- Light switches should be flameproof and should be on the outside of the store.
- The building should be ventilated at high and low levels and have at least two exits.

- Naked flames and spark producing materials should be clearly prohibited, including smoking.

- The storeroom should be clearly marked with red and white squares and 'Highly Flammable' signage.

In addition to these requirements, there are specifically storage regulations for LPG. LPG must be stored in the open and usually in a locked cage. It should be stored off the floor and protected from direct sunlight and frost or snow. The storage of LPG is governed by the Highly Flammable Liquids and Liquefied Petroleum Gases Regulations. Note that these regulations apply when 50 or more litres are stored, and permission must be obtained from the District Inspector of Factories.

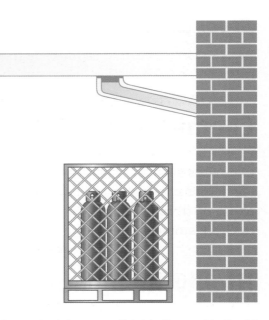

Figure 5.12 Storage of highly flammable liquids

Glass

Glass should be stored vertically in racks. The conditions for glass storage should be:

- clean – storing glass in dirty or dusty conditions can cause discoloration

- dry – if moisture is allowed between the sheets of glass it can make them stick together, which may make them difficult to handle and more likely to break.

If only a small number of sheets of glass are to be stored, they can be leant against a stable surface, as shown in the photo.

Storage of glass

FAQ

What happens if there is a delivery of timber but there is no room in the wood store?

It is probably best to remove some of the old stock from the wood store and either store it flat on timber cross-bearers or on edge in racks. This timber should be used first and as soon as possible. The new timber can now be stored in the wood store.

What should I do if I notice a leakage in the LPG store?

Leaking LPG should be treated as a very dangerous situation. Don't turn on any lights or ignite any naked flames, for example a cigarette lighter. Any kind of spark could ignite the LPG. Report the situation immediately and don't attempt to clear up the spillage yourself.

On the job: Unloading timber

Damon and Molly are helping to unload some timber that has just been delivered. They are carrying it from the lorry to the joinery shop and placing it in the wood store. While Damon is walking through the joinery shop with some timber, he trips over a hammer someone has left on the floor. Meanwhile, Molly is in the store with a sheet of plywood. She can't see any room on the shelves for the sheet of wood so she leans it against the wall. She has heard that it is OK to store sheet material on edge, so it should be fine.

Before Damon and Molly started bringing the timber into the shop, what should they have done to avoid Damon's accident? What do you think of Molly's storage of the plywood sheet? Would you have done the same thing?

Knowledge check

1. List some of the basic common sense things you can do to stay safe when loading and handling materials.

2. What is carcassing timber and how should it be stored?

3. Why is it important to ensure that timber is not exposed to moisture during storage?

4. What might happen if you store sheet materials against a wall?

5. What does the term 'shelf life' mean?

6. Name the four risks associated with adhesives.

7. What is the maximum height allowed for brick packs?

8. Describe how slabs should be stored.

9. Why should pipes be stacked on timber cross-bearers?

10. What is the most common fine aggregate?

11. How can you prevent aggregates from being contaminated?

12. How should bags of plaster and cement be stored?

13. What is LPG?

14. What do the labels on abrasive papers tell you?

15. Describe the characteristics of a special storeroom for highly flammable liquids.

16. What are the ideal conditions for glass storage and why?

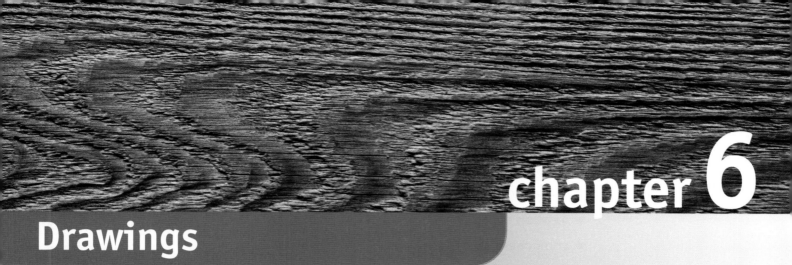

chapter 6

Drawings

OVERVIEW

Drawings are the best way of communicating detailed and often complex information from the designer to all those concerned with a job or project. Drawings are part of the legal contract between client and contractor and mistakes, either in design or interpretation of the design, can be costly. Details relating to drawings must follow guidelines by the British Standards Institute: *BS 1192 Construction Drawing Practice*.

This chapter will help you to understand the basic principles involved in producing, using and reading drawings correctly.

This chapter will cover the following topics:

- Types of drawing
- Drawing equipment
- Scales, symbols and abbreviations
- Datum points
- Types of projection
- Contract documents.

These topics can be found in the following modules:

CC 2002K	CC 2003K
CC 2002S	CC 2003S

Types of drawing

Working drawings

Working drawings are scale drawings showing plans, elevations, sections, details and location of a proposed construction. They can be classified as:

- location drawings
- component range drawings
- assembly or detail drawings.

Location drawings

Location drawings include block plans and site plans.

Block plans identify the proposed site by giving a bird's eye view of the site in relation to the surrounding area. An example is shown in Figure 6.1.

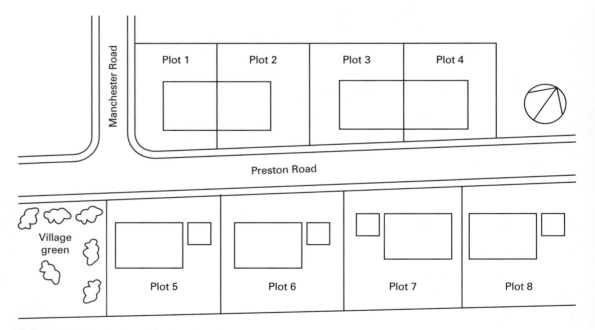

Figure 6.1 Block plan showing location

Site plans give the position of the proposed building and the general layout of the roads, services, drainage etc. on site. An example is shown in Figure 6.2.

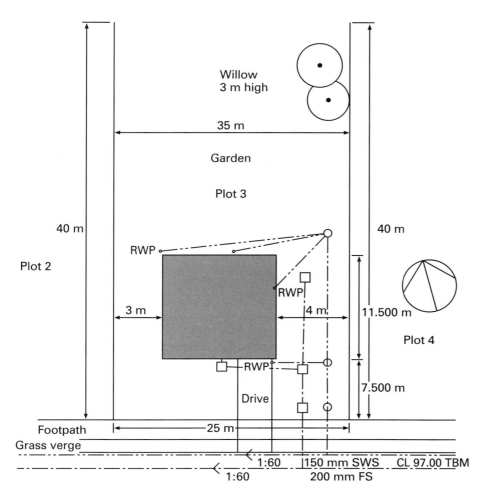

Figure 6.2 Site plan

Component range drawings

Component range drawings show the basic sizes and reference system of a standard range of components produced by a manufacturer. This helps in selecting components suitable for a task and available off-the-shelf. An example is shown in Figure 6.3.

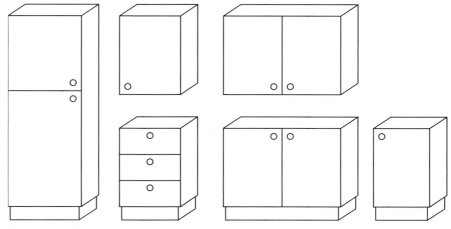

Figure 6.3 Component range drawing

Assembly or detail drawings

Assembly or detail drawings give all the information required to manufacture a given component. They show how things are put together and what the finished item will look like. An example is shown in Figure 6.4.

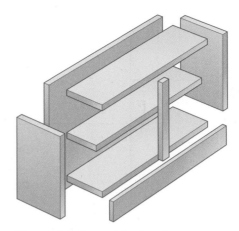

Figure 6.4 Assembly drawing

Title panels

Every drawing must have a title panel, which is normally located in the bottom right-hand corner of each drawing sheet. See Figure 6.5 for an example. The information contained in the panel is relevant to that drawing only and contains such information as:

- drawing title
- scale used
- draughtsman's name
- drawing number/ project number
- company name
- job/project title
- date of drawing
- revision notes
- projection type.

Remember

It is important to check the date of a drawing to make sure the most up-to-date version is being used, as revisions to drawings can be frequent

ARCHITECTS	CLIENT
Peterson, Thompson Associates	Carillion Development
237 Cumberland Way	
Ipswich	**JOB TITLE**
IP3 7FT	Appleford Drive
Tel: 01234 567891	Felixstowe
Fax: 09876 543210	4 bed detached
Email: enquiries@pta.co.uk	
DRAWING TITLE	**SCALE:** 1:50
Plan – garage	**DRAWING NO:** 2205-06
DATE: 27.08.2008	**DRAWN BY:** RW

Figure 6.5 Typical title panel

Drawing equipment

A good quality set of drawing equipment is required when producing drawings. It should include:

- set squares (A)
- protractors (B)
- compasses (C)
- dividers
- scale rule (D)
- pencils (E)
- eraser (F)
- drawing board (G)
- T-square (H).

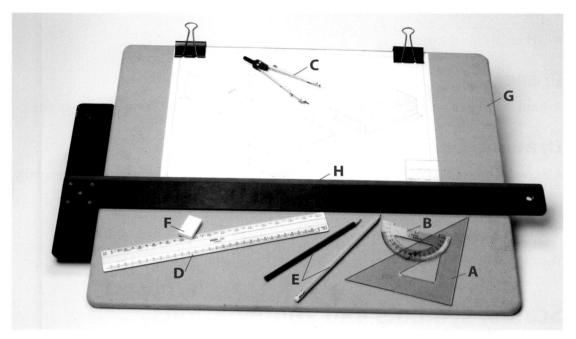

Drawing equipment

Set squares

Two set squares are required, one a 45° set square and the other a 60°/30° set square. These are used to draw vertical and inclined lines. A 45° set square (A) is shown in the photograph.

Protractors

Protractors (B) are used for setting out and measuring angles.

Compass and dividers

Compasses (C) are used to draw circles and arcs. Dividers (not shown) are used for transferring measurements and dividing lines.

Did you know?

Set squares, protractors and rules should be occasionally washed in warm soapy water

Scale rule

A scale rule that contains the following scales is to be recommended:

1:5/1:50   1:10/1:100   1:20/1:200   1:250/1:2500

An example (D) is shown in the photo.

Pencils

Two pencils (E) are required:

- HB for printing and sketching
- 2H or 3H for drawing.

Eraser

A vinyl or rubber eraser (F) is required for alterations or corrections to pencil lines.

Drawing board

A drawing boards (G) is made from a smooth flat surface material, with edges truly square and parallel.

T-square

The T-square (H) is used mainly for drawing horizontal lines.

Scales, symbols and abbreviations

Scales in common use

In order to draw a building on a drawing sheet, the building must be reduced in size. This is called a scale drawing.

The preferred scales for use in building drawings are shown in Table 6.1.

These scales mean that, for example, on a block plan drawn to 1:2500, one millimetre on the plan would represent 2500 mm (or 2.5 m) on the actual building. Some other examples are:

- On a scale of 1:50, 10 mm represents 500 mm.
- On a scale of 1:100, 10 mm represents 1000 mm (1.0 m).
- On a scale of 1:200, 30 mm represents 6000 mm (6.0 m).

Remember

A scale is merely a convenient way of reducing a drawing in size

Why not try these for yourself?

- On a scale of 1:50, 40 mm represents:…………
- On a scale of 1:200, 70 mm represents:…………
- On a scale of 1:500, 40 mm represents:………….

The use of scales can be easily mastered with a little practice.

Variations caused through printing or copying will affect the accuracy of drawings. Hence, although measurements can be read from drawings using a rule with common scales marked, it is recommended that you work to written instructions and measurements wherever possible.

A rule marked with scales used in drawings or maps is illustrated in Figure 6.6.

Type of drawing	Scales
Block plans	1:2500, 1:1250
Site plans	1:500, 1:200
General location drawings	1: 200, 1:100, 1:50
Range drawings	1:100, 1:50, 1:20
Detail drawings	1:10, 1:5, 1:1
Assembly drawings	1:20, 1:10, 1:5

Table 6.1 Preferred scales for building drawings

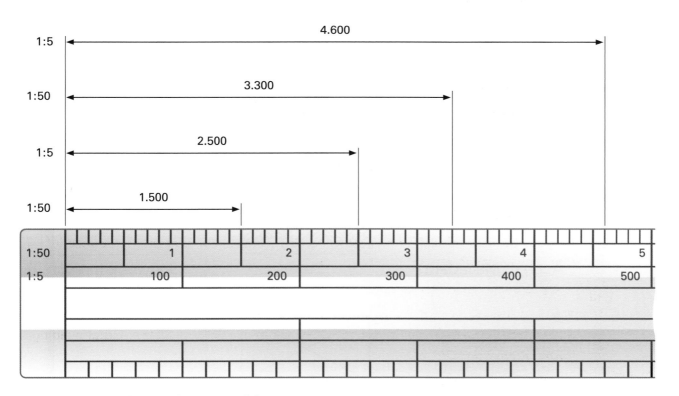

Figure 6.6 Rule with scales for maps and drawings

Symbols and abbreviations

The use of symbols and abbreviations in the building industry enables the maximum amount of information to be included on a drawing sheet in a clear way. Figure 6.7 shows some recommended drawing symbols for a range of building materials.

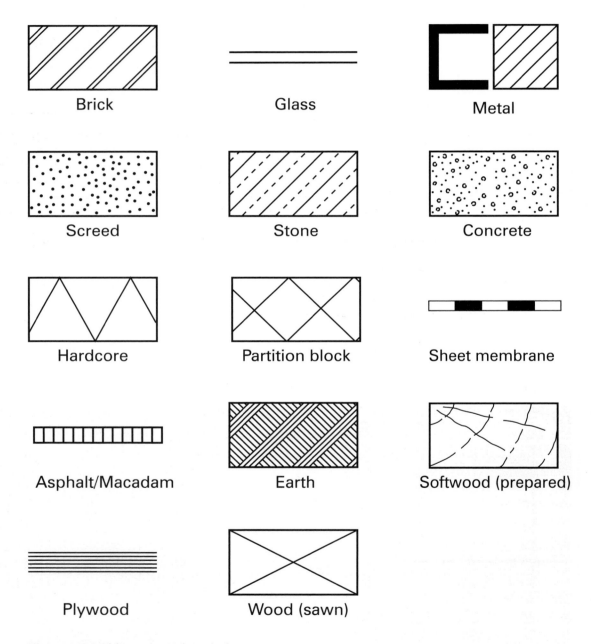

Figure 6.7 Building material symbols

Figure 6.8 illustrates the recommended methods for indicating types of doors and windows and their direction of opening.

Figure 6.9 shows some of the most frequently used graphical symbols, which are recommended in the *British Standard BS 1192*.

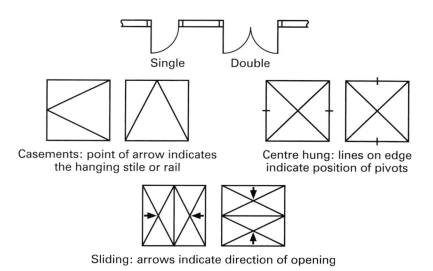

Figure 6.8 Doors and windows, type and direction of opening

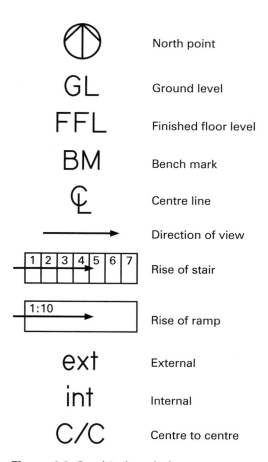

Figure 6.9 Graphical symbols

Table 6.2 lists some standard abbreviations used on drawings.

Item	Abbreviation	Item	Abbreviation
Airbrick	AB	Cement	ct
Asbestos	abs	Column	col
Bitumen	bit	Concrete	conc
Boarding	bdg	Cupboard	cpd
Brickwork	bwk	Damp proof course	DPC
Building	bldg	Damp proof membrane	DPM
Cast iron	CI	Drawing	dwg

Table 6.2 Standard abbreviations used on drawings (*continued overleaf*)

Item	Abbreviation	Item	Abbreviation
Foundation	fnd	Polyvinyl chloride	PVC
Hardboard	hdbd	Reinforced concrete	RC
Hardcore	hc	Satin anodised aluminium	SAA
Hardwood	hwd	Satin chrome	SC
Insulation	insul	Softwood	swd
Joist	jst	Stainless steel	SS
Mild steel	MS	Tongue and groove	T&G
Plasterboard	pbd	Wrought iron	WI
Polyvinyl acetate	PVA		

Table 6.2 Standard abbreviations used on drawings (continued)

FAQ

Why not just write the full words on a drawing?

This would take up too much space and clutter the drawing, making it difficult to read.

Datum points

The need to apply levels is required at the beginning of the construction process and continues right up to the completion of the building. The whole country is mapped in detail and the Ordnance Survey place datum points (bench marks) at suitable locations from which all other levels can be taken.

Ordnance bench mark (OBM)

OBMs are found cut into locations such as walls of churches or public buildings. The height of the OBM can be found on the relevant Ordnance Survey map or by contacting the local authority planning office. Figure 6.10 shows the normal symbol used, though it can appear as shown in Figure 6.11.

Figure 6.10 Ordnance bench mark

Site datum

It is necessary to have a reference point on site to which all levels can be related. This is known as the site datum. The site datum is usually positioned at a convenient height, such as finished floor level (FFL).

The site datum itself must be set in relation to some known point, preferably an OBM and must be positioned where it cannot be moved.

Figure 6.11 shows a site datum and OBM, illustrating the height relationship between them.

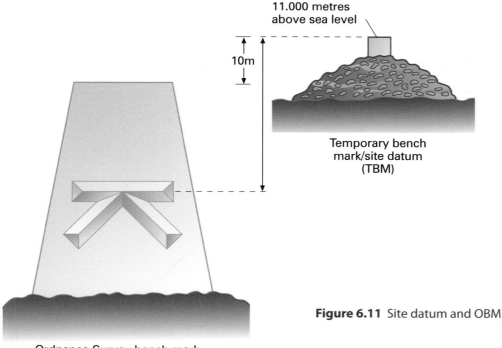

11.000 metres above sea level

10m

Temporary bench mark/site datum (TBM)

Ordnance Survey bench mark (O.S.B.M.) 10.000 metres above sea level

Figure 6.11 Site datum and OBM

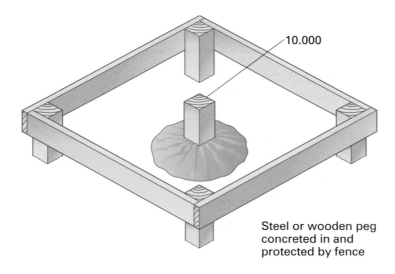

10.000

Steel or wooden peg concreted in and protected by fence

Figure 6.12 Datum peg suitably protected

If no suitable position can be found a datum peg may be used, its accurate height transferred by surveyors from an OBM, as with the site datum. It is normally a piece of timber or steel rod positioned accurately to the required level and then set in concrete. However, it must be adequately protected and is generally surrounded by a small fence for protection, as shown in Figure 6.12.

Temporary bench mark (TBM)

When an OBM cannot be conveniently found near a site it is usual for a temporary bench mark (TBM) to be set up at a height suitable for the site. Its accurate height is transferred by surveyors from the nearest convenient OBM.

All other site datum points can now be set up from this TBM using datum points, which are shown on the site drawings. Figure 6.13 shows datum points on drawings.

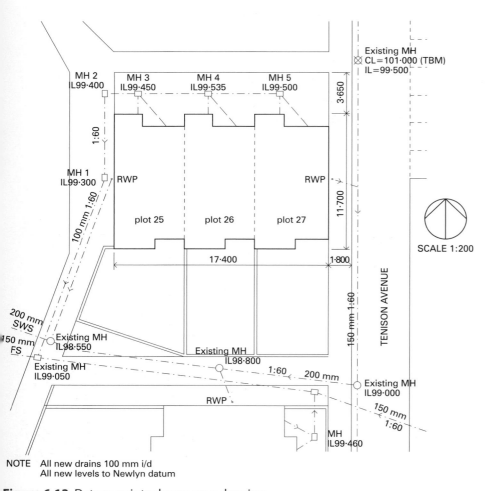

NOTE   All new drains 100 mm i/d
All new levels to Newlyn datum

Figure 6.13 Datum points shown on a drawing

Types of projection

Building, engineering and similar drawings aim to give as much information as possible in a way that is easy to understand. They frequently combine several views on a single drawing.

These may be elevations (the view we would see if we stood in front or to the side of the finished building) or plan (the view we would have if we were looking down on it). The view we see depends on where we are looking from. There are then different ways of 'projecting' what we would see onto the drawings.

The two main methods of projection, used on standard building drawings, are orthographic and isometric.

Orthographic projection

Orthographic projection works as if parallel lines were drawn from every point on a model of the building on to a sheet of paper held up behind it (an elevation view), or laid out underneath it (plan view).

There are then different ways that we can display the views on a drawing. The method most commonly used in the building industry, for detailed construction drawings, is called 'third angle projection'. In this the front elevation is roughly central. The plan view is drawn directly below the front elevation and all other elevations are drawn in line with the front elevation. An example is shown in Figure 6.14.

Front Elevation

Side Elevation

Figure 6.14 Orthographic projection

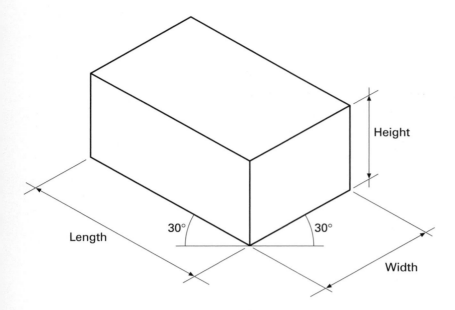

Figure 6.15 Isometric projection of rectangular box

Isometric projection

In isometric views, the object is drawn at an angle where one corner of the object is closest to the viewer. Vertical lines remain vertical but horizontal lines are drawn at an angle of 30° to the horizontal. This can be seen in Figure 6.15, which shows a simple rectangular box.

Figures 6.16 and 6.17 show the method of drawing these using a T–square and set square.

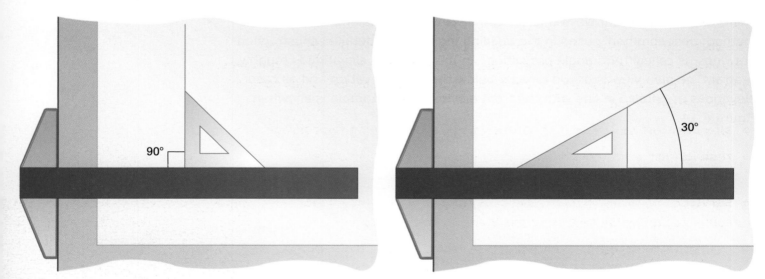

Figure 6.16 Drawing vertical lines

Figure 6.17 Drawing horizontal lines

Contract documents

Contract documents are also vital to a construction project. They are created by a team of specialists – the architect, structural engineer, services engineer and quantity surveyor – who first look at the draft of drawings from the architect and client. Just which contract documents this team goes on to produce will vary depending on the size and type of work being done, but will usually include:

- plans and drawings
- specification
- schedules
- bill of quantities
- conditions of contract.

Plans and drawings have already been covered, so here we will start with the specification.

Specification

The specification or 'spec' is a document produced alongside the plans and drawings and is used to show information that cannot be shown on the drawings. Specifications are almost always used, except in the case of very small contracts. A specification should contain:

- **site description** – a brief description of the site including the address
- **restrictions** – what restrictions apply such as working hours or limited access
- **services** – what services are available, what services need to be connected and what type of connection should be used
- **materials description** – including type, sizes, quality, moisture content etc.
- **workmanship** – including methods of fixing, quality of work and finish.

A good 'spec' helps avoid confusion when dealing with subcontractors or suppliers

The specification may also name subcontractors or suppliers, or give details such as how the site should be cleared, and so on.

Schedules

A schedule is used to record repeated design information that applies to a range of components or fittings. Schedules are mainly used on bigger sites where there are multiples of several types of house (4-bedroom, 3-bedroom, 3-bedroom with dormers, etc.), each type having different components and fittings. The schedule avoids the wrong component or fitting being put in the wrong house. Schedules can also be used on smaller jobs such as a block of flats with 200 windows, where there are six different types of window.

The need for a specification depends on the complexity of the job and the number of repeated designs that there are. Schedules are mainly used to record repeated design information for:

- doors
- windows
- ironmongery
- joinery fitments
- sanitary components
- heating components and radiators
- kitchens.

A schedule is usually used in conjunction with a range drawing and a floor plan.

The following are basic examples of these documents, using a window as an example:

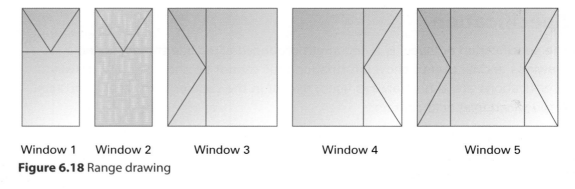

Window 1   Window 2   Window 3   Window 4   Window 5

Figure 6.18 Range drawing

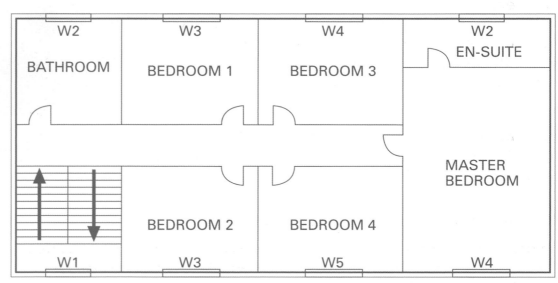

Figure 6.19 Floor plan

WINDOW SCHEDULE		
WINDOW	LOCATIONS	NOTES
Window 1	Stairwell	
Window 2	Bathroom En-suite	Obscure glass
Window 3	Bedroom 1 Bedroom 2	
Window 4	Bedroom 3 Master bedroom	
Window 5	Bedroom 4	

Figure 6.20 Schedule for a window

The schedule shows that there are five types of window, each differing in size and appearance; the range drawing shows what each type of window looks like; and the floor plan shows which window goes where. For example, the bathroom window is a type two window, which is 1200 × 600 × 50 cm with a top-opening sash and obscure glass.

Bill of quantities

The bill of quantities is produced by the quantity surveyor. It gives a complete description of everything that is required to do the job, including labour, materials and any items or components, drawing on information from the drawings, specification and schedule. The same single bill of quantities is sent out to all prospective contractors so they can submit a tender based on the same information – this helps the client select the best contractor for the job.

Every item needed should be listed on the bill of quantities

All bills of quantities contain the following information:

- **preliminaries** – general information such as the names of the client and architect, details of the work and descriptions of the site

- **preambles** – similar to the specification, outlining the quality and description of materials and workmanship, etc.

- **measured quantities** – a description of how each task or material is measured with measurements in metres (linear and square), hours, litres, kilograms or simply the number of components required

- **provisional quantities** – approximate amounts where items or components cannot be measured accurately
- **cost** – the amount of money that will be charged per unit of quantity.

The bill of quantities may also contain:

- any costs that may result from using subcontractors or specialists
- a sum of money for work that has not been finally detailed
- a sum of money to cover contingencies for unforeseen work.

This is an extract from a bill of quantities that might be sent to prospective contractors, who would then complete the cost section and return it as their tender.

Item ref No	Description	Quantity	Unit	Rate £	Cost £
A1	Treated 50 × 225 mm sawn carcass	200	M		
A2	Treated 75 × 225 mm sawn carcass	50	M		
B1	50 mm galvanised steel joist hangers	20	N/A		
B2	75 mm galvanised steel joist hangers	7	N/A		
C1	Supply and fit the above floor joists as described in the preambles				

Figure 6.21 Extract from a bill of quantities

To ensure that all contractors interpret and understand the bill of quantities consistently, the Royal Institution of Chartered Surveyors and the Building Employers' Confederation produce a document called the *Standard Method of Measurement of Building Works* (SMM). This provides a uniform basis for measuring building work, for example stating that carcassing timber is measured by the metre whereas plasterboard is measured in square metres.

Conditions of contract

Almost all building work is carried out under a contract. A small job with a single client (e.g. a loft conversion) will have a basic contract stating that the contractor will do the work to the client's satisfaction, and that the client will pay the contractor the agreed sum of money once the work is finished. Larger contracts with clients such as the Government will have additional clauses, terms or stipulations, which may include any of the following.

Variations

A variation is a modification of the original drawing or specification. The architect or client must give the contractor written confirmation of the variation, then the contractor submits a price for the variation to the quantity surveyor (or client, on a small job). Once the price is accepted, the variation work can be completed.

Interim payment

An **interim** payment schedule may be written into the contract, meaning that the client pays for the work in instalments. The client may pay an amount each month, linked to how far the job has progressed, or may make regular payments regardless of how far the job has progressed.

Final payment

Here the client makes a one-off full payment once the job has been completed to the specification. A final payment scheme may also have additional clauses included, such as:

- **Retention** – This is when the client holds a small percentage of the full payment back for a specified period (usually six months). It may take some time for any defects to show, such as cracks in plaster. If the contractor fixes the defects, they will receive the retention payment; if they don't fix them, the retention payment can be used to hire another contractor to do so.
- **Penalty clause** – This is usually introduced in contracts with a tight deadline, where the building must be finished and ready to operate on time. If the project overruns, the client will be unable to trade in the premises and will lose money, so the contractor will have to compensate the client for lost revenue.

Did you know?

On a poorly run contract, a penalty clause can be very costly and could incur a substantial payment. In an extreme case, the contractor may end up making a loss instead of a profit on the project

On the job: Drawing plans

James is about to draw a kitchen plan. What should James consider when creating the drawing in terms of openings like doors and windows, and services such as water, electricity and gas supplies? Once James has created an outline of the kitchen, where should he start drawing? What sort of things should James have already discussed with the client (think about things like appliances, an extractor, electrical points, the client's budget etc.)?

Knowledge check

1. Briefly explain why drawings are used in the construction industry.

2. What do the following abbreviations stand for: DPC; hwd; fnd; DPM?

3. Sketch the graphical symbols which represent the following: brickwork; metal; sawn timber; hardcore.

4. Can you name the main types of projection that are used in building drawings?

5. What does a block plan show?

6. What are dividers used for?

7. What type of information could be found in a drawing's title panel?

8. Name two ways in which you can find out the height of a OBM?

9. In isometric projection, at what angle are horizontal lines drawn?

10. What type of information can be found in specifications?

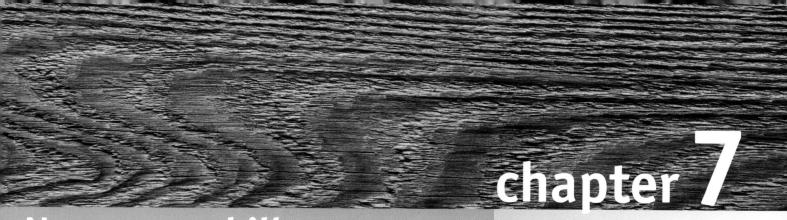

chapter 7

Numeracy skills

OVERVIEW

Throughout your career in the construction industry you will have to make use of numbers and calculations in order to plan and carry out work. You will therefore need to make sure you are confident dealing with numbers, which may mean that you have to develop and improve your maths and numeracy skills.

Although you may often use a calculator to do calculations, you may find that a calculator is not always available and you may have to work something out on paper or in your head. This chapter will help you refresh and practise your skills.

This chapter will cover the following topics:

- Numbers
- Calculations
- Measures.

These topics can be found in the following modules:

CC 2002K   CC 2002S

Numbers

Place value

0, 1, 2, 3, 4, 5, 6, 7, 8 and 9 are the ten digits we can work with. We can write any number you can think of, however huge, using any combination of these ten digits. In a number, the value of each digit depends upon its place value. Table 7.1 is a place value table and shows how the digit 2 has a different value, depending on its position.

Millions	Hundred thousands	Ten thousands	Thousands	Hundreds	Tens	Units	Value
2	9	4	1	3	7	8	2 million
	2	5	3	1	0	7	2 hundred thousand
	7	2	5	6	6	4	2 × ten thousand = 20 thousand
		6	2	4	9	2	2 thousands
		5	6	2	9	1	2 hundreds
			8	4	2	7	2 tens = 20
				1	6	2	2 units

Table 7.1 A place value table for the digit 2

Positive numbers

A positive number is a number that is greater than zero. If we make a number line, positive numbers are all the numbers to the right of zero.

0 1 2 3 4 5 6 7 8 9 10 11 12 13 ...

Positive numbers

Negative numbers

A negative number is a number that is less than zero. If we make another number line, negative numbers are all the numbers to the left of zero.

... −13 −12 −11 −10 −9 −8 −7 −6 −5 −4 −3 −2 −1 0

←——————————————————————
Negative numbers

Zero is neither positive nor negative.

Decimal numbers

Most of the time, the numbers we use are whole numbers. For example, we might buy six apples, or two loaves of bread or one car. However, sometimes we need to use numbers that are less than whole numbers, for example, we might eat one and a quarter sandwiches, two and a half cakes and drink three-quarters of a cup of tea. You can use decimals to show fractions or parts of quantities. Table 7.2 shows the value of the digits to the right of a decimal point.

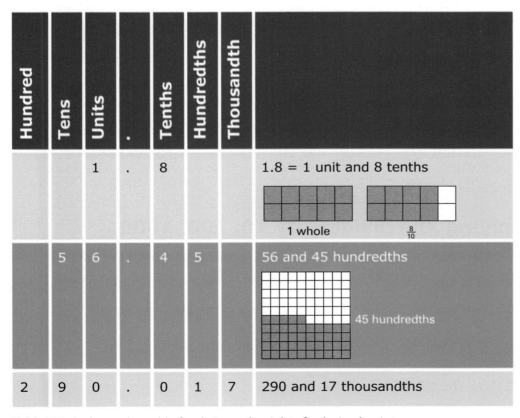

Hundred	Tens	Units	.	Tenths	Hundredths	Thousandth	
		1	.	8			1.8 = 1 unit and 8 tenths
	5	6	.	4	5		56 and 45 hundredths
2	9	0	.	0	1	7	290 and 17 thousandths

Table 7.2 A place value table for digits to the right of a decimal point

Did you know?

Knowing about place value helps you to read numbers and to put numbers and quantities in order of size

Rounding to a number of decimal places

To round a number to a given number of decimal places, look at the digit in the place value position after the one you want.

- If it is 5 or more, round up.
- If it is less than 5 round down.

For example, say we wanted to round 4.634 to two decimal places, the digit in the third decimal place is 4, so we round down. Therefore, 4.634 rounded to 2 decimal places (d.p.) is 4.63.

If we look at the number 16.127, the digit in the third decimal place is 7, so we round up. Therefore, 16.127 rounded to 2 d.p. is 16.13.

Rounding to a number of significant figures

The most significant figure in a number is the digit with the highest place value. To round a number to a given number of significant figures, look at the digit in the place value position after the one you want.

- If it is 5 or more round up.
- If it is less than 5 round down.

For example, say we wanted to write 80 597 to one significant figure, the most significant figure is 8. The second significant figure is 0, so we round down. Therefore, 80 597 to 1 significant figure (s.f.) is 80 000.

If we wanted to write 80 597 to two significant figures, the first two significant figures are 8 and 0. The third significant figure is 5, so we round up. Therefore, 80 597 to 2 s.f. is 81 000.

Multiplying and dividing by 10, 100, 1000...

- To multiply a number by 10, move the digits one place value to the left.
- To multiply a number by 100, move the digits two place values to the left.
- To multiply a number by 1000, move the digits three place values to the left.

For example:

	5			3.25
5 × 10 =	50		3.25 × 10 =	32.5
5 × 100 =	500		3.25 × 100 =	325
5 × 1000 =	5000		3.25 × 1000 =	3250

Remember

If a calculation results in an answer with a lot of decimal places, such as 34.568 923, you can round to 1 or 2 decimal places to make it simpler

- To divide a number by 10, move the digits one place value to the right.
- To divide a number by 100, move the digits two place values to the right.
- To divide a number by 1000, move the digits three place values to the right.

For example:

$80\,000 \div 10 = 8000$	$473.6 \div 10 = 47.36$
$80\,000 \div 100 = 800$	$473.6 \div 100 = 4.736$
$80\,000 \div 1000 = 80$	$473.6 \div 1000 = 0.4736$

Converting decimals to fractions

You can use place value to convert a decimal to a fraction. For example:

0.3 is 3 tenths which is $\frac{3}{10}$

0.25 is 25 hundredths which is $\frac{25}{100}$

$\frac{25}{100}$ simplifies to $\frac{1}{4}$ (by dividing the top and bottom numbers by 25)

Table 7.3 shows some useful fraction/decimal equivalents.

See page 137 for more on simplifying fractions.

Decimal	0.1	0.25	0.333333	0.5	0.75	0.01
Fraction	$\frac{1}{10}$	$\frac{1}{4}$	$\frac{1}{3}$	$\frac{1}{2}$	$\frac{3}{4}$	$\frac{1}{100}$

Table 7.3 Useful fraction/decimal equivalents

Multiples

Multiples are the numbers you get when you multiply any number by other numbers. For example:

- the multiples of 3 are 3, 6, 9, 12, 15, 18, 21, 24, 27, 30 and so on
- the multiples of 4 are 4, 8, 12, 16, 20, 24, 28, 32, 36, 40 and so on
- the multiples of 5 are 5, 10, 15, 20, 25, 30, 35, 40, 45, 50 and so on.

Did you know?

Knowing how to multiply and divide by 10, 100, 1000 etc. is useful for converting metric units of measurement (see page 145) and finding percentages (see page 140)

Did you know?

Knowing how to convert between fractions and decimals helps with working out parts of quantities and calculating percentages (see page 140)

Remember

The multiples of a number are the numbers in its 'times' table (multiplication table)

Common multiples

Here are the multiples of 3 and 5:

- Multiples of 3: 3, 6, 9, 12, 15, 18, 21, 24, 27, 30, 33, 36...
- Multiples of 5: 5, 10, 15, 20, 25, 30, 35...

3 and 5 have the multiples 15 and 30 in common. 15 and 30 are common multiples of 3 and 5. The lowest common multiple of 3 and 5 is 15.

Factors

The factors of a number are the whole numbers that divide into it exactly. For example:

- The factors of 18 are 1, 2, 3, 6, 9 and 18
- The factors of 30 are 1, 2, 3, 5, 6, 10, 15 and 30
- 5 is a factor of 5, 10, 15, 20...
- 7 is a factor of 7, 14, 21, 28...

Common factors

Here are the factors of 28 and 36:

- The factors of 28 are 1, 2, 4, 7, 14, 28
- The factors of 36 are 1, 2, 3, 4, 6, 9, 12, 18, 36

From these lists you can see that 28 and 36 have the factors 1, 2 and 4 in common. 1, 2 and 4 are the common factors of 28 and 36. 4 is the highest common factor of 28 and 36.

Fractions

Fractions describe parts of a whole, for example a half of a pie, a third of a can of cola or a quarter of a cake.

In a fraction:

$$\frac{3}{4}$$

the top number is called the **numerator**

the bottom number is called the **denominator**

The denominator shows how many *equal parts* the whole is divided into. The numerator shows how many of those parts you have.

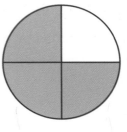

$\frac{3}{4}$

Finding a fraction of a quantity

To find a fraction of a quantity you divide by the denominator and multiply by the numerator. For example:

- to find $\frac{1}{2}$ of 500 m, divide by 2 (500 m ÷ 2 = 250 m)

- to find $\frac{2}{5}$ of £40, divide by 5 (£40 ÷ 5 = £8) and multiply by 2 (£8 × 2 = £16)

Equivalent fractions

Two fractions are equivalent if they have the same value. For example:

$$\frac{1}{2} = \frac{2}{4} = \frac{3}{6} = \frac{4}{8}$$

 $\frac{1}{2}$  $\frac{2}{4}$ $\frac{3}{6}$ $\frac{4}{8}$

$$\frac{2}{3} = \frac{4}{6} = \frac{6}{9} = \frac{8}{12}$$

 $\frac{2}{3}$ $\frac{4}{6}$ $\frac{6}{9}$  $\frac{8}{12}$

Simplifying fractions

To simplify a fraction, write it as an equivalent fraction with smaller numbers in the numerator and denominator.

For example, $\frac{8}{12}$ simplifies to $\frac{2}{3}$.

When a fraction cannot be simplified any more, it is in its simplest form, or its lowest terms.

For example:

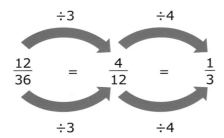

$$\frac{12}{36} = \frac{4}{12} = \frac{1}{3}$$

$\frac{12}{36} = \frac{1}{3}$ in its simplest form.

Multiplying fractions

To multiply a fraction by a whole number, multiply the numerator by the whole number.

For example:

$$\frac{2}{9} \times 4 = \frac{8}{9}$$

To multiply a fraction by another fraction, multiply the numerators and the denominators. For example:

$$\frac{2}{3} \times \frac{5}{8} = \frac{2 \times 5}{3 \times 8} = \frac{10}{24}$$

Give the answer in its simplest form:

$$\frac{10}{24} = \frac{5}{12} \qquad \text{(dividing numerator and denominator by 2)}$$

Dividing fractions

To divide one fraction by another, you invert (turn upside down) the fraction you are dividing by, and multiply.

For example:

$$\frac{3}{4} \div \frac{2}{3} = \frac{3}{4} \times \frac{3}{2} = \frac{9}{8} = 1\frac{1}{8}$$

To divide a fraction by a whole number, or to divide a whole number by a fraction, write the whole number as a fraction with denominator 1, and use the same method.

$$\frac{4}{5} \div 3 = \frac{4}{5} \div \frac{3}{1} = \frac{4}{5} \times \frac{1}{3} = \frac{4}{15}$$

$$6 \div \frac{3}{4} = \frac{6}{1} \div \frac{3}{4} = \frac{6}{1} \times \frac{4}{3} = \frac{24}{3} = 8$$

Adding fractions

To add fractions with the same denominator, add the numerators.

For example:

$$\frac{1}{3} + \frac{1}{3} = \frac{2}{3}$$

$$\frac{1}{5} + \frac{3}{5} = \frac{4}{5}$$

To add fractions with different denominators, first write the fractions as equivalent fractions with the same denominator. Use the lowest common multiple of the two denominators. You can use any common multiple as the denominator, but using the lowest common multiple keeps the numbers smaller and the calculations simpler.

For example:

$$\frac{1}{2} + \frac{1}{3}$$

The denominators are not the same. The lowest common multiple of 2 and 3 is 6.

To write $\frac{1}{2}$ as an equivalent fraction with denominator 6, multiply numerator and denominator by 3, to give $\frac{3}{6}$.

To write $\frac{1}{3}$ as an equivalent fraction with denominator 6, multiply numerator and denominator by 2, to give $\frac{2}{6}$

The calculation is now $\frac{3}{6} + \frac{2}{6} = \frac{5}{6}$

A mixed number has a whole number and a fraction part, for example $3\frac{1}{4}$.

To add mixed numbers, add together the whole number parts and then the fractions.

For example, if we wanted to add together $1\frac{1}{2}$ and $2\frac{1}{3}$:

Add the whole numbers: $1 + 2 = 3$

Now add the fractions: $\frac{1}{2} + \frac{1}{3} = \frac{3}{6} + \frac{2}{6} = \frac{5}{6}$

Combine the two answers: $1\frac{1}{2} + 2\frac{1}{3} = 3\frac{5}{6}$

Subtracting fractions

To subtract fractions with the same denominator, subtract the numerators.

For example: $\frac{7}{8} - \frac{3}{8} = \frac{4}{8} = \frac{1}{2}$ 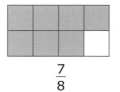 $-$ 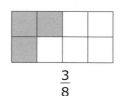 $=$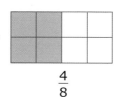

$$\frac{7}{8} \qquad\qquad \frac{3}{8} \qquad\qquad \frac{4}{8}$$

To subtract mixed numbers, first write them as improper (top heavy) fractions with a common denominator

For example: $2\frac{1}{3} - \frac{1}{2} = \frac{7}{3} - \frac{1}{2} = \frac{14}{6} - \frac{3}{6} = 1\frac{5}{6}$

Percentages

Percentages are another way of showing parts of a quantity. Percentage means 'number of parts per hundred'. The symbol % means per cent. For example:

- 1% means 1 out of a hundred or $\frac{1}{100}$

- 10% means 10 out of a hundred or $\frac{10}{100}$

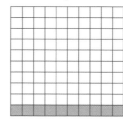

10%

- 84% means 84 out of a hundred or $\frac{84}{100}$

100% means the whole quantity.

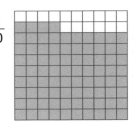

84%

Finding a percentage of a quantity

To find a percentage of a quantity, find 1% first, by dividing by 100, then multiply by the number you need. For example:

20% of £45

$$1\% \text{ of } £45 = \frac{£45}{100} = £45 \div 100 = £0.45$$

So   20% of £45 = 20 × £0.45 = £9

Percentage change

A number can be increased or decreased by a percentage. Wages are often increased by a percentage (e.g. a 4% rise in wages). Items are often reduced by a percentage in sales (e.g. 10% off, 20% reduction, etc.). For example:

A set of paintbrushes costs £12.99. In the sale there is 15% off.

(a) How much money do you save by buying the paintbrushes in the sale?

(b) What is the sale price of the paintbrushes?

(a) Work out 15% of £12.99

1% of £12.99 = £0.1299

So   15% of £12.99 = 15 × £0.1299 = £1.9485 = £1.95 to the nearest penny

You save £1.95

(b) The sale price is

£12.99 – £1.95 = £11.04

Ratio

A **ratio** describes a relationship between quantities. You can read a ratio as a 'for every' statement. For example:

- Green paint is made by mixing blue and yellow in the ratio 1 : 2. This means, *for every* 1 litre of blue paint you need 2 litres of yellow paint.

- A labourer and a bricklayer agree to share their bonus in the ratio 2 : 5. The bonus is £42. The ratio 2:5 means that for every 2 parts the labourer

Did you know?

Percentages are also used for:

- paying deposits – a deposit is a percentage of the whole price (e.g. 20% deposit)
- paying interest – interest is a percentage of money. It is repaid on top of the money (e.g. a loan from a bank of £1000 at 15% interest)
- profit – when charging a client for work you have carried out, you will need to add on to your costs a percentage for your profit
- Value Added Tax (VAT) – VAT is a government tax added to many items or goods that we buy (the standard VAT rate is 17.5%)

receives, the bricklayer receives 5 parts. So the bonus needs to be split into 2 + 5 = 7 parts. One part can be calculated: £42 ÷ 7 = £6. Therefore, the labourer receives two parts = 2 × £6 = £12 and the bricklayer receives five parts = 5 × £6 = £30.

Calculations

Did you know?

Scales for scale drawings can also be given as ratios (see page 146)

Addition

When adding numbers using a written method, write digits with the same place value in the same column. For example, to work out 26 + 896 + 1213 write the calculation:

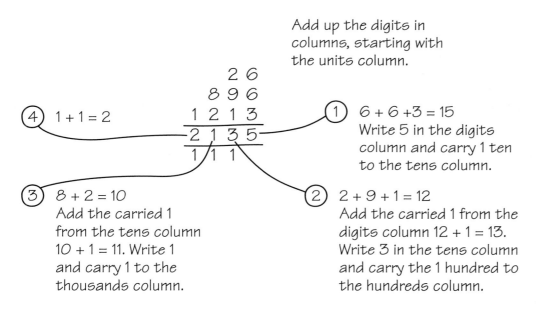

Add up the digits in columns, starting with the units column.

④ 1 + 1 = 2

① 6 + 6 + 3 = 15
Write 5 in the digits column and carry 1 ten to the tens column.

③ 8 + 2 = 10
Add the carried 1 from the tens column 10 + 1 = 11. Write 1 and carry 1 to the thousands column.

② 2 + 9 + 1 = 12
Add the carried 1 from the digits column 12 + 1 = 13. Write 3 in the tens column and carry the 1 hundred to the hundreds column.

```
   2 6
   8 9 6
 1 2 1 3
 2 1 3 5
 1 1 1
```

To add decimals, write the numbers with the decimal points in line:

```
  4.56
 10.2
  0.32
 15.08
```

In a problem, these words mean you need to add:

- What is the *total* of 43 and 2457?   (43 + 2457)

- What is the *sum* of 56 and 345?   (56 + 345)

- *Increase* 3467 by 521.   (3467 + 521)

Subtraction

When subtracting numbers using a written method, write digits with the same place value in the same column.

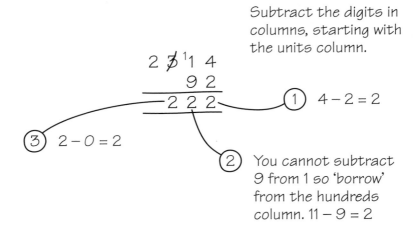

Subtract the digits in columns, starting with the units column.

① $4 - 2 = 2$

③ $2 - 0 = 2$

② You cannot subtract 9 from 1 so 'borrow' from the hundreds column. $11 - 9 = 2$

In a problem, these words mean you need to subtract:

- *Find the difference* between 200 and 45.　　　$(200 - 45)$
- *Decrease* 64 by 9.　　　$(64 - 9)$
- *How much greater than* 98 is 110?　　　$(110 - 98)$

Multiplication

Knowing the multiplication tables up to 10×10 helps with multiplying single digit numbers. You can use multiplication facts you know to work out other multiplication calculations. For example:

$$20 \times 12$$

You know that $20 = 2 \times 10$

So　　$20 \times 12 = 2 \times 10 \times 12$

$$= 2 \times 12 \times 10$$

$$= 24 \times 10 = 240$$

To multiply larger numbers you can write the calculation in columns or use the grid method. Both methods work by splitting the calculation into smaller ones.

In columns

```
     2 5
 ×   3 6
   1 5 0        6 × 25 ⎫ Add these to
   7 5 0       30 × 25 ⎭ find 36 × 25
   9 0 0
```

The grid method

36 × 25 =

×	20	5
30	600	150
6	120	30

```
30 ×  20 = 600
30 ×   5 = 150
 6 ×  20 = 120
 6 ×   5 =  30
           900
```

Division

Division is the opposite of multiplication. Knowing the multiplication tables up to 10 × 10 helps with division. Each multiplication fact gives two related division facts. For example:

$$4 \times 6 = 24 \qquad 24 \div 6 = 4 \qquad 24 \div 4 = 6$$

To divide by a single digit number, use short division.

161 ÷ 7 =

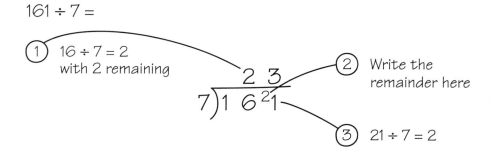

① 16 ÷ 7 = 2
   with 2 remaining

② Write the remainder here

③ 21 ÷ 7 = 2

To divide by 10 or more, use long division.

$$
\begin{array}{r}
2\,4\,6 \\
12\,\overline{)\,2\,9\,5\,2} \\
2\,4 \\
\hline
5\,5 \\
4\,8 \\
\hline
7\,2 \\
7\,2 \\
\hline
0
\end{array}
$$

$2 \times 12 = 24$
$29 - 24 = \quad 5$. Bring down the next 5
$12 \times 4 = 48$
$55 - 48 = \quad 7$. Bring down the 2
$12 \times 6 = 72$

Estimating

Sometimes an accurate answer to a calculation is not required. You can estimate an approximate answer by rounding all the values in the calculation to 1 significant figure (see page 134). For example:

Estimate the answer to the calculation 4.9×3.1

4.9 rounds to 5 to 1 s.f.

3.1 rounds to 3 to 1 s.f.

A sensible estimate is $5 \times 3 = 15$

Measures

Units of measurement

The metric units of measurement are shown in Table 7.4.

Length	millimetres (mm), centimetres (cm), metres (m), kilometres (km)
Mass (weight)	grams (g), kilograms (kg), tonnes (t)
Capacity (the amount a container holds)	millilitres (ml), centilitres (cl), litres (l)

Table 7.4 Units of measurement

Did you know?

You can use values rounded to 1 s.f. to estimate approximate areas and prices

Metric units are all based on 10, 100, 1000 which makes it easy to convert between units.

milli means one thousandth $1 \text{ mm} = \dfrac{1}{1000} \text{ m}$ $1 \text{ ml} = \dfrac{1}{1000} \text{ litre}$

centi means one hundredth $1 \text{ cm} = \dfrac{1}{100} \text{ m}$ $1 \text{ cl} = \dfrac{1}{100} \text{ litre}$

kilo means one thousand $1 \text{ kg} = 1000 \text{ g}$ $1 \text{ km} = 1000 \text{ m}$

Table 7.5 shows some useful metric conversions.

Length	Mass	Capacity
1 cm = 10 mm 1 m = 100 cm = 1000 mm 1 km = 1000 m	1 kg = 1000 g 1 tonne = 1000 kg	1 l = 100 cl = 1000 ml

Table 7.5 Useful metric conversions

Remember

To convert from a smaller unit to a larger one, divide; to convert from a larger unit to a smaller one, multiply

To convert 2657 mm to metres: $2657 \div 1000 = 2.657 \text{ m}$

To convert 0.75 tonnes to kg: $0.75 \times 1000 = 750 \text{ kg}$

For calculations involving measurements, you need to convert all the measurements into the same unit. For example, a plasterer measures the lengths of cornice required for a room. He writes down the measurements as 175 cm, 2 metres, 225 cm, 1.5 m. In order to work out the total length of cornice needed, we first need to write all the lengths in the same units:

175 cm   2 metres = 2 × 100 = 200 cm   1.5 m = 1.5 × 100 = 150 cm   225 cm

So the total length is:

$$175 + 200 + 225 + 150 = 750 \text{ cm}$$

Imperial units

In the UK we still use some imperial units of measurement (see Table 7.6).

Length	inches, feet, yards, miles
Mass (weight)	ounces, pounds, stones
Capacity (the amount a container holds)	pints, gallons

Table 7.6 Some imperial units of measurement

To convert from imperial to metric units, use the approximate conversions shown in Table 7.7.

Length	Mass	Capacity
1 inch = 2.5 cm 1 foot = 30 cm 5 miles = 8 km	2.2 pounds = 1 kg 1 ounce = 25 g	1.75 pints = 1 litre 1 gallon = 4.5 litres

Table 7.7 Converting imperial measurements to metric

For example, if a wall is 32 feet long, what is its approximate length in metres?

1 foot = 30 cm

So   32 feet = 32 × 30 cm = 960 cm = 9.6 m.

Scale drawings

Building plans are drawn to scale. Each length on the plan is in proportion to the real length. On a drawing to a scale of 1 cm represents 10 m:

- a length of 5 cm represents an actual length of 5 × 10 = 50 m
- a length of 12 cm represents an actual length of 12 × 10 = 120 m
- an actual length of 34 m is represented by a line 34 ÷ 10 = 3.4 cm long.

Scales are often given as ratios. For example:

- a scale of 1 : 100 means that 1 cm on the drawing represents an actual length of 100 cm (or 1 metre)
- a scale of 1 : 20 000 means that 1 cm on the drawing represents an actual length of 20 000 cm = 20 m.

Table 7.8 shows some common scales used in the construction industry.

1 : 5	1 cm represents 5 cm	5 times smaller than actual size
1 : 10	1 cm represents 10 cm	10 times smaller than actual size
1 : 20	1 cm represents 20 cm	20 times smaller than actual size
1 : 50	1 cm represents 50 cm	50 times smaller than actual size
1 : 100	1 cm represents 100 cm = 1 m	100 times smaller than actual size
1 : 1250	1 cm represents 1250 cm = 12.5 m	1250 times smaller than actual size

Table 7.8 Common scales used in the construction industry

Did you know?

A useful rhyme to help you remember the pints to litre conversion is 'a litre of water's a pint and three quarters'

Remember

You can use a scale to:

- work out the actual measurement from a plan
- work out how long to draw a line on the plan to represent an actual measurement

Let's look at the following examples:

(a) A plan is drawn to a scale of 1 : 20. On the plan, a wall is 4.5 cm long. How long is the actual wall?

      1 cm on the plan = actual length 20 cm

So   4.5 cm on the plan = actual length 4.5 × 20 = 90 cm or 0.9 m

(b) A window is 3 m tall. How tall is it on the plan?

      3 m = 300 cm

      an actual length of 20 cm is 1 cm on the plan

      an actual length of 5 × 20 = 100 cm is 5 × 1 cm on the plan

      an actual length of 3 × 100 cm is 3 × 5 cm on the plan.

The window is 15 cm tall on the plan.

To make scale drawings, architects use a scale rule. The different scales on the ruler give the equivalent actual length measurements for different lengths in cm, for each scale.

Perimeter of shapes with straight sides

The perimeter of a shape is the distance all around the outside of the shape. To find the perimeter of a shape, measure all the sides and then add the lengths together. For example:

The perimeter of this room is

4.5 + 3.2 + 4.5 + 3.2 = 15.4 m

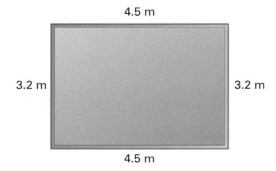

4.5 m

3.2 m   3.2 m

4.5 m

Did you know?

You can use the perimeter to work out the length of picture rail to go all round a room

Area of shapes with straight sides

The area of a 2-D (flat) shape is the amount of space it covers. Area is measured in square units, such as square centimetres (cm²) and square metres (m²).

This rectangle is drawn on squared paper. Each square has an area 1 cm².

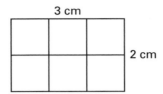

You can find the area by counting the squares. Area = 6 squares = 6 cm²

You can also calculate the area by multiplying the number of squares in a row by the number of rows: 3 × 2 = 6

The area of a rectangle with length *l* and width *w* is

$$A = l \times w$$

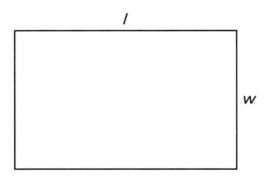

For example, if the length of a rectangular room is 3.6 metres and the width is 2.7 metres, the area is

$$A = 3.6 \times 2.7 = 9.72 \text{ m}^2$$

Area of a triangle

The area of a triangle is given by the formula:

$$A = \tfrac{1}{2} \times h \times b$$

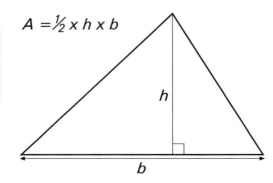

$A = \dfrac{1}{2} \times h \times b$ where *h* is the **perpendicular** height and *b* is the length of the base. The perpendicular height is drawn to meet the base at right angles (90°).

Definition

Perpendicular – at right angles to

For example, say we wanted to find (a) the area and (b) the perimeter of this triangle

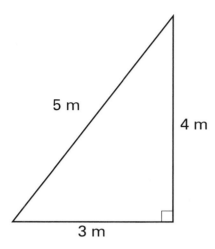

(a) Perpendicular height = 4 m, base = 3 m

$$A = \frac{1}{2} \times h \times b = \frac{1}{2} \times 4 \times 3 = 6 \text{ m}^2$$

(b) Perimeter = 5 + 4 + 3 = 12 m

Pythagoras' theorem

You can use Pythagoras' theorem to find unknown lengths in right-angled triangles. In a right-angled triangle:

- one angle is 90° (a right angle)
- the longest side is opposite the right angle and is called the **hypotenuse**.

Pythagoras' theorem says that for any right-angled triangle with sides a and b and hypotenuse c,

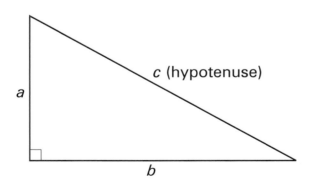

$$c^2 = a^2 + b^2$$

Definition

Hypotenuse – the longest side on a right angled triangle

For example, if we wanted to find the length of the hypotenuse of this triangle

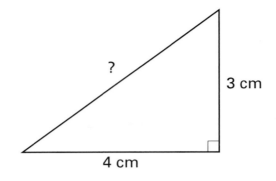

$$c^2 = a^2 + b^2$$

In the theorem, c is the hypotenuse.

$$c^2 = 3^2 + 4^2$$
$$= 9 + 16$$
$$= 25$$
$$c = \sqrt{25} = 5$$

The hypotenuse is 5 cm long.

a^2 is 'a squared' and is equal to $a \times a$. 3^2 is '3 squared' and is equal to 3×3. The opposite or inverse of squaring is finding the square root. $\sqrt{25}$ means 'the square root of 25': $5 \times 5 = 25$, so $\sqrt{25} = 5$

Learning these squares and square roots will help with Pythagoras' theorem calculations. Table 7.9 shows some square roots you will often find useful.

$1^2 = 1 \times 1 = 1$	$\sqrt{1} = 1$
$2^2 = 2 \times 2 = 4$	$\sqrt{4} = 2$
$3^2 = 3 \times 3 = 9$	$\sqrt{9} = 3$
$4^2 = 4 \times 4 = 16$	$\sqrt{16} = 4$
$5^2 = 5 \times 5 = 25$	$\sqrt{25} = 5$
$6^2 = 6 \times 6 = 36$	$\sqrt{36} = 6$
$7^2 = 7 \times 7 = 49$	$\sqrt{49} = 7$
$8^2 = 8 \times 8 = 64$	$\sqrt{64} = 8$
$9^2 = 9 \times 9 = 81$	$\sqrt{81} = 9$
$10^2 = 10 \times 10 = 100$	$\sqrt{100} = 10$

Table 7.9 Useful square roots

Using Pythagoras' theorem to find the shorter side of a triangle

You can rearrange Pythagoras' theorem like this:

$$c^2 = a^2 + b^2$$

$$a^2 = c^2 - b^2$$

For example, say we wanted to find the length of side *a* in this right-angled triangle

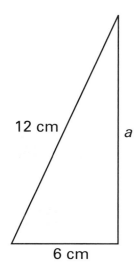

$$a^2 = c^2 - b^2$$
$$= 12^2 - 6^2$$
$$= 144 - 36 = 108$$
$$a = \sqrt{108} = 10.3923... \quad \text{Using the } \sqrt{} \text{ key on a calculator}$$
$$= 10.4 \text{ cm (to 1 decimal place)}$$

You can also use Pythagoras' theorem to find the perpendicular height of a triangle. For example, if we wanted to find the area of this triangle we would need to find the perpendicular height:

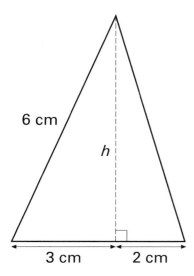

Using Pythagoras $h^2 = 6^2 - 3^2$
$$= 36 - 9 = 27$$
$$h = \sqrt{27} = 5.196... = 5.2 \text{ cm (to 1 d.p.)}$$

Area $= \dfrac{1}{2} \times b \times h$

$\quad\;\; = \dfrac{1}{2} \times 5 \times 5.2$ Base length = 3 + 2 cm

$\quad\;\; = 13 \text{ cm}^2$

Areas of composite shapes

Composite shapes are made up of simple shapes such as rectangles and squares. To find the area, divide up the shape and find the area of each part separately. For example, to work out the area of this L-shaped room:

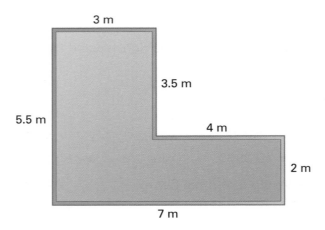

First divide it into two rectangles, A and B:

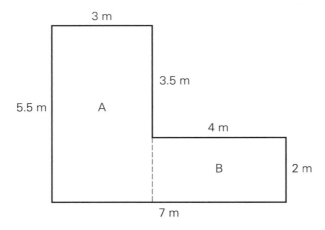

Area of rectangle A = 3 × 5.5 = 16.5 m²

Area of rectangle B = 4 × 2 = 8 m²

Total area of room = 16.5 + 8 = 24.5 m²

You could divide the rectangle in the example above into two different rectangles, C and D, like this (see page 153):

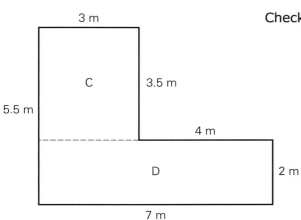

Check that you get the same total area.

Some shapes can be divided into rectangles and triangles. For example, to find the area of this wooden floor:

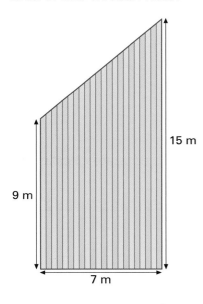

Divide the floor into a right-angled triangle A and a rectangle B

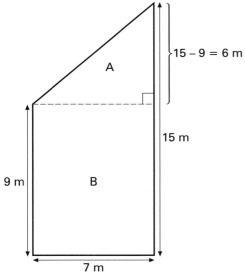

Triangle A has vertical height 6 m and base 7 m

Area triangle A $= \frac{1}{2} \times b \times h$

$= \frac{1}{2} \times 7 \times 6 = 21$ m²

Area rectangle B $= 9 \times 7 = 63$ m²

Total area $= 21 + 63 = 84$ m².

Circumference of a circle

The circumference of a circle is its perimeter – the distance all the way around the outside. The formula for the circumference of a circle of radius r is

$$C = 2\pi r$$

The radius is the distance from the centre of a circle to its outside edge. The diameter is the distance across the circle through the centre (diameter = 2 × radius).

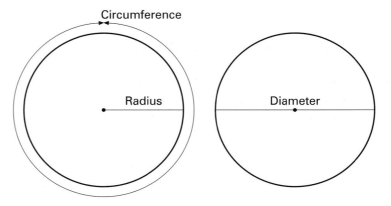

$\pi = 3.141\,592\,654\ldots$

To estimate the circumference of a circle, use $\pi = 3$. For more accurate calculations use $\pi = 3.14$, or the π key on a calculator.

For example, to estimate the circumference of this circular pond, with radius 2 m:

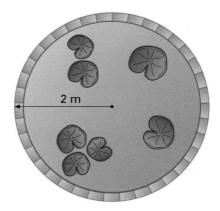

For an estimate, use $\pi = 3$

$$\text{Circumference} = 2\pi r = 2 \times \pi \times r = 2 \times 3 \times 2 = 12 \text{ m}$$

Calculate the circumference of a circular patio with radius 3.5 m.

$$\text{Circumference} = 2\pi r = 2 \times \pi \times r = 2 \times 3.14 \times 3.5$$

$$= 21.98 \text{ m}$$

$$= 22 \text{ m to the nearest metre}$$

Area of a circle

The formula for the area of a circle of radius r is

$$\text{Area} = \pi r^2$$

We can calculate the area of a circle with radius 3.25 m

$$\text{Area} = \pi r^2$$

$$= \pi \times r^2 = 3.14 \times 3.25 \times 3.25 = 33.166\,25$$

$$= 33 \text{ m}^2 \text{ to the nearest metre.}$$

Did you know?

If you are given the diameter of the circle, you halve the diameter to find the radius

Part circles and composite shapes

You can use the formulae for circumference and area of a circle to calculate perimeters and areas of parts of circles, and shapes made from parts of circles. For example, we can work out the perimeter and area of this semicircular window.

The diameter of the semicircle is 1.3 m, so the radius is 1.3 ÷ 2 = 0.65 m. The length of the curved side is half the circumference of the circle with radius 0.65 m.

$$\text{Length of curved side} = \frac{1}{2} \times 2\pi r = \frac{1}{2} \times 2 \times \pi \times r$$

$$= \frac{1}{2} \times 2 \times 3.14 \times 0.65 = 2.041 \text{ m}$$

Circumference of the semicircle = curved side + straight side

$$= 2.041 + 1.3 = 3.341\text{m}$$

$$= 3.34 \text{ m (to the nearest cm)}$$

Area of semicircle = half the area of the circle with radius 0.65m

$$= \frac{1}{2} \times \pi r^2 = \tfrac{1}{2} \times \pi \times r^2$$

$$= \frac{1}{2} \times 3.14 \times 0.65 \times 0.65 = 0.663\,325 \text{ m}^2$$

$$= 0.66 \text{ m}^2 \text{ (to 2 d.p.)}$$

To find the area of a quarter circle, use $\frac{1}{4}\pi r^2$.

To find the perimeter of a quarter circle, work out $\frac{1}{4}$ circumference + 2 × radius.

To find the area of a composite shape including parts of circles, divide it into circles and simple shapes and find the areas separately.

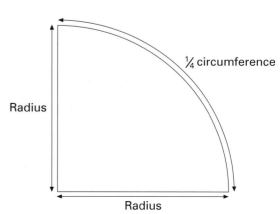

¼ circumference

Radius

Radius

Units of area

Area is measured in **square units** such as mm², cm², m².

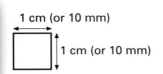

1 cm (or 10 mm)

1 cm (or 10 mm)

The area of this square is 1 cm² or 10 × 10 = 100 mm²

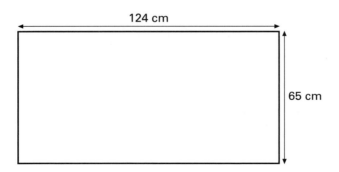

1 m (or 100 cm)

1 m (or 100 cm)

The area of this square is 1 m² or 100 × 100 = 10 000 cm²

We can work out the area of this rectangle

(a) in cm²

(b) in m²

124 cm

65 cm

(a) $A = l \times w = 124 \times 65 = 8060$ cm²

(b) 8060 ÷ 10 000 = 0.806 m²

Volume

Volume is the amount of space taken up by a 3-D or solid shape. Volume is measured in cube units, such as cubic centimetres (cm³) and cubic metres (m³).

A cuboid is a 3-D shape whose faces are all rectangles.

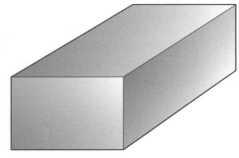

A cube is a 3-D shape whose faces are all squares.

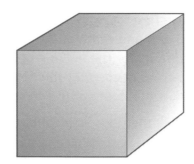

This cuboid is made of 1cm³ cubes.

You can find the volume by counting the cubes.
Volume = number of cubes = 36 cm³.

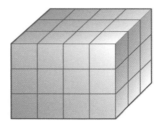

You can also calculate the volume by splitting the solid into equal rectangular layers.

Each layer has 4 × 3 cubes.

There are three layers, so the total number of cubes is 3 × 4 × 3 = 36 cm³.

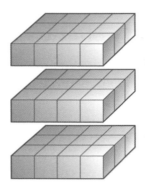

The volume of a cuboid with length *l*, width *w* and height *h* is

$$V = l \times w \times h$$

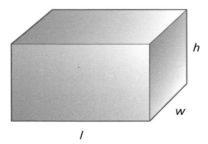

For a cube, length = width = height, so the volume of a cube with side *l* is

$$V = l^3$$

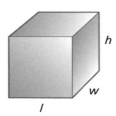

You can find the volume of concrete needed for a rectangular floor by measuring the length and width of the floor, and the depth of the concrete required and using the formula for the volume of a cuboid. For example, we can work out the volume of concrete needed for the floor of a rectangular room with length 3.7 m and width 2.9 m, if the depth of the concrete is to be 0.15 m.

Visualise the floor as a cuboid, like this

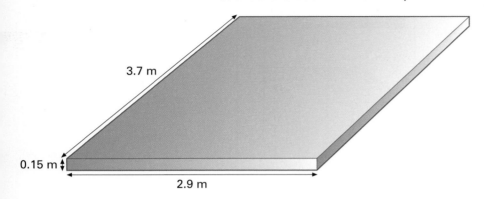

The depth of the concrete (0.15 m) is the height of the cuboid.

$$V = l \times w \times h$$
$$= 3.7 \times 2.9 \times 0.15$$
$$= 1.6095 \text{ m}^3$$
$$= 1.61 \text{ m}^3 \text{ (to 2 d.p.)}$$

Units of volume

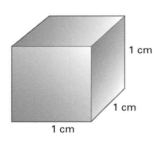

Volume is measured in **cube units** such as mm^3, cm^3, m^3. The volume of this cube is 1 cm^3 or $10 \times 10 \times 10 = 1000$ mm^3.

The volume of this cube is 1 m^3 or $100 \times 100 \times 100 = 1\,000\,000$ cm^3.

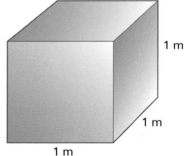

A cuboid is a 3-D shape with rectangular faces (like a box). The formula for the volume of a cuboid is $V = l \times w \times h$. For example, we can calculate the volume of this cuboid:

(a) in cm^3

(b) in m^3

(a) $V = l \times w \times h = 56 \times 84 \times 221 = 1\,039\,584$ cm^3

(b) $1\,039\,584 \div 1\,000\,000 = 1.039584$ $m^2 =$
    1.04 m^3 (to 2 d.p.)

Remember

$1 \text{ cm}^3 = 1000 \text{ mm}^3$

$1 \text{ m}^3 = 1\,000\,000 \text{ cm}^3$

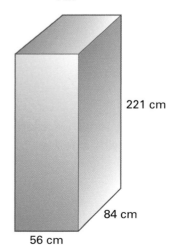

Rounding to a sensible degree of accuracy

Sometimes measurement calculations give an answer to a large number of decimal places. It is sensible to round the answer to a measurement that is practical. For example, Ahmed has a piece of wood 190 cm long. He wants to cut it into 7 equal lengths.

He works out $190 \div 7 = 14.28578142$ cm

You cannot measure 0.28578142 of a centimetre!

It is sensible to round to 14.3 cm (to 1 d.p.), which is 14 cm and 3 mm and can be measured.

For more on rounding to a number of decimal places, see page 134.

Let's look at another example:

A plasterer calculates the total area of walls in a room as 36 m^2.

Plasterboard sheets have area 2.88 m^2.

The number of plasterboard sheets needed is $36 \div 2.88 = 12.5$

If the plasterer buys 12 sheets he will not have enough.

If he buys 13 sheets he will have half a sheet left over.

In this case it is sensible to round up, and buy 13 sheets.

You may find there are times when you need to round down.

In some situations it is most sensible to round down. How many 2 metre lengths can be cut from 7 metres of pipe?

$7 \div 2 = 3.5$

You can cut three 2 metre lengths. The rest (0.5 of a 2 metre length) is wasted.

So in this case it is sensible to round down: you can only cut three 2 metre lengths.

chapter 8

Hand tools and portable power tools

OVERVIEW

Whatever type of work is involved, the quality of the finished product depends upon the skill of the operator in selecting and using the correct tools to cut, shape and assemble the materials for the task. There is a wide range of tools available and this chapter looks at the main hand and portable power tools that the carpenter and joiner should be able to use and maintain properly. We will also look at the importance of safety and the precautions you should take when using tools. Issues relating to the protection and maintenance of tools will also be addressed.

This chapter will cover the following topics:

- Hand tools
- Portable power tools.

These topics can be found in the following modules:

CC 1001K   CC 1001S

Hand tools

Although portable power tools are increasingly being used, and can complete many tasks much faster than hand tools, the good carpenter and joiner will still need to use a wide range of hand tools.

It is essential to have a basic set of hand tools. It is also good practice to extend your tool kit by purchasing good tools, as and when needed for a particular job, so building up a set of high-quality tools over time.

In this section we will look at each type of hand tool in turn, including where and how to use them. We will also address general safety measures and recommendations. Finally we will look at how to protect and maintain tools, so that they give long and efficient service.

Safety first

The following are general safety rules when using any hand tools:

- Wear the correct **PPE** for the task, particularly good quality safety glasses if there is any danger of fast-flying objects.
- Do not wear loose clothing or jewellery that could catch on tools.
- Secure all work down before working on it.
- Keep hands well away from the sharp edges of tools, especially saw teeth, chisels and drill bits.
- Never use a tool to do a job for which it was not designed.
- Make sure that tools are kept sharp and stored properly when not in use.
- Never force the tool – the tool should do the work.
- Never place any part of your body in front of the cutting edge.
- Keep work areas clean.

Summary of hand tool types

Hand tools can be grouped in the following categories, each of which is covered separately below, with illustrations:

- measuring and marking out tools
- sawing tools
- cutting tools
- planing tools
- shaping tools

Definition

PPE – personal protective equipment, which might include a helmet, safety glasses and gloves

- drilling or boring tools
- percussion tools
- screwdrivers
- holding and clamping tools
- lever tools
- tools for levelling
- jigs and guides.

Measuring and marking out tools

The main tools for measuring and marking out are:

- folding rules
- retractable steel tape measures
- metal steel rules
- pencils
- marking knife

- tri-square
- sliding bevel
- mitre square
- combination square
- gauges.

Folding rules

Folding rules are used in the joiner's shop or on site. They are normally one metre long when unfolded and made of wood or plastic. They can show both metric and imperial units.

Folding rule

Retractable steel tape measures

Retractable steel tape measures, often referred to as spring tapes, are available in a variety of lengths. They are useful for setting out large areas or marking long lengths of timber and other materials. They have a hook at right angles at the start of the tape to hold over the edge of the material. On better tapes this should slide, so that it is out of the way when not measuring from an edge.

Retractable steel tape measure

Remember

Imperial measurements (e.g. yards, feet, inches) have been replaced by metric (e.g. metres, millimetres) but most older items will have been constructed in imperial

Remember

Check the sliding hook regularly for signs of wear

Metal steel rules

Metal steel rules, often referred to as bar rules, are used for fine, accurate measurement work. They are generally 300 mm or 600 mm long and can also serve as a short straight edge for marking out. The rule can also be used on its edge for greater accuracy.

They may become discoloured over time. If so, give them a gentle rub with very fine emery paper and a light oil. If they become too rusty, replace them.

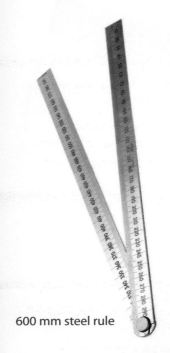

600 mm steel rule

Pencils

Pencils are an important part of a tool kit. They can be used for marking out exact measurements, both across and along the grain. They must be sharpened regularly, normally to a chisel-shaped point, which can be kept sharp by rubbing on fine emery paper. A chisel edge will draw more accurately along a marking out tool, like a steel rule, than a rounded point.

A variety of pencils

Pencils are graded by the softness or hardness of the lead. B grades are soft, H grades hard, with HB as the medium grade. Increasing hardness is indicated by a number in front of the H. Harder leads give a finer line but are often more difficult to rub out. A good compromise for most carpentry work is 2H.

Did you know?

The 'lead' in pencils is actually graphite, a naturally occurring form of compressed carbon. If graphite is pressed very much harder, it becomes diamond

Marking knife

Marking knives are used for marking across the grain and can be much more accurate than a pencil. They also provide a slight indentation for saw teeth to key into.

Marking knife

Tri-square

Tri-squares are used to mark and test angles at 90° and check that surfaces are at right angles to each other.

They should be regularly checked for accuracy. To do this, place the square against any straight-edged spare timber and mark a line at right angles. Turn the square over and draw another line from the same point. If the tool is accurate the two lines should be on top of each other.

Tri-square

Sliding bevel

The sliding bevel is an adjustable tri-square, used for marking and testing angles other than 90°. When in use, the blade is set at the required angle then locked by either a thumbscrew or set screw in the stock.

Sliding bevel

Mitre square

Remember

Do not over-tighten thumbscrews on a sliding bevel as they may snap

Mitre square

The blade of a mitre square is set into the stock at an angle of 45° and is used for marking out a mitre cut.

Combination square

A combination square does the job of a tri-square, mitre square and spirit level all in one. It is used for checking right angles, 45° angles and also that items are level.

Combination square

Remember

All squares should be checked for accuracy on a regular basis

Gauges

Gauges are instruments used to check that an item meets standard measurements. They are also used to mark critical dimensions, such as length and thickness.

Marking gauge

A marking gauge is used for marking lines parallel to the edge or end of the wood. The parts of a marking gauge include stem, stock, spur (or point) and thumbscrew. A marking gauge has only one spur or point.

Marking gauge

Mortise gauge

A mortise gauge is used for marking the double lines required when setting out mortise and tenon joints, hence the name. It has one fixed and one adjustable spur or point. Figure 8.1 shows setting of the adjustable point to match the width of a chisel.

Mortise gauge

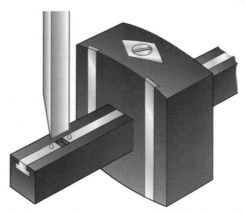

Figure 8.1 Setting mortise gauge to chisel blade width

165

Cutting gauge

The cutting gauge is very similar to the marking gauge but has a blade in place of the spur. This is used to cut deep lines in the timber, particularly across the grain, to give a clean, precise cut (e.g. for marking the shoulders of tenons).

Callipers and dividers

Callipers and dividers enable accurate checking of widths and gaps. They can have a simple friction joint or knurled rod and thread. The latter are more accurate for repetitive work, as the width setting can be maintained.

Callipers are designed for either internal and external gaps. Although some come with a graduated scale it is usually better to check measurements against a steel rule.

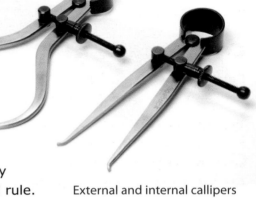

External and internal callipers

Sawing tools

There is a wide range of tools that cut using saw teeth. They can be classified as:

- hand saws, including rip saw, cross cut and panel saw
- back saws, so named because they have reinforcing metal along the back edge, including tenon, dovetail and bead saw
- saws for cutting circles or curves, including bow saw, coping saw, fret saw, pad, compass or keyhole saw.

They vary in the way they cut depending on the shape and size of the teeth, also on the angle the teeth are bent outwards from the line of the saw blade. This angle is called the set of the teeth. Cross cut teeth are designed to act like tiny knives that sever wood fibres while rip saw teeth are shaped to act like small chisels.

Saws are still commonly categorised by the number of 'teeth per inch' (TPI), but this is now taken to mean the number of teeth every 25 mm, as we have used below.

Many modern saws are designed to be used until blunt and then replaced. On older saws teeth need to be regularly maintained, which involves sharpening and setting. This is described in the section on tool maintenance later in this chapter.

Did you know?

The cut that the saw makes is called a saw kerf

Rip saw

Rip saws are usually used for cutting with the grain. Typically they are 650 mm to 700 mm long with 3–4 teeth per 25 mm. The teeth are filed at 90° across the blade, shaped like chisels and thus may be described as a gang of cutting chisels in a row. The saw should be at 60° to the work to cut most efficiently.

When starting to cut with a saw, make the first cut by drawing the saw backwards, as this avoids the saw jumping out from the mark, which can cause injury. Use your thumb as a guide as you start to cut.

Cross cut saw

Cross cut saws, as the name implies, are used for cutting across the grain, but can also be used for short rips on light timber.

Typically they are 650 mm long with 5–8 teeth per 25 mm. The teeth are bevelled (i.e. filed at an angle) across the saw to produce a knife-like edge on the forward and back edge of the tooth. The angle of this bevel is between 60° and 75°.

The cutting edge should ideally be at 45° to the work when sawing across the grain.

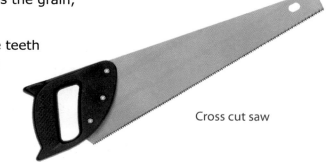

Cross cut saw

Panel saw

Panel saws are used for cutting plywood, large tenons and most fine work (e.g. on polished materials). They are obtainable up to 600 mm long. The teeth are practically the same as the cross cut saw, but usually there are 7–12 teeth per 25 mm.

Panel saw

Tenon saw

The tenon saw has teeth in a similar pattern to a cross cut saw but with 12–14 teeth per 25 mm. It is used for cutting joints and general bench work. There is a reinforcing strip along the top of the blade made from steel or brass to keep the saw rigid. Tenon saws are typically 300 mm to 350 mm long.

There is no special virtue in brass or steel for the reinforcement, except that brass is kept clean more easily.

Tenon saw

Dovetail saw

The dovetail saw is a smaller edition of the tenon saw with the same tooth pattern, but the teeth are finer, 18–24 per 25 mm, and it has a thinner blade. It is used for dovetails and other fine work. Dovetail saws are obtainable up to 200 mm in length.

Bead saw

Even finer than the dovetail saw is the bead saw or 'gents saw' as it is sometimes known. They have 15–25 teeth per 25 mm and are designed for very delicate work. They are usually 150 mm to 200 mm long.

Bow saw

The bow saw, also known as a frame saw because of its construction, has a thin blade and is used for cutting circular or curved work. The blade is held in tension by a cord twisted tourniquet fashion, though modern versions are fitted with an adjustable steel rod. The teeth are of the cross cut saw pattern and blades are from 200 mm to 400 mm in length.

Bow saw

Coping saw

The coping saw has a very narrow blade held in tension by the springing of the frame and is also used for cutting curves, especially internal and external shapes. The teeth are of the rip saw pattern, typically 14 teeth per 25 mm and blades can be 150 mm long.

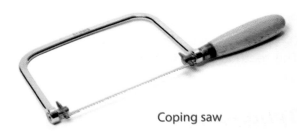

Coping saw

For internal shapes the blade can be released from the frame and reconnected and tensioned after pushing the blade through a hole drilled though part of the wood to be removed.

Fret Saw

Fret saws are very similar to coping saws but have a deep throat in the frame and smaller, finer blades, up to 32 teeth per 25 mm. They are used for cutting very tight curves and blades are generally no more than 125 mm long.

Fret saw

Pad, compass or key hole saws

Pad saws have no frame, so can be used where coping or fret saws cannot reach. They tend to have a single handle with interchangeable blades to carry out a range of tasks, especially for cutting key holes and large internal shaped work. The handle may be angled to the blade. Key hole saw handles are usually in line with the blade.

Blades are tapered with a slot, which fits into the saw handle and is then secured with two screws. Blades for cutting wood may come in lengths from 125 mm to 375 mm and have teeth set in the rip saw or cross cut saw pattern. Other blades will cut plastic, metal, etc.

Pad saw

Key hole saw

Hack saw

A hack saw is a framed saw that is used to cut metal components such as pipes. A hack saw can cut through copper, brass and steel.

Hack saw

Remember

Most modern saws require little maintenance, but should have teeth covered for protection when not in use and be wiped down with a clean, lightly oiled rag

Flooring saw

Flooring saws are specially designed to cut through floorboards in situ. They have a curved blade, which means they can cut into a board without having to drill through first, and also cause less damage to neighbouring boards. There are also teeth on an angled front edge, which allow cutting into skirting boards.

Blades are short and stiff, generally no more than 320 mm, with 8 teeth per 25 mm.

Flooring saw

Cutting tools

There is a wide range of tools designed to cut material using a sharp blade rather than saw teeth, including knives, scissors, chisels, gouges and axes. Chisels, gouges and axes are covered in more detail below.

Retractable knife

Chisels

As with saws, chisels are available in a variety of shapes and sizes. The ones that you are most likely to use are described below. Like all edge tools, chisels work best when they are very sharp. This ensures less effort needed to cut, which enables greater accuracy.

Firmer chisel

A firmer chisel is one of the strongest chisels. It is used for general purpose wood cutting and designed to be used with a mallet, if required.

The blade is generally about 100 mm long when new, rectangular in cross-section and tapering slightly from the bolster to the cutting edge. Size is normally by blade width, which may range from 6 mm to 50 mm.

Selection of firmer chisels

Bevelled-edge chisel

Bevelled-edge chisels are variations on the standard firmer chisel and come in similar sizes. The two long edges are bevelled, which makes them lighter but not as strong. Hence, they should not be used with a mallet, except for very light taps.

They are used for short paring and other fine work. The bevelled edge also helps when cleaning out corners that are less than 90°.

Bevelled-edge chisels

Paring chisel

Paring chisels may be rectangular in cross-section or have bevelled edges. They are longer, normally around 175 mm, more slender than the firmer chisels and are, thus, a more delicate tool. Their main use is for cutting deep grooves and long housings. They should not be hit with a mallet.

Paring chisel

Mortise chisel

The mortise chisel is designed for heavy duty work, with a thick stiff blade and generally shorter than the firmer and paring chisel. It may also have a slight bevel on all edges to allow easy withdrawal from the work. Sizes are typically 6 mm to 50 mm wide.

The handle is longer than on other chisels, with a wide curved end designed to take blows from a mallet. It is sometimes encircled with a metal ring to give extra strength and a leather washer inserted between the shoulder of the blade and the handle to absorb mallet blows. The handles can either be held by a simple tang or in a socket.

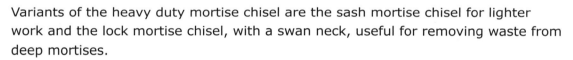

Mortise chisel

Variants of the heavy duty mortise chisel are the sash mortise chisel for lighter work and the lock mortise chisel, with a swan neck, useful for removing waste from deep mortises.

Gouges

Gouges are a type of chisel with a curved cross-section to the blade. They come in two main types, similar to chisels, called firmer and paring.

Firmer gouge

Firmer gouges come with two blade types: out-cannel and in-cannel.

- Out-cannel gouges have the cutting edge ground on the outside, so that they can be used to make concave cuts (i.e. cuts into the surface).

- In-cannel gouges have the cutting edge ground on the inside, so that they can be used to make convex cuts (i.e. leaving a protruding bulge on the surface).

Blade length is about 100 mm when new and they are designed to be used with a mallet. Blade widths typically range from 6 mm to 25 mm, though larger ones are available. Both are usually ground square across, though the out-cannel blade is sometimes rounded for cutting deep hollows.

Figures 8.2 and 8.3 show the two gouges being used to make convex and concave cuts.

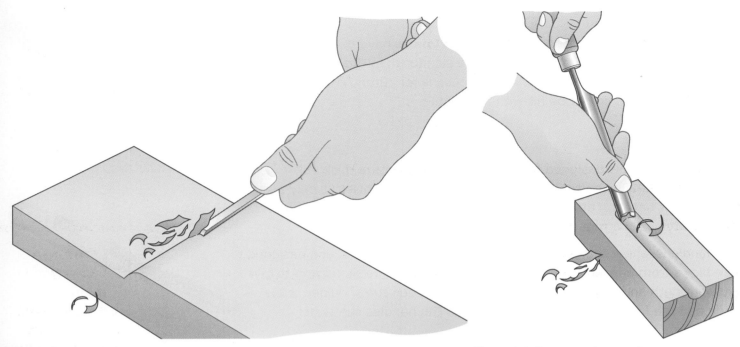

Figure 8.2 In-cannel gouge in use

Figure 8.3 Out-cannel gouge in use

Paring gouge

Similar to paring chisels, and sometimes known as scribing gouges, paring gouges are longer and thinner than firmer gouges and used for finer work. They are only ground on the inside for in-cannel cuts. They are not designed to be used with a mallet and may have a double bend in the neck to keep the handle clear of the work surface.

Axes

When an axe is used correctly, it is highly effective at removing waste wood and cutting wedges. However, it must be kept sharp, and the edge protected when not in use.

Axes vary in the shape of the head and its weight. The head is fitted on to a wooden shaft, preferably hickory, and held by wedges in a similar manner to a hammer. Axes designed to be used with one hand are typically around 1 kg in weight. Felling axes, with longer handles and designed to be used with two hands, may be 2.75 kg or heavier.

Hand axe

Planing tools

Planing tools are used to cut thin layers of wood, leaving a flat surface. They come in many forms, each developed to carry out a specific function or job, from levelling a surface to cutting bevels, rebates or grooves.

The most widely used are as follows:

- smoothing plane
- jack plane
- jointer or trying plane
- rebate plane

- plough plane
- shoulder plane
- bull nose plane
- block plane.

The first three are often referred to as bench planes. The different parts of a smoothing plane are labelled in the photo. Other planes have similar component parts.

Smoothing plane

The smoothing plane is the shortest of the bench planes and is used for final finishing or cleaning up, bevelling and chamfering. It can be used to follow the grain and can even be used with one hand.

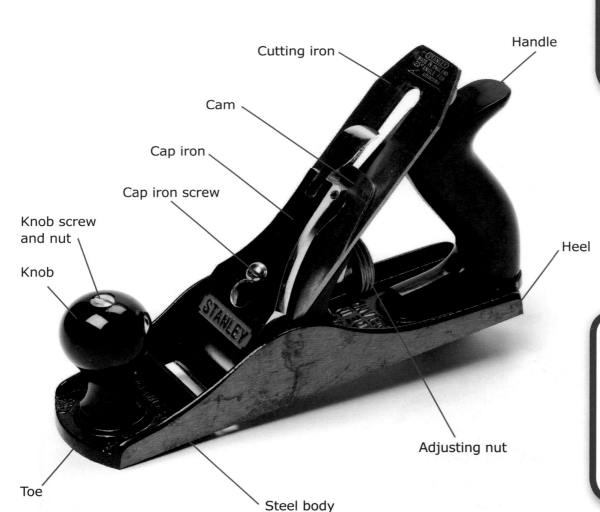

Smoothing plane

Jack plane

Jack plane

The jack plane is a medium length bench plane, generally 350 mm to 380 mm, with similar blade width to the smoothing plane. It is used for rapid and accurate removal of waste material, such as dressing doors for hanging.

Jointer or trying plane

The jointer, try or trying plane (sometimes also known as the long plane) is the longest bench plane, ranging from 450 mm to 600 mm. It is generally used for smoothing long edges of timber. The length helps to make the surface as level as possible. Typical blade width is 60 mm.

The sequence of operations for squaring off a length of timber is:

1. One surface is planed flat.

2. Then plane one edge square to this dressed face.

3. The wood can then be dressed on all round to the required width and thickness.

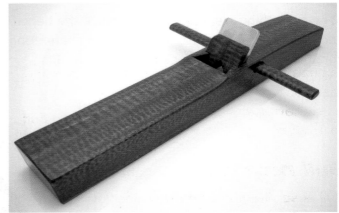

Jointer plane

Rebate plane

Rebate planes, as the name suggests, are used for making and cleaning out rebates. The simplest is known as a bench rebate plane. It is similar to the other bench planes described above but has the cutting blade extending across the full width of the sole and at an angle of 55° to 60°. It is usually 37 mm wide and at right angles across the sole, but can be obtained with the blade set at a skew.

More complex versions are called 'rebate and fillister planes', which have removable guide fences and depth gauges to assist in making accurate cuts. This avoids having to clamp a batten to the wood to act as a guide, which is advised with a standard bench rebate plane.

Rebate plane

Plough plane or combination plane

Plough planes also have guide fences and depth gauges. They are usually used for cutting grooves along the grain of the timber, as well as rebates, typically from 3 mm to 12 mm wide. However, they come with a multiple choice of square and shaped cutters adapted for shaping **tongues**, bends, **beads**, **mouldings** etc. The combination plane is essentially the same but may have additional capabilities.

Plough plane

Shoulder plane

Shoulder planes are similar to rebate planes but, as the name implies, they are primarily for shaping the shoulders of tenons etc. They are often much narrower with blade widths down to 15 mm.

Shoulder plane

Bull nose plane

Bull nose planes are similar to shoulder planes but have the cutting edge very close to the front of the plane. Hence, they are particularly useful for cleaning out the corners of stopped housings or rebates. Some have a removable front end so that the plane can be worked right into the corners. They are obtainable with cutter widths from 10 mm to 28 mm.

Bull nose plane

Compass plane

Compass planes have a flexible sole to enable them to be used on concave or convex surfaces. There is screw adjustment to change the radius of the curve, within limits. Blade width is typically about 45 mm.

Block plane

Block plane

Block planes come in a variety of designs but their key difference from other planes is the low angle at which the cutting blade is set, typically 20° but can be as low as 12°. This enables them to cut end grain.

Did you know?

Block planes are called this because they were used to cut butchers blocks, which traditionally used end grain timber

Shaping tools

Even though work on curved surfaces can be carried out with planes and chisels, the following tools are used for shaping curved surfaces and edges:

- spokeshaves
- rasps and files
- Surforms®.

Round face spokeshave

Spokeshave

Spokeshaves perform a similar role to a smoothing plane, except that they can do this on curved surfaces. They have a steel cutter set in either wood or metal handles and are designed to be used with two hands, pushing the tool along the wood to remove shavings.

There are two main types, a flat face for convex surfaces and a round face for concave surfaces. Cutter widths are typically around 50 mm but they can be found up to 100 mm.

Flat face spokeshave

Rasps and files

Rasps and files have a hard, rough surface of miniature teeth designed to smooth wood or metal by removing very small shavings. Rasps have teeth formed individually, whereas files have teeth formed by cutting grooves in the surface in patterns.

Files, in particular, come in a wide range of shapes and sizes, with rectangular, triangular, oval or circular cross-sections, and combinations of these. They are generally classified by the number of teeth per 25 mm and the pattern of cuts used to create them. These may range from 26 teeth per 25 mm on coarse files to 60 teeth on smooth ones, and the grade may differ between faces on the same file. Generally the longer the file the coarser it is.

Files

Rasps are mainly flat, round or half round and the teeth coarse. They are mostly used for initial shaping of wood, able to remove a lot of material quickly but leaving a rough surface which then needs to be smoothed with a file or spokeshave.

Rasp

Surforms®

Surforms® differ from rasps because the teeth are punched right through the metal, enabling timber shavings to pass through. Hence, the tool is less inclined to clog up. They can be flat or round, the latter being useful for enlarging holes and shaping cuts.

Surform®

Drilling or boring tools

There are countless reasons why holes may be required in timber, for example to insert screws, dowels or other fittings. Hand tools used for these tasks can be broken down into:

- bradawls
- hand drills
- carpenter's brace.

Many of these tasks, traditionally completed with hand tools, have been taken over by the power drill. In particular, the hand drill and carpenter's brace are much less used today. However, hand tools for making holes still play an important part in the workplace and the good carpenter and joiner should be able to use them all.

Bradawl

The bradawl is used for making pilot holes for small screws or a centre mark for drilling. They come in various cross-sections, and with sharp points, tips like a sharpened screwdriver blade or even spiral cutters. The points work by pushing aside the wood fibres when pressure is applied, but this can cause the wood to split. A sharp screwdriver tip can reduce this risk by cutting the fibres initially, the hole then being further deepened by a point.

Bradawl

Hand drill or wheel brace

Hand drills, or wheel braces, are useful for boring screw holes up to 6 mm in diameter. A selection of twist bits is required, usually with a round shank as used with power drills.

Hand drill

Carpenter's brace

The carpenter's brace has been in use for several hundred years and can hold a wide variety of bits for boring holes and other tasks. Most types of brace have a ratchet mechanism so that the tool can be used in restricted spaces.

The chuck of the brace is designed to grip bits with square shanks, called a tang. However, some braces can accept rounded shanks that are designed for power tools. The main types of bits are described on the page opposite.

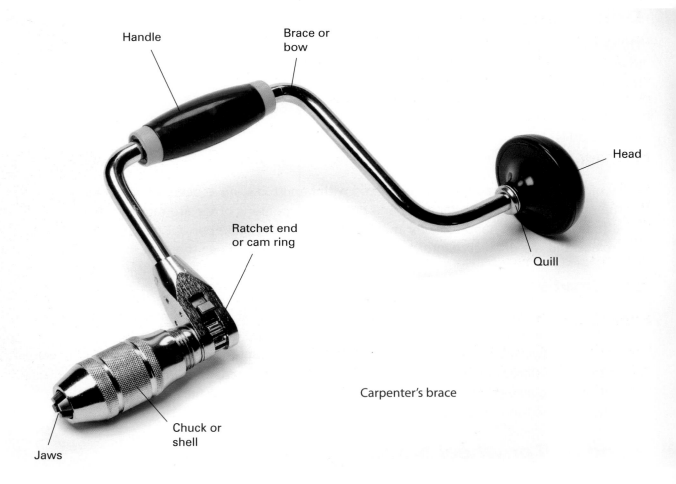

Carpenter's brace

Centre bit

A centre bit is a short, fast cutting bit. It has a lead screw and a single cutting spur. It is used to bore accurate but shallow holes. To go right through the wood it is normal to drill into wood until the point comes through, then use this as the centre mark to drill from the other side.

Twist bit

Twist bits have a lead screw and two cutting spurs. The shank has a helical twist or spiral that clears the wood from the hole. Irwin bits have a single spiral, the Jennings a double spiral. Both are used for boring holes from 4 mm upwards. The Jennings bits are claimed to be more accurate and stronger because of the double spiral, so are favoured for longer bits, but, in practice, there is little to choose between them.

Irwin bit

Jennings bit

Forstner bit

Forstner bits are used for high-quality, accurate work where a flat bottom to a hole is required. They have no threaded point and need a pilot hole made by other cutters to start and then steady pressure to make them cut. They can range in size from 10 mm to 50 mm.

Expansion bit

Expansion bits are similar to centre bits but with an adjustable cutting spur. They are only designed to cut a shallow hole.

Forstner bits

Expansion bit

Safety tip

Never use an expansion bit in any power tool

Countersink bit

Countersink bits are used to form a recess for a screw head. They come with different cutter shapes. Snail countersinks are used in braces, rose countersinks in hand drills or power drills.

Snail countersink

Rose countersink

Screwdriver bit

A carpenter's brace with a screwdriver bit can be very useful for removing stubborn and worn screws, as considerable pressure can be applied and the brace handle gives excellent leverage. They are often more controllable than a power drill and less likely to jump out, which can damage the screw further. They can also be used to insert screws.

Screwdriver bits

Percussion tools

Percussion tools can be grouped into:

- hammers
- mallets
- punches.

Hammers

Hammers are available in various types and sizes and are essential to the craftsman. The head has a protruding part, often called the bell, with a flat face for striking the work, an eye for fitting over the end of the shaft and a protruding part at the back, the pein, most often shaped as a claw, wedge or ball. Wooden shafts, made from a hardwood like ash or hickory, are still preferred for most work, as they absorb the shock. The head is secured to the shaft by driving a wedge or wedges into slots in the head of the shaft.

Claw hammer

The claw hammer is used to drive nails into timber, with the claw available to withdraw bent or unwanted nails. It should be of the best quality to perform safely and efficiently. They come in various weights up to about 570 g. Increasingly they have steel shafts and integral heads, as much more leverage can be applied when using the claw without danger of loosening the head.

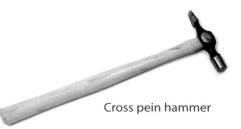

Claw hammer

It is best to place a spare scrap of timber or hardboard under the claw as packing to protect the surface of the timber when using it. You can also get additional leverage and withdraw a nail straighter, thus causing less damage to the timber, by using thicker packing under the claw.

Cross pein hammer

The cross pein hammer has a wedge-shaped pein. The pein can have various patterns and the hammers come typically with head weights from 200 g to 570 g. The Warrington pattern with a tapered pein is favoured by bench joiners for lighter work.

Cross pein hammer

Mallets

Carpenter's mallets have a much larger head than hammers, and are usually made from hardwood (beech is the commonest) with an ash or hickory shaft. The head can be rectangular in section or round.

They are primarily for use with chisels but also for a variety of 'persuading' purposes, such as knocking components or material into place without causing damage.

Mallet

There are also soft-faced mallets with either a rubber head or tightly rolled rawhide, glued and often loaded with lead. These are useful for material that wooden mallets may damage.

Punches

Punches come in a wide range of shapes and sizes with different shaped tips. The type most frequently employed by a carpenter and joiner is the nail punch.

Nail punches have a square head, a knurled gripping section and a tapered point with a hollow tip. They are used to punch nail heads below the surface of the wood. The punches are available in various sizes to suit the different diameters of nails available. Smaller ones are called pin punches. The hollow tip prevents the punch slipping off the nail.

Nail punch

Pin punch

Push pin nail drivers can be used with pins or small nails when a hammer would be impractical. They have a built-in spring mechanism to propel the nail into the wood and can be activated one-handed.

Push pin nail driver

Screwdrivers

Screwdrivers are an important part of any tool kit. There are many types and styles but they all serve a similar purpose, that is inserting and withdrawing screws. Hence, they come with different tips to match the size and type of screws available, the most common of which for wood are:

- slotted head
- Phillips cross head
- Posidrive, similar to a Phillips head but with an additional square hole in the centre for added grip by the screwdriver.

Slotted screwdriver

Phillips screwdriver

Posidrive screwdriver

Screwdrivers also vary by the means used to drive screws. They can be grouped as:

- basic screwdrivers with no moving parts
- ratchet screwdrivers
- pump screwdrivers.

Basic screwdriver

Basic screwdrivers are available in various sizes from 50 mm to 400 mm. The handle is made of hardwood or unbreakable plastic. Traditional wooden handles are generally bulbous to give a good grip. Plastic ones are often moulded with flutes.

Screwdrivers to fit slotted screws may have tips that flare out, flared but then ground so that they are narrower towards the point, or parallel (i.e. the same width as the shank). Those made to fit other screw types are generally parallel.

Various basic screwdrivers

Ratchet screwdriver

Ratchet screwdrivers have a ratchet mechanism built into the handle so that when turned one way the blade locks and acts like a basic screwdriver. When turned the other way the ratchet allows the handle to turn but leaves the blade in place.

There is a means to reverse the ratchet. Hence, screws can be driven in or withdrawn without changing grip on the handle. They often have interchangeable tips for different screw types, and can also be locked in one position so that they act like a basic screwdriver until the screw is sufficiently deep in the wood for the ratchet to work.

Ratchet type screwdriver

Pump action screwdrivers

Pump action screwdrivers have spiral grooves down the length of the screwdriver shank. There is then a cylinder as an extension of the handle, which can move over the grooves against an internal spring.

When pressure is applied to the handle it tries to follow the grooves and, if the handle is held still, the blade is forced to turn. When pressure is released the spring in the handle allows it to return to its starting point, leaving the blade in place and ready to pump the handle down again without changing grip.

This is similar to a ratchet screwdriver and, hence, these screwdrivers are often known as spiral ratchet screwdrivers. Like them, the mechanism can also be locked in place to operate like a basic screwdriver, and normally have a chuck to take different bits.

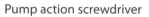

Pump action screwdriver

The pump action allows screws to be driven and removed quickly. Therefore they are useful in repetitive operations, though it is generally advisable to drill a pilot hole unless screwing into very soft wood.

Holding and clamping tools

Joiners frequently need to grip items and hold them steady while using tools, or while fixing them in position with nails, screws or glue. These include simple tools like pliers and pincers, as well as carpenters' vices attached to fixed or portable work benches, and a range of clamps (or cramps).

Pliers

Pliers are a universal gripping device used to grip small nuts and bolts. They can also be used to bend and cut, as well as to strip wire.

Pliers

Pincers

Pincers are the simplest gripping tool and are an essential wood worker's tool. They are primarily used for removing nails but have many other uses. A good quality pair should give many years of service.

Pincers

When using pincers for removing nails they should be operated similarly to a claw hammer. Scraps of wood should be used to avoid damage to the timber and thicker pieces used to get additional leverage. There is often a small claw on one handle, which is useful for removing tacks.

Clamps

Clamps, or cramps, have been developed to hold almost any shape of wood or other material in almost any position. Some are spring-operated, others have screw threads to pull jaws onto the material, others may use web straps that can be ratcheted tight, often around a frame. They may incorporate quick-release mechanisms.

The two you are likely to use most frequently are described below.

Sash clamp

Sash clamps are designed to hold large pieces of wood or frames together, usually while gluing.

A steel bar, either flat or T-section, has holes drilled along its length. An end shoe, flat to fit over the edge of the work to be clamped, is able to be moved a limited distance down the bar by using a screw mechanism driven by a tommy bar.

Sash clamp

> **Did you know?**
>
> Wooden packers can be used on shoes to avoid damage to the work

A second shoe, the tail shoe, is designed to fit over the other edge of the work. It can be locked in place by a steel pin through any of the holes in the bar.

With the end shoe withdrawn towards the end, the clamp is placed over the work and the tail shoe moved to give a loose fit. It is then locked in place with the pin. Pressure is then applied by screwing down the end shoe. The clamp must be square across the job to get even pressure.

G clamp

G clamps take their name from their shape, which resembles the letter 'G'. They come in various sizes and forms, and may include a quick-release mechanism but all have a fixed end and an adjustable end.

G clamp

They are very versatile and particularly useful for clamping work pieces to the bench to allow operations to be carried out, such as routing or sawing. Again, pieces of scrap material can be used as packers to prevent damage to the work.

Lever tools

Lever tools are known by various names (e.g. wrecking bar, crowbar, nail bar or tommy bar) and are designed for prising items apart, levering up heavy objects or pulling large nails.

Wrecking bar

Tools for levelling

The most common levelling tool on site is the spirit level. It has a metal body into which are fitted one or more curved glass or plastic tubes known as 'vials'. These contain a liquid and a bubble of air. They work on the principle that a bubble of gas, enclosed in a glass tube containing a liquid, will always rise to the highest possible point within the tube. There are usually marks on the glass either side of the centre at the width of the bubble. When the bubble is positioned between them the tool is level.

Levels vary in length from 200 mm to 1 m or more and are used for marking and testing level surfaces or, in some cases, marking and testing vertical surfaces (plumbing). Good quality levels have adjustable tubes that can be reset should any inaccuracy develop.

Check for accuracy regularly by spanning something like a door frame and marking two level points, one at each end. Reverse the tool against the marks and see if the bubble is still level. If not, reset or replace the tool.

Safety tip

Care must be taken when prising and removing large nails and the correct PPE worn

Spirit level

A range of electronic levels and plumb bobs are becoming available, based on lasers. They are often mounted on a stand and self-levelling.

Laser levels work by shining a beam of light at whatever you wish and can be used to transfer levels. Rotary laser levels are used often in the construction of suspended ceilings and floors and they work in the same way except the laser spins very fast and projects a line around the room which you can work to.

Laser level

Jigs and guides

There are other items of equipment, on the bench and on site, which help in carrying out accurate work. Those designed to assist a specific task are generally referred to as **jigs**, some of which you can make yourself or they can be provided by the manufacturer (e.g. jigs to help fit kitchen units). Other items can provide more general support. Two of the commonest are guides to assist when sawing, described below.

Definition

Jig – any device made to hold a specific piece of work and help guide the tools being used

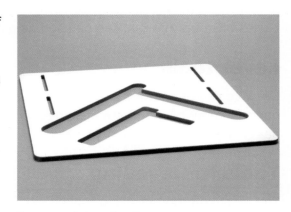

Router stair-housing jig

Mitre boxes and blocks

Mitre boxes or blocks are used to guide the saw when cutting angles, particularly mitres. Usually they are made to suit the job and replaced when worn, though purpose-built versions can be purchased.

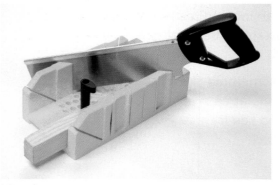

Mitre box

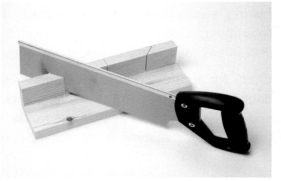

Mitre block

Bench hook

Bench hook

A bench hook is a very simple device for holding the timber on a bench when sawing.

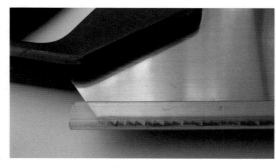
Saw teeth covered for protection

Tool protection

When stored in a tool bag, tool box or cupboard all tools should be prevented from moving about, otherwise cutting edges and sharp points can be damaged. Some useful tips for protecting tools are given below.

- Planes should have blades retracted before being stored.

- Protective covers can be made or purchased for saw teeth.

- Most chisels are provided with plastic end covers when purchased and should be used. A strong, purpose-made chisel roll is also useful when storing in a tool bag or box.

Chisel with plastic end cover

- The sharp points on twist bits should always be protected as, unlike chisels, reshaping is difficult.

Maintenance of hand tools

Good-quality hand tools will last a lifetime if well looked after. All metal tools need to be kept free of rust, which can be achieved by rubbing them over occasionally with an oily cloth to preserve a light film of oil.

Chisel roll

Saw setting and sharpening

Most modern saws are of the hard-point type and cannot be re-sharpened. For these, replacement is the only option. Older saws do require periodic maintenance, the frequency depending on the amount of work they do.

There are four stages in returning a saw to tip-top condition:

1. topping
2. shaping
3. setting
4. sharpening.

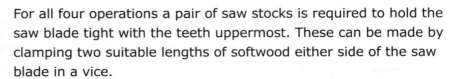

Saw blade held in saw stocks

For all four operations a pair of saw stocks is required to hold the saw blade tight with the teeth uppermost. These can be made by clamping two suitable lengths of softwood either side of the saw blade in a vice.

There are then various saw files available, designed for the different tasks.

Topping a saw blade

Topping or levelling the teeth is completed by running a flat saw file along the top of the teeth. A simple jig to help with this can be made of hardwood, with a slot cut in it to hold the file at right angles. The wood can then rub against the blade to ensure that the file remains flat on the teeth.

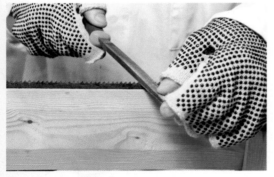

Shaping a saw blade

Shaping is undertaken with a saw file designed for the task. This restores the teeth to their original shape and size. The aim is to have all teeth with the same angle, or pitch.

Setting saw teeth

Teeth are then set individually to the required angle to give the saw blade clearance in its kerf. This is easiest with a tool called a saw set, which is used to bend the tips of the teeth to the correct angle.

The tool is operated by pre-selecting the required points per 25 mm for the saw blade being sharpened. Starting at one end, the tool is then positioned over a tooth and the handles squeezed. This will automatically set the angle of the tooth tip correctly. The procedure is repeated for each alternate tooth and then repeated from the opposite side for the teeth angled in the other direction.

Sharpening puts a cutting edge on the teeth. It is completed with a triangular file, holding the file at an angle to suit the type of saw blade and lightly filing each alternative 'V' between the teeth three times. When arriving at the end of the saw the blade is turned through 180° and the process repeated for the other teeth.

Rip saw teeth are filed at right angles across the blade to create a chisel effect. Cross cut saw teeth are normally filed with the file at approximately 70°, sharpening both the front of one tooth and the back of the next at the same time.

Saw set tool

Planes and chisels

New plane or chisel blades will come with a bevel of 25° on the cutting edge. This has to be sharpened (honed) to an angle of 30° before use and regularly re-sharpened to this angle. See Figure 8.4.

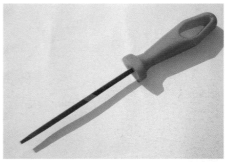

File used for sharpening saw teeth

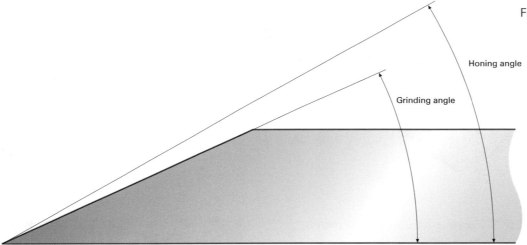

Honing angle

Grinding angle

Figure 8.4 Honing and grinding angles

Sharpening is best done by hand using an oilstone. These come in coarse, medium or fine grades, often as a combination with two of them on opposite sides. Tool sharpening should always be completed on a fine grade, the others only used for reshaping. Stones can be cleaned with a brush and paraffin and should be re-levelled occasionally by rubbing on a sheet of glass with carborundum paste. The stone should always be kept oiled during sharpening.

Combination oilstone

The following procedures will enable you to sharpen the blades of planes and chisels.

Step 1 Position the blade. Holding the blade in both hands, position its grinding angle flat on the stone. This means the blade will be at an angle of 25°. Then raise the back end up a further 5°, so that the tool is now at the correct sharpened angle of 30°.

Step 2 Grind an initial burr. Slowly move the blade forwards and backwards until a small burr has formed at the cutting edge. Make sure that you use the entire oilstone surface, to avoid hollowing the stone.

Step 3 Form a wire edge. Remove the burr by holding the blade perfectly flat and drawing the blade towards you once or twice to form a wire edge on the end of the blade.

Step 4 Remove the wire edge. The wire edge can be removed by drawing the blade over a piece of waste timber. Now inspect the sharpened edge and, if you can still see a dull white line, then repeat Steps 2 to 4.

Two additional tips specifically for sharpening planes are:

1. Hold the blade at a slight angle to the line of the stone. This helps to ensure that the whole of the cutting edge makes contact with the stone.

2. To ensure that the cap iron fits neatly with the blade, also use the oilstone to flatten and straighten its edge.

Sharpening twist bits

The life of a twist bit is shortened every time it is sharpened, so only sharpen when necessary. Twist bits can come with or without spurs. See Figure 8.5 for a twist bit with spurs. The spurs are the first to become dull, followed by the cutters. If the screw point becomes blunt the twist bit must be replaced.

Twist bits can be sharpened with various shaped files, the shape depending on the type of twist bit.

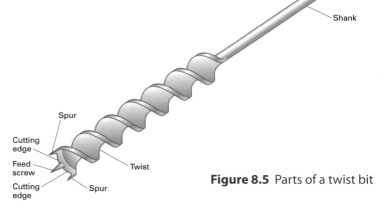

Figure 8.5 Parts of a twist bit

Sharpening a twist bit with a triangular file

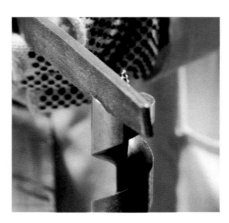

Sharpening a twist bit with a flat file

Portable power tools

By the end of this section you will have the basic knowledge to begin to use portable powered hand tools safely but, as with any tool, you will need practice to become competent in their use.

These tools are covered by legislation and you must fully comply with the following:

- *The Provision and Use of Working Equipment Regulations 1998*
- *The Abrasive Wheels Regulations 1974*
- *The Protection of Eye Regulations 1974*
- *The Health and Safety at Work Act 1974.*

Summary of power tool types

The following power tools are described in this section:

- power drills
- powered planer
- routers
- portable circular saws
- chop saws
- jig saws
- sanders
- cordless tools.

First we will cover:

- safety issues common to all power tools
- power supplies
- power tool maintenance.

Safety first

If used correctly, powered hand tools can save time, money and effort. Skill in using them comes only through training and experience – it cannot be picked up on site. All the safety precautions listed at the start of this chapter for hand tools apply but there are additional risks with power tools, especially portable ones.

Although each type of power tool has its own safe working procedures the following basic rules apply to *any* portable powered tools.

Remember

You must receive training and obtain permission before using any power tool and an instructor must be present while you carry out the operation

- Only use powered hand tools if you have been authorised to do so and have been taught how to use them correctly.

- Never use a power tool without permission.

- If you are unfamiliar with the equipment, read the instruction booklet and practise using the tool. Go ahead with the job only when you are sure you can do so safely and be certain of producing good results.

- Always select the correct tool for the work in hand. Check that it is in good condition and that any blade or cutter has been fitted correctly. *If in doubt, ask someone with experience.*

- Ensure that the tool and power supply are compatible. Do not mix voltages; for example 110 V tools must only be used on 110 V supplies.

Extension lead on drum

- Ensure that plugs, wires, extension leads etc. are in good condition. If not, do not use the tool.

- Ensure that any extension leads are fully uncoiled before use, as they can easily overheat.

- Never carry any tool by its lead.

- Always use fitted safety guards correctly, as these have been designed with your safety in mind. Never use a tool without them.

- Do not let others touch tools or extension cords while in use and disconnect tools when finished with.

- Never put a tool down until all moving parts have stopped and, before making any adjustments, disconnect a tool from its power supply or switch off at source. Just switching off at the tool is not good enough.

- Should a tool not operate, do not tamper with it but report the matter immediately to your supervisor. Repairs should only be carried out by a competent person.

- Wear goggles when there is any danger of flying particles. A dust mask or respirator, ear protectors and safety helmet may also be necessary.

- Report any accident (or 'near miss') to your supervisor, whether or not it results in injury to yourself or others.

- If you are injured, even when it is a minor wound or scratch, get immediate first aid treatment.

- Guard against electric shock. Hence, never allow electrical equipment to become damp or wet and never use electrical equipment in damp or wet conditions.

Safety tip

Tools cannot be careless but you can, so do the job safely – it is quicker in the long run

Transformer used to reduce voltage from 230 V to 110 V

Power supplies

For tools that need to be connected to a power supply this can be 230 V (normal domestic mains supply) or 110 V (as a reduced voltage via a transformer). The use of tools at 230 V is not recommended, as any electrical shock from the tool at this voltage can be fatal. Only tools using a supply of 110 V, reduced from 230 V through a transformer, can be used on construction sites.

All power tools are now made with double insulation and should be stamped with a double insulation sign and a kite mark. See Figures 8.6 and 8.7.

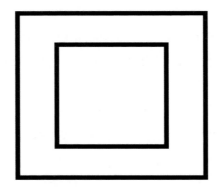

Figure 8.6 Double insulation sign

Figure 8.7 BS Kitemark

Power tool maintenance

Most modern power tools are designed to operate over a long period of time with minimum maintenance. However, regular attention to powered hand tools improves safety and ensures that they work efficiently.

In particular they should be cleaned regularly (especially important with cutting and abrasive tools because the dust produced can easily damage the motor). This may mean a daily clean for tools in constant use.

Information for cleaning and maintenance of power tools can be found in the instruction manual for each machine. This will clearly state which actions should and can be undertaken by you and which must be undertaken by authorised repair agents. Not only are they trained, and have the necessary specialist tools to complete the work correctly, but unauthorised repairs will invalidate any guarantee.

Regular visual checks must be made to spot damage to the tool, leads and plugs. Make sure that the ventilation slots are clear. Any damaged parts should be replaced before use, but only attempt this if it is a user-authorised task and you have been trained to carry it out.

Power drills

Power drills are highly versatile and can be used with different bits, not just to drill holes in a range of materials but to drive or remove screws, cut circular holes with a saw cutter, or buff and polish. Various designs of chuck are available for rapid interchange of bits, the chuck size being matched to the power of the drill.

The simplest power drills just operate at a single speed. However, most drills today come with additional capabilities, which may include one or more of the following:

- dual speed, the slower speed for use when boring into brickwork or masonry using a tipped drill bit
- variable speed, controlled by trigger pressure
- reverse action, that is able to operate clockwise or anticlockwise and mainly used for driving or withdrawing screws

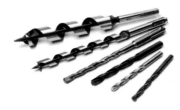

- percussion or hammer action, primarily for brickwork or masonry
- torque control, particularly useful to avoid over tightening screws, nuts or bolts.

Choosing the most suitable drill and drill bit for the job depends on:

- the type of material that needs to be drilled
- size of holes required or task to be completed
- how often the drill will be used.

Some of the common types of power drill are described below.

Palm grip drill

The most common and versatile of the electric drills, palm grip drills can be used to carry out work on timber, steel, alloy and masonry. Dust should be cleaned from bore holes at regular intervals.

They are available with single or twin speeds but may have other options as listed above.

Back handle drill

The back handle drill is a heavy-duty version of the palm grip drill. They have a larger chuck, allowing holes of greater diameter to be bored. They can also withstand long periods of heavy pressure but, as with the palm grip drill, dust should be cleaned from bore holes at regular intervals.

They are available in single-, two- and four-speed models but may also have other options as listed above. A handle can be attached to give greater control during use.

Rotary percussion (hammer) drills

Hammer drills have a percussive hammer effect which enables them to bore into hard materials. This percussive action is optional and is brought into operation by means of a switch. It is particularly effective for drilling into masonry or concrete. A depth gauge can be fitted to the side handle to provide accuracy where required.

110 V hammer drill

Good practice when using drills

- Always tighten the chuck securely.

- Take advantage of speed selection. Start drilling into hard materials at slow speed and increase gradually.

- Slow down the rate of speed just prior to breaking through the surface, particularly metals, to avoid snatching and twisting.

- Keep drill vents clear to maintain adequate ventilation.

- Use sharp drill bits at all times.

- Apply as much pressure as possible, consistent with the size of the drill. This is more important than speed, as it is essential to keep the edge of the drill biting into the material rather than let it rub on the bottom of the hole.

- Especially when using larger drill sizes, make a pilot hole with a smaller drill. For example, for a 13 mm hole use a 10 mm or 12 mm drill and then open this out with a 13 mm drill.

- Check carbon brushes regularly. Excessive sparking indicates excessive wear or a possible short circuit. Report this to a supervisor.

- Make sure all plug connections are secure and correct.

- Keep all cables clear of the cutting area during use.

- Keep holes clear of dust. Drills should be withdrawn from holes at intervals, as an accumulation of dust not only causes overheating but also tends to blunt the drill bit.

Bad practice when using drills

- Never use a drill designed to operate with a 3-wire source (i.e. live, neutral and earth) on a 2-wire supply (i.e. live and neutral only). Always connect an earth.

- Do not use a drill bit with a bent spindle.

- Do not exceed the manufacturer's recommended maximum capacities for drill sizes on appropriate materials.

- Do not use high-speed steel (HSS) bits without cooling or lubrication.

- Never use a hole saw cutter without the pilot cutter.

- Do not cool the hot point of a drill by dipping it in water, as this will crack the tip.

Powered planers

Powered planers are invaluable for removing large amounts of waste wood, especially on site. With large models it is possible to remove up to 3 mm in one pass. The machine has an adjustable depth gauge to allow the greater depths needed for rebating to be achieved.

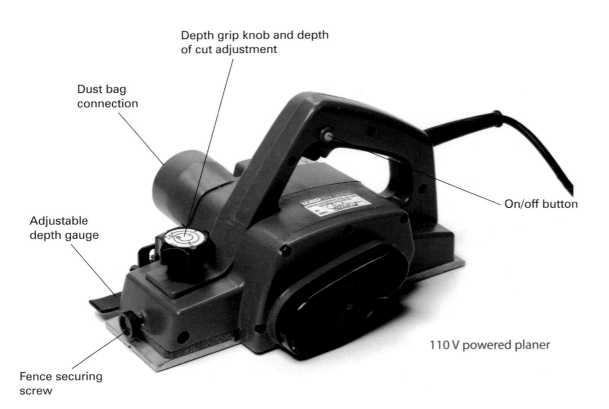

Depth grip knob and depth of cut adjustment

Dust bag connection

On/off button

Adjustable depth gauge

Fence securing screw

110 V powered planer

The sequence of operations is similar to using a hand plane. One surface is planed square and then one edge is planed square to this dressed face. The wood can then be dressed to the required width and thickness.

Remember:

- Check that the cutters are sharp to prevent overloading the motor and producing a poor finish.
- Secure material to be planed in a vice or on a workbench against a stop.
- Make adjustments to the planer before connecting to the power supply.
- Do not put the machine down until the cutters have stopped rotating.
- Wear ear, eye and nose/mouth protection.
- All machines should be checked on a regular basis by a qualified electrical engineer but the operator should check the visual condition of the planer, voltage, power cable and plug.

Routers

The router is a very versatile machine. Like a power drill it has a chuck able to take different sizes of bit, the size depending on the power of the machine. However, the variety of bits and cutters that can be fitted is very wide and many different operations can be completed, including:

- cutting straight, curved and moulded grooves
- cutting slots and recesses
- rebating
- beading and moulding
- dovetailing
- laminate trimming.

Routers used by carpenters and joiners on site are most likely to be:

- heavy-duty routers
- heavy-duty plunge routers.

With the heavy-duty router the cutter projects from the base and great care must be taken to feed it gradually into the material to prevent it from snatching.

Most routers now are of the plunge type, where the cutter is brought downwards into contact with the material to be cut and then withdrawn out of the material when downward pressure is released. This is made possible by two spring-loaded plungers, which are fitted to each side of the machine.

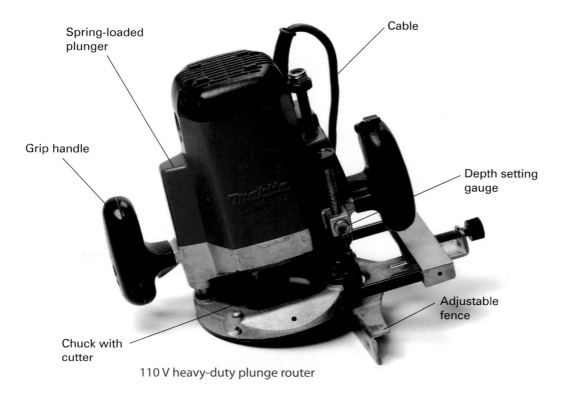

Spring-loaded plunger

Cable

Grip handle

Depth setting gauge

Adjustable fence

Chuck with cutter

110 V heavy-duty plunge router

Both types are capable of producing excellent finishes, mainly due to the very high speed of the machine, but these speeds demand great control and concentration from the operator.

For general work router bits and cutters are made of high-speed steel (HSS), but carbide-tipped bits and cutters offer a longer cutting life. Edging bits have a guide pin or roller which allows the cutter to follow an accurate path without biting into the material.

Routers come with various items as standard, including guides and fences, but some useful additional accessories are:

- trammel point and arm, used for cutting circles
- dovetail kit for producing dovetail joints
- bench or table stand that converts the router into a small spindle moulder
- trimming attachment to cut veneered edging.

Router bits for grooving and trimming

Router bits for edging

Remember:

- Secure the work piece before commencing work.

- Make sure the cutter/bit is free to rotate.

- Allow the machine to reach its maximum speed before making the first trial cut (into waste material).

- Make any cut from left to right.

- Move the router quickly enough to make a continuous cut but never overload the motor (listen to the drone) by pushing too fast, which will blunt the cutter and burn the wood.

- At the end of each operation switch off the motor. The cutter/bit should be freed from the work and allowed to stop revolving before being left unattended.

- Goggles or safety glasses must always be worn as well as other appropriate items of PPE.

- All machines should be checked on a regular basis by a qualified electrical engineer but the operator should check the visual condition of the router, voltage, power cable and plug.

Portable circular saws

Portable circular saws are mainly used for cross cutting and ripping, but can also be used for bevel cuts, grooves and rebates. They can be used to cut a wide variety of materials:

- timber, including softwood and hardwood

- manufactured boards, including plywood, chipboard, blockboard, laminboard and fibre board

- plastic laminates.

We will refer mainly to timber below, but the same principles apply to all materials.

Before use the saw should be adjusted so that, when cutting normally, the blade will only just break through the underside of the timber. This is achieved by releasing a locking device, which controls the movement of the base plate in relation to the amount of blade exposed.

A detachable fence is supplied for ripping. For bevel cutting the base tilts up to 45° on a lockable quadrant arm. The telescopic saw guard, covering the exposed blade, will automatically spring back when the cut is complete.

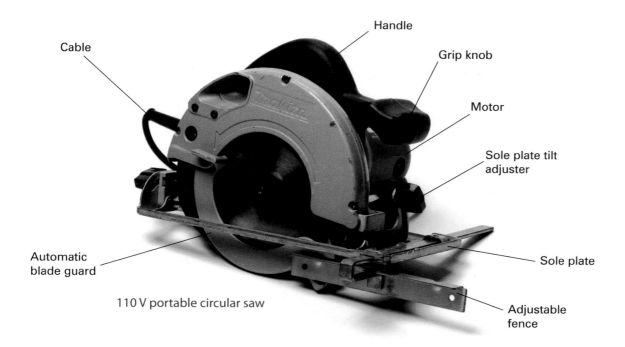

Cable

Handle

Grip knob

Motor

Sole plate tilt adjuster

Automatic blade guard

Sole plate

Adjustable fence

110 V portable circular saw

Remember:

- Wear appropriate PPE.
- Any timber being cut should be securely held, clamped or fixed, making sure that fixings are clear of the saw cut.
- Ensure the power cable is clear of the cutting action.
- Always use both hands on the saw handles, as this reduces the risk of the free hand making contact with the cutting edge of the blade.
- The saw should be allowed to reach maximum speed before starting to cut, and should not be stopped or restarted in the timber.
- At the finish of the cut, keep the saw suspended away from the body until the blade stops revolving.
- Disconnect the machine before making any adjustments or when not in use. It is not sufficient just to switch off the machine.
- Do not overload the saw by forcing it into the material.
- Keep saw blades sharp.
- Always work in safe, dry conditions.
- All machines should be checked on a regular basis by a qualified electrical engineer but the operator should check the visual condition of the saw, voltage, power cable and plug.

Chop saws

Chop saws have a circular blade rotating in a housing and a fixed bed on to which items to be cut can be fixed. The blade can be pulled down on to the work to cut it.

They can be fitted with different blades, depending on the material to be cut and the task. These can then be set to cut:

- square at 90° in both directions
- mitres at any angle up to 45°
- bevels up to 45°
- compound bevels up to 45° (i.e. two angles in one cut, both face side and face edge of the material).

Remember

When cutting angles or bevels, the machine's effective cutting depth will be reduced

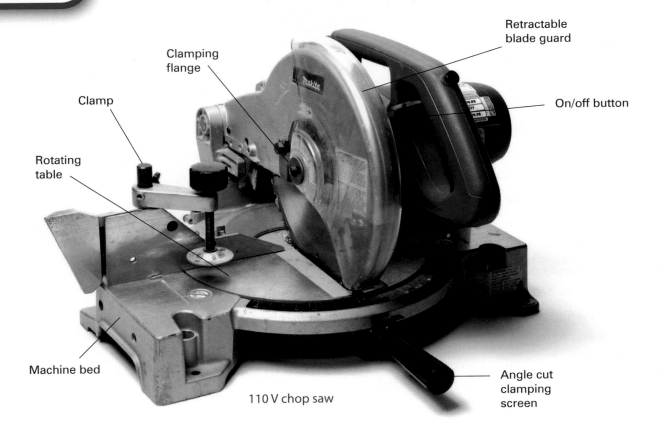

110 V chop saw

Chop saws are capable of causing serious injury and should be securely fixed to a workbench, or purpose-made stand, and at a height that is comfortable for the operator. The built-in bed tends to be short, so side extension tables or trestles must be used to support longer lengths of timber when cutting.

Remember:

- Because of the high risk of injury when using this type of machine, safety must be a priority. READ and understand the manufacturer's instructions before using the machine.

- Chop saws are available for 230 V or 110 V, so be sure to connect the machine to the correct voltage.

- Ensure that cables are clear of the machine's moving parts, and are not causing obstruction to the materials to be cut, or in a position that may cause a person to trip.

- Check the machine for damage, wear and operating functions before use.

- Complete all machine settings (blade changes, angles, stops, guards, fences etc.) when disconnected from the power supply. Blade changing must be carried out by a competent person and in conjunction with the manufacturer's instructions.

- Secure the work piece firmly to the machine bed. Hands should not be used to hold material in place while cutting.

- Concentrate on what you are doing. Keep an eye on the work being cut and *never* allow your hands to be closer to the blade than 150 mm.

- Do not force the machine to cut but allow it to cut at its own speed.

- Always use the recommended blade (blade size, tooth type, HSS, TCT etc.). Do not use blunt, wrongly set, or buckled blades.

- Wear protective clothing and equipment suitable for the type of work.

Jig saws

Powered jig saws have a reciprocating blade, which moves up and down at high speed to cut timber or other material. A range of interchangeable saw blades is available.

Jig saws are mainly used for cutting slots and curves but can also be used for straight cutting, usually with the aid of a guide attachment or fence to aid accuracy. Care must be taken to select the correct blade and speed setting, for example fast for wood, slow for metals.

Cutting can start at one edge of the material or, if cutting slots or holes, the saw blade can be inserted through a pre-drilled hole in the material to be removed.

Remember:

- Always change or adjust the blades with the machine disconnected.

- Always allow the machine to run with ease, never force it around a curve.

- Select the correct blade for the task.

On/off switch

Speed selector

Guard

Blade

Base plate

110 V jig saw

- Always position the machine before cutting; and avoid re-entry with an activated blade.

- Stop and allow the blade to become stationary before withdrawing it from the work piece.

- Use lubricants when cutting metals; oils for mild steel and paraffin for cutting aluminium.

- Never allow a cable to be in front of the cutter during use.

- All machines should be checked on a regular basis by a qualified electrical engineer but the operator should check the visual condition of the saw, voltage, power cable and plug.

Sanders

There are several types of powered sander available, designed for specific tasks. The two that you are most likely to use in carpentry and joinery work are the belt and orbital sanders.

All sanders produce large quantities of dust, which can be harmful if breathed in. Always use any dust collection facility available with the machine and, if in doubt, wear a mask or respirator.

Belt sanders

Belt sanders are designed for fairly heavy-duty work and, using a coarse abrasive, can quickly remove large amounts of wood or other material like old paint. However, they can also be used to achieve a fine finish, depending on the abrasive belt used.

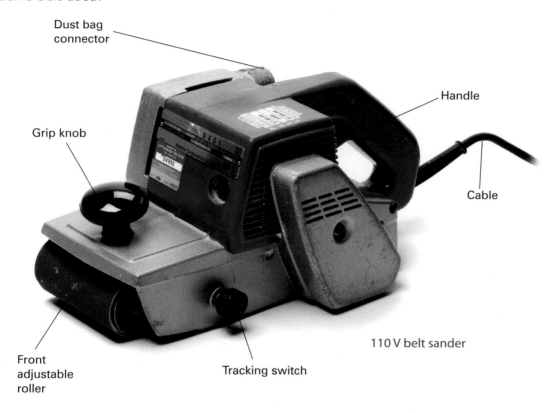

Dust bag connector

Handle

Grip knob

Cable

Front adjustable roller

Tracking switch

110 V belt sander

The fineness or coarseness of the grit used on the abrasive belt is graded according to size, as shown in Table 8.2.

Belt sanders tend to produce large amounts of dust. To compensate for this a dust bag is fitted, which collects the dust while at the same time allowing air to escape.

Grit size	Grade
20 grit and less	Very coarse
24 to 30 grit	Coarse
40 to 60 grit	Medium
80 to 150 grit	Fine
150 grit and over	Very fine

Table 8.1 Grades of abrasive belt

Orbital sanders

Orbital sanders use abrasive sheets, held to the base of the machine by spring clips situated at the front and rear of the sander. When the machine is switched on the base moves in 3 mm diameter orbits and, with care, a good quality finish can be produced.

A wide range of grit sizes is available but, unlike the belt sander, they should be used for finishing only and not used for removing waste material. They do not have any built-in dust collection facility, so a mask or respirator should be used.

Handle

On/off switch

Clip holding abrasive sheet

110 V orbital sander

Felt or rubber pad

Cordless tools

An increasing range of battery operated (i.e. cordless) tools is becoming available, including drills, jig saws, sanders, screwdrivers etc. They are particularly convenient in situations where the mains power supply is not easily accessible, or even not available. There is no cable to cause problems when working some distance from a power point or hanging down when working at height.

Cordless tools usually have the same chuck options and can carry out all the functions of tools connected to mains power. However, the heavier the task the more rapidly batteries will discharge. Battery power is indicated by the voltage rating, for example 10 V, 12 V etc. – the higher voltage tools are more suited for heavy-duty work.

Cordless drill

Batteries are rechargeable and this generally only takes a couple of hours. However, it pays to carry a spare, which can be used while waiting for the spent one to recharge. Battery life is considerably shortened if this is not done correctly, so always use the charger purchased with the tool.

Be careful, when the battery is not connected to either tool or charger, that nothing is allowed to touch the terminals, especially anything metal, as this will cause a short and probably destroy the battery.

Powered screwdrivers

Powered screwdrivers are really a specialised, simple powered drill, only available in a cordless form and specifically designed to drive or remove screws. They generally do not have removable batteries, so the whole tool is inserted into a purpose-built charger.

They are only designed to operate at slow speeds and should have variable torque control so that screws are not over-tightened. The chuck can take a full range of screwdriver bits.

Their shape enables them to be used in areas that would be difficult with a power drill.

Cordless screwdriver

Nail guns

There are three main types of nail guns in use in the construction industry today and these are cartridge powered nail guns, air powered nail guns and portable nail guns.

Cartridge powered nail guns are primarily used for fixing timber into concrete or steel and work by ramming a piston along the barrel of the gun at high speed and pressure forcing the nail into the materials. When the trigger is pulled a mechanical pin fires into the cartridge, which causes a reaction and forces the piston along the barrel. Depending on what you are fixing into, the cartridges available are colour coded to allow you to pick the correct cartridge. The nails used are special steel nails and you must use the nails specified by the manufacturer. Manufactures may use different colour coding so check the instructions prior to use.

Cartridge powered nail gun

Air powered nail guns work in a similar way with a piston firing the nail into the materials, except these are generally used for timber to timber fixing. Air powered nail guns are operated by a pneumatic process and an air compressor is required to give the force needed to insert the nail. Again special nails supplied by the manufacturer must be used and usually come in a cartridge or strip form. As these guns use a compressor for power they are mainly found in workshops and are not the preferred portable option.

Air powered nail gun

Portable nail gun

Portable nail guns also use the piston method for fixing into materials but are usually powered by gas and battery. When the gun is pushed into place the pressure lets a small amount of gas into a chamber to mix with the air. Pressing the trigger produces a spark which causes combustion and forces the piston and the nail into the work materials. The gun comes with a charger and battery, and it is better to buy at least one spare battery so that you can continue work while the other charges. Nails come in strip form, which must be bought from the manufacturer. Nails usually come with sufficient gas cartridges to fit the amount of nails supplied. There are two main types of portable guns available: one is generally used for first fixing, and the other is a slightly smaller model used for finish work.

Nail gun safety

Nail guns are very dangerous. They cannot be fired unless they are pressed into the work. The pressure allows the trigger to be pulled. When using nail guns keep both hands away from the work area as the nail may hit a weak spot and shoot straight through or even hit a very hard knot and shoot out from the side or elsewhere. Safety glasses as well as the usual PPE must be worn at all times when operating a nail gun.

FAQ

Can I use a 110 V tool with a 230 V power supply?

Yes and no. You cannot plug a 110 V tool into a 230 V socket simply because the plug and socket are different. If you use a transformer, you can use a 110 V power tool with a 230 V power supply as the transformer will 'knock down' the supply.

Do I always have to turn off a power tool to make an adjustment?

Yes, of course. Not only must you turn the machine off, you must also make sure it is removed from the power supply. If you don't, when you are making the adjustment the tool may be accidentally turned on and you could be very seriously injured.

Why do I have to fully unwind a reel-up extension cable when I use one with a power tool? Surely the unwound cable could cause an accident.

An extension cable left wound around the reel can overheat and cause a fire. Unwind the cable fully but, to prevent an accident, make sure the cable is not lying in the area where people will be walking.

On the job: Using power tools

Will is about to use a jig saw. What safety checks would you advise Will to carry out before he uses the jig saw? Can you think of any safety checks Will should carry out whilst he is using the jig saw? Finally, when Will has finished with the jig saw, is there anything you would recommend that he does?

Knowledge check

1. Name three types of rules which could be used by carpenters/joiners. State the advantages of each.

2. What are the following used for: sliding bevel; tri-square; combination square?

3. State the uses of the following: smoothing plane; rebate plane; plough plane.

4. Name four types of bit that can be used with a ratchet brace.

5. Name and state the uses of five different types of saw.

6. Sketch the following: bevelled edge chisel; mortise chisel; paring chisel.

7. What do the kite mark and double square symbol indicate?

8. Name three portable power tools and state a task for which each can be used.

9. What type of work is a percussion drill best suited for?

10. What is the advantage of using a two-speed drill?

11. Why is it necessary to secure material being cut with a portable saw?

12. How are cordless tools powered?

13. What three things should always be checked by the operator before using a portable power tool?

14. Why is it important to maintain and clean power tools regularly?

15. State three points to remember when using a portable circular saw.

16. How are abrasive sanding belts graded?

chapter 9

Timber technology

OVERVIEW

This chapter is designed to give you an overview of the different types of timber materials you can expect to be working with during your working life as a carpenter and joiner. It will also help to build your knowledge and understanding of the way in which timber is processed, preserved and protected.

This chapter will cover the following topics:

- Classification of timber
- Identification of timber
- Conversion of timber
- Seasoning of timber
- Timber defects
- Timber decay
- Preservation and protection
- Manufactured boards.

These topics can be found in the following modules:

CC 2003K   CC 2003S

Classification of timber

Timber is classified as either hardwood or softwood. This can sometimes be confusing, as not all hardwoods are physically hard or soft woods soft. For example, the balsa tree is classed as a hardwood although it is very soft and light. The wood of a yew tree, classified as softwood timber, is harder than most hardwoods.

Hardwood and softwood refer to the **botanical** differences and not to the strength of the timber. Generally speaking, hardwood trees are **deciduous**, broad-leafed, with an encased seed. Softwood trees are usually **evergreen** with needles and seeds held in cones.

Identification of timber

One of the best ways of becoming familiar with, and being able to identify, a reasonably wide range of timber species, is to form a personal collection. Small, matchbox-size samples are ideal for this.

The key to learning and memorising timber species is colour. Once timbers have been grouped together, identification becomes a process of elimination. Tables 9.1 (softwoods) and 9.2 (hardwoods) outline the general appearance, distribution and uses of timber species most commonly used.

Definition

Botanical – the classification of trees based upon scientific study

Deciduous – the name given to a type of tree that sheds its leaves every year

Evergreen – a type of tree that keeps its leaves all year round

Name	Main sources	Description	Main uses
Douglas fir	Canada, USA	Pinkish-brown timber with distinct grain and a clean, sweet smell	Good quality joinery, strip flooring, stairs and plywood veneers
Larch	Europe, including UK	Strong, resinous reddish-brown timber with a good straight grain	Gates, fences etc.
Pitch pine	Southern USA	Hard, tough, heavy timber, very distinct grain, generally pink to brown; easily recognised by its resinous, turpentine smell	Shipbuilding, polished softwood joinery and church furniture
Redwood (commonly known as Pine)	Europe	Pinkish-white timber with distinct orange/red grain; clean pine smell	General purpose joinery

Table 9.1 Commonly used softwoods (continued opposite)

Name	Main sources	Description	Main uses
Whitewood (also known as European Spruce)	Europe	Similar to redwood but paler and lacks the resinous pine smell	Joists, rafters, floorboards etc.
Western red cedar	Canada, USA	Light, soft, spongy timber with a woolly texture; pink to reddish-brown, darkening to grey when exposed to the weather	Externally for good quality timber buildings, saunas etc.

Table 9.1 Commonly used softwoods *(continued)*

Name	Main sources	Description	Main uses
Ash	Europe	Creamy, white timber with occasional dark streaks (black heart), generally straight grained but can be quite coarse	Furniture, boat building, sports equipment etc.
Oak	Europe	English oak is generally acknowledged to be the toughest, whilst Polish or Slovenian oak is usually the least tough and is easily worked. The timber is a pale yellowish-brown, with a distinct darker figure of 'silver grain'	High-class joinery, panelling, doors etc.
Beech	Europe	Cream to pale brown timber which darkens a little on exposure to air and has an attractive dark fleck running through it	Furniture, kitchen utensils, wood block floors etc.
Mahogany	Africa	Reddish-brown timber with grey tinges; very attractive grain with a tendency to woolliness	High-class joinery, furniture, boat building and plywood veneers
Mahogany	Spain, Cuba	Superb timber, usually variable and deep in colour, from yellowish-brown to deep rich red, with curling grain	As for African Mahogany but considered to be superior

Table 9.2 Commonly used hardwoods *(continued overleaf)*

Name	Main sources	Description	Main uses
Maple	Canada, North East USA	Medium dark, reddish-brown timber with a close, even texture, similar to sycamore to which it is closely related	Hardwood strip flooring etc.
Sapele	West Africa	Medium dark, reddish-brown timber with a pronounced stripe; when carefully cleaned it will give a superb finish	Furniture, veneers etc.
Teak	India, Java, Thailand	Outstanding timber to use, this golden brown timber, often with a distinct dark figure, has a greasy, oily texture	High-class joinery, furniture, boat building etc.
Walnut	Europe	Purplish-brown with very attractive dark figure, sometimes with a curl or burr; valuable timber much sought after by cabinet makers and veneer manufacturers	Furniture, veneers etc.

Table 9.2 Commonly used hardwoods *(continued)*

Conversion of timber

Conversion is the sawing of a log into board or planks ready for use by the carpenter or joiner. How the timber is converted directly affects its usefulness.

Softwoods are nearly always sawn in their country of origin, while hardwoods are often imported in log form and converted by the timber merchant, sometimes to customer requirements.

Methods of conversion

There are four main methods of conversion:

- through and through
- tangential
- quarter
- boxed heart.

Through and through sawn timber

Also known as flat slab or slash sawing, through and through sawing is the simplest and most economical method of converting timber. See Figure 9.1. Although there is very little wastage with this method, the majority of the boards produced are prone to a large amount of shrinkage and distortion.

Tangential sawn timber

The tangential sawn timber method of conversion is used to provide floor joists and beams, as it gives the strongest timber. See Figure 9.2. It is also used on pitch pine and douglas fir for decorative purposes to produce 'flame figuring' or 'fiery grain'.

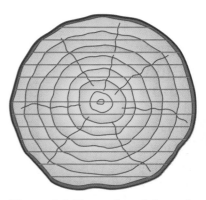

Figure 9.1 Through and through sawn timber

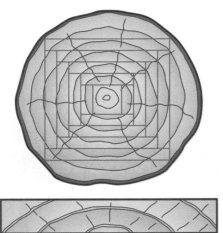

Figure 9.2 Tangential sawn timber

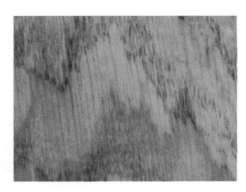

Flame figuring

Quarter sawn timber

Quarter sawn timber, shown in Figure 9.3, produces the best quality timber. However, it is also the most expensive, both in time involved and material wastage. It produces the greatest quantity of 'rift' or 'radial sawn' boards, which are generally superior for joinery purposes.

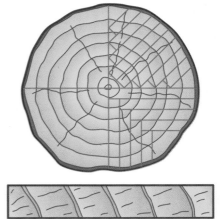

Figure 9.3 Quarter sawn timber

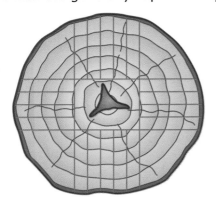

Figure 9.4 Boxed heart sawn timber

Boxed heart sawn timber

Boxed heart sawn timber is a type of radial sawing that is used when the heart of the tree is rotten or badly shaken. It is sometimes known as floor boarding sawing, as the boards are ideal for this purpose because they are hard-wearing and do not distort. See Figure 9.4.

Seasoning of timber

Timber from newly felled trees contains a high proportion of moisture in the form of sap, which is made up of water and minerals drawn up from the soil. Most of this water has to be removed by some form of drying, which is called seasoning.

The main reasons for seasoning are:

- to make sure that shrinkage occurs before the timber is used
- to make sure that the moisture content of the timber is below the 'dry rot' safety line of 20 per cent (discussed later in this chapter)
- to make sure that dry timber is used
- dry timber is stronger
- seasoned timber is less likely to split or distort
- wet timber will not accept glue, paint or polish.

The timber should be dried to a moisture content that is similar to the surrounding atmosphere in which it will be used. See Table 9.3.

Timber location	Moisture content (per cent)
Carcasing timber (joists etc.)	18–20
External joinery	16–18
Internal timber where there is a partial intermittent heating system	14–16
Internal timber where there is a continuous heating system	10–12
Internal timber placed directly over, or near, sources of heat	7–10

Table 9.3 Moisture content table

There are two main methods of seasoning timber:

1. natural, normally called 'air seasoning'
2. artificial, normally called 'kiln seasoning'.

Air seasoning

For air drying the timber is stacked in a pile in open-sided, covered sheds, which protect the timber from rain but still allow a free circulation of air. A moisture content of 18–20 per cent can be achieved in a period of 2–12 months.

Kiln seasoning

Most timber that is used is kiln seasoned. If done correctly the moisture content of the timber can be reduced without causing any timber defects.

Depending on the size of the timber, the length of time the timber needs to stay in the kiln varies between two days and six weeks.

There are two types of kiln in general use:

1. compartment kiln
2. progressive kiln.

A compartment kiln is shown in Figure 9.5. It is usually a brick or concrete structure, in which the timber remains stationary during the drying process.

The drying of the timber depends on three factors:

- air circulation supplied by fans
- heat, usually supplied by heating coils
- humidity, which is raised by steam sprays.

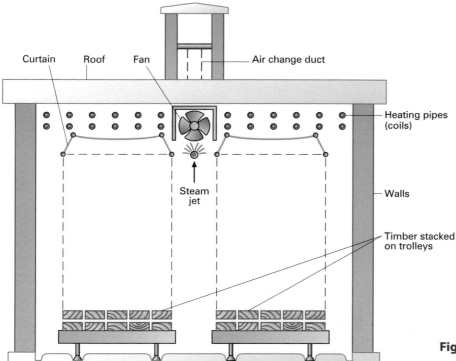

Figure 9.5 Compartment kiln

In a progressive kiln the timber is stacked on trolleys, which pass slowly through a long chamber, with gentle changes to heat and humidity as it moves from one end of the chamber to the other.

There are three clear advantages of using kiln, as opposed to air, drying:

1. the speed at which the seasoning can be completed

2. the facility to dry timber to any desired moisture content

3. the sterilising effect of the heated air upon fungi and insects in the timber, lessening the likelihood of fungal or insect attack.

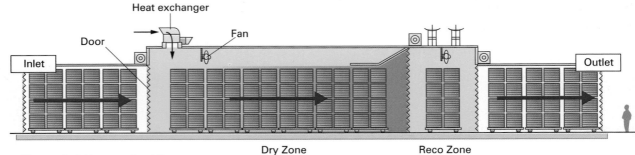

Figure 9.6 Progressive kiln

Timber defects

Defects are faults that are found in timber. Some present a serious structural weakness in the timber, others do little more than spoil its appearance.

Defects can be divided into two groups:

1. seasoning defects

2. natural defects.

Seasoning defects

Seasoning defects can be further divided into:

- bowing
- springing
- winding (or twist)
- cupping
- shaking
- collapse
- case hardening.

Bowing

Bowing is usually caused by poor stacking during seasoning. It is a serious defect, causing good timber to be suitable only for use in short lengths. See Figure 9.7.

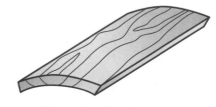

Figure 9.7 Bowing

Springing

Springing is an edgeways curvature of a board. It is usually caused by the release of internal stresses during seasoning. See Figure 9.8.

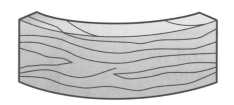

Figure 9.8 Springing

Winding (or twist)

Winding, also known as twist, is very serious as it restricts the use of the timber to short lengths. It is caused by poor seasoning and poor stacking. See Figure 9.9.

Figure 9.9 Winding (or twist)

Cupping

Cupping is very common in flat sawn boards. It occurs through shrinkage of the timber when drying. See Figure 9.10.

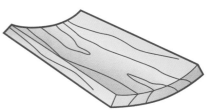

Figure 9.10 Cupping

Shaking

Shaking is caused by the board being dried too rapidly. It is particularly common at the ends of boards, spreading along the grain. See Figure 9.11.

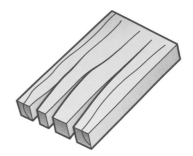

Figure 9.11 Shaking

Collapse

Collapse is a rare defect caused by too rapid drying in the early stages of seasoning. The moisture is drawn out too rapidly causing dehydrated cells to collapse. See Figure 9.12.

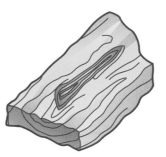

Figure 9.12 Collapse

Case hardening

Case hardening is caused by too rapid drying, resulting in the outside cells of the timber drying and hardening, sealing off the moisture in the central part of the board. See Figure 9.13.

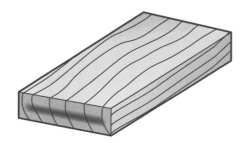

Figure 9.13 Case hardening

Natural defects

Natural defects can be further divided into:

- heart shakes
- cup shakes
- star shakes
- knots.

Heart shakes

Heart shakes are usually the result of disease or over-maturity of the tree. The shakes radiate from the centre of the log and are caused by internal shrinkage. See Figure 9.14.

Cup or ring shakes

Cup shakes, also known as ring shakes, are caused by a separation of the annual rings and are usually due to a lack of nutrient or twisting of the tree in high winds. In bad cases economic conversion of the log is very difficult. See Figure 9.15.

Star shakes

Star shakes are radial cracks which occur around the outside of the log. They are caused by shrinkage at the outside of the log whilst the middle remains stable. This is usually because the log has been left too long before conversion. See Figure 9.16.

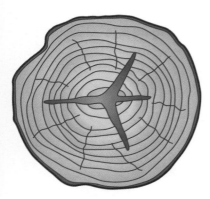

Figure 9.14 Heart shakes

Figure 9.15 Cup shakes

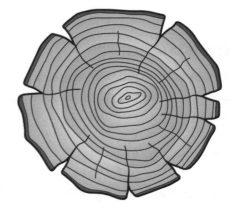

Figure 9.16 Star shakes

Knots

Knots mainly occur in softwood and mark the origin of a branch in the tree. Knots are termed either 'dead' or 'live' depending on the condition of the branch which caused it. Dead knots are often loose. Small live knots are no real problem. Small dead knots and large knots, either dead or live, are a serious structural weakness.

Types of knot are shown in Figures 9.17 to 9.20.

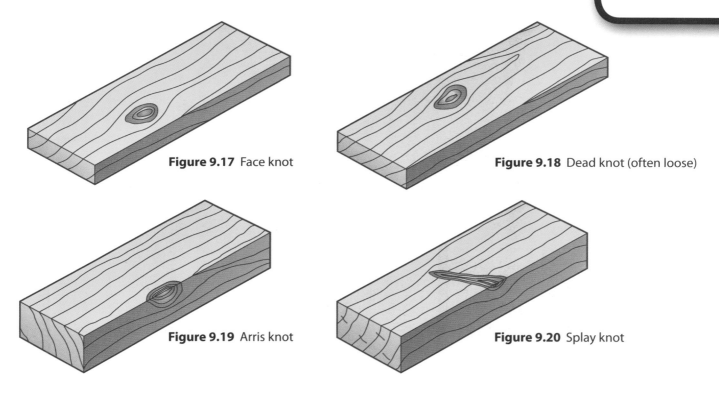

Figure 9.17 Face knot

Figure 9.18 Dead knot (often loose)

Figure 9.19 Arris knot

Figure 9.20 Splay knot

Timber decay

The decay of timber is caused by one or both of the following:

- wood-destroying fungi
- wood-boring insects.

Wood-destroying fungi

There many forms of fungi which, under suitable conditions, will attack timber until it is destroyed. They mainly fall into two groups:

1. dry rot
2. wet rot.

Both types can be a serious hazard to constructional timbers. Figure 9.21 shows some possible causes of dry and wet rot.

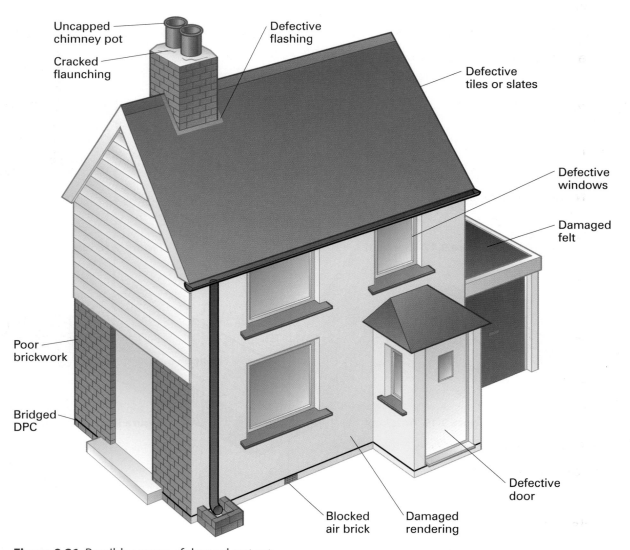

Uncapped
chimney pot

Cracked
flaunching

Defective
flashing

Defective
tiles or slates

Defective
windows

Damaged
felt

Poor
brickwork

Bridged
DPC

Blocked
air brick

Damaged
rendering

Defective
door

Figure 9.21 Possible causes of dry and wet rot

Dry rot

This is the most common and most serious of wood-destroying fungi. The fungus does most damage to softwoods but also attacks hardwoods, particularly when they are close to softwoods already infected. If left undetected or untreated it can destroy much of the timber in a building.

Dry rot is so called because of the dry, crumbly appearance of the infected timber. It is, however, excessive moisture that is the main cause of the decay. If the wood is kept dry and well ventilated there should be little chance of dry rot occurring.

The main conditions for an attack of dry rot are:

- damp timber, with a moisture content above 20 per cent (known as the **dry rot safety line**)
- poor, or no, ventilation (i.e. no circulation of fresh air).

The attack of dry rot takes place in three stages, which are shown in Figures 9.22 to 9.24:

- The spores (seeds) of the fungus germinate and send out hyphae (roots) which bore into the timber.
- The hyphae branch out and spread through the timber. A fruiting body now starts to grow.
- The fruiting body, which resembles a large, fleshy pancake, starts to ripen. When fully ripened it discharges millions of red spores into the air. The spores attach to fresh timber and the cycle starts again.

Definition

Dry rot safety line – when the moisture content of damp timber reaches 20 per cent. Dry rot is likely if the moisture content exceeds this

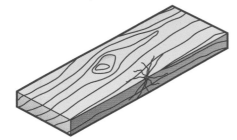

Figure 9.22 Dry rot stage one

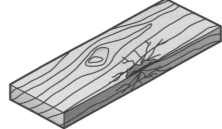

Figure 9.23 Dry rot stage two

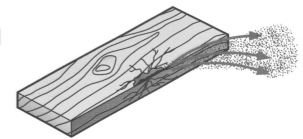

Figure 9.24 Dry rot stage three

Identification of dry rot

Dry rot can be identified by:

- an unpleasant, musty smell
- visible distortion of infected timber; warped, sunken and/or shrinkage cracks
- probing to test the timber for softening or crumbling
- the appearance of fruiting bodies
- presence of fine, orange-red dust on the floorboards and other parts of the structure
- presence of whitish-grey strands on the surface of the timber.

Did you know?

Fungal fruiting bodies are the sign of advanced decay in timber

Eradication of dry rot

Dry rot is eradicated by carrying out the following actions:

- Eliminate all possible sources of dampness, such as blocked air bricks, bridged damp proof course, leaking pipes etc.

- Determine the extent of the attack.
- Remove all infected timber.
- Clean and treat surrounding walls, floors etc. with a suitable fungicide.
- Treat any remaining timber with a preservative.
- Replace rotted timber with new treated timber.
- Monitor completed work for signs of further attack.

Wet rot

This is a general name given to another type of wood-destroying fungus. The conditions where wet rot is found are usually wet rather than damp. Although wet rot is capable of destroying timber it is not as serious a problem as dry rot and, if the source of wetness is found, the wet rot can be halted.

The most likely places to find wet rot are:

- badly maintained external joinery, where water has penetrated
- ends of rafters and floor joists
- fences and gate posts
- under leaking sinks or baths etc.

Identification of wet rot

The signs to look for to identify wet rot are:

- timber becomes darker in colour with cracks along the grain
- decay usually occurs internally leaving a thin layer of relatively sound timber on the outside
- localised areas of decay close to wetness
- a musty, damp smell.

Eradication of wet rot

As wet rot is not as serious a problem as dry rot, less extreme measures are normally involved. It is usually sufficient to remove the rotted timber, treat the remaining timber with a fungicide and replace any rotted timber with treated timber. Lastly, if possible, the source of any wetness should be rectified.

Wood-boring insects

Many insects attack or eat wood, causing structural damage. There are four main types of insects found in the UK, all classed as beetles.

Common Furniture Beetle

This wood-boring insect can damage both softwoods and hardwoods. The larvae of the beetle bore through the wood, digesting the cellulose. After about three years, the beetle forms a pupal chamber near the surface, where it changes into an adult beetle. In the summer, the beetle bites its way out to the surface, forming the characteristic round flight hole, measuring about 1.5 mm in diameter. After mating, the females lay their eggs (up to 80) in cracks, crevices or old flight-holes. The eggs hatch and a new generation begins a fresh life cycle.

Death Watch Beetle

This wood-boring insect is related to the Common Furniture Beetle, but is much larger, with a flight hole of about 3 mm in diameter, usually found in decaying oak. The female lays up to 200 eggs. While generally attacking hardwoods only, this wood-boring insect has been known to feed on decaying softwood timbers. The Death Watch Beetle is well known for making a tapping sound, caused by the head of the male during the mating season.

Powder Post Beetle

This beetle gets its name from the way it can reduce timber to a fine powder. Powder Post beetles generally attack timber with a high moisture content. As with all other beetles, the female lays eggs and the larvae do the damage.

House Longhorn Beetle

This wood-destroying insect attacks seasoned softwoods, laying its eggs in the cracks and crevices of wood. In Great Britain, this insect is found mainly in Surrey and Hampshire.

As well as beetles, weevils and wood wasps can also cause a problem.

Wood-boring Weevil

This is a wood-boring insect similar in appearance and size to the Common Furniture Beetle. It differs in that it will only attack timber that is already decayed by wood-rotting fungi. There are over 50,000 species of weevil, all of which have

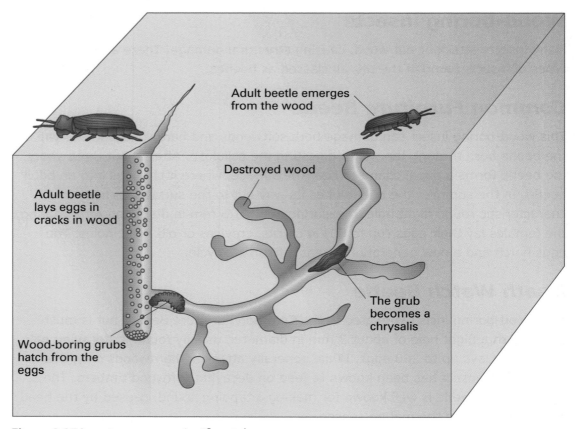

Figure 9.25 Insects can cause significant damage

long snouts. The Wood-boring Weevil is prolific, and is known to have up to two complete life cycles in one year. Its presence may therefore be accompanied by serious structural collapse of timber.

Wood Wasp

These insects are also known as horntails. They are wasp-like insects and may be seen or heard buzzing and making cosmetic damage around the home. The body is 2–5 cm long, and the wasp is multi-coloured, blue or black, with wings of matching colour. Walls, wood floors, doors and other wooden surfaces may have holes that are approximately 3 mm in diameter. These holes may also appear on items covering wooden structures, such as carpeting, wallpaper, and linoleum.
Wood Wasps lay their eggs in forest trees that are either damaged or weak. It can take up to five years for adult Wood Wasps to emerge, so often they are found in trees that have been felled for construction purposes – often even after the wood has been stored. However, most Wood Wasp holes will appear within the first two years after use of the cut wood. The good news is that these insects only lay their eggs in weakened forest trees, and will not re-infest buildings or household structures.

Eradication of insects

If the insect attack is confined to a small area the infected timbers can be cut out and replaced. If the attack is over a large area it is better to pass the work to a specialist firm.

Preservation and protection

Not only do we need to protect timber from fungal and insect attack but also, where timber is exposed, it needs protecting from the weather. Preservation extends the life of timber and greatly reduces the cost of maintenance.

Types of preservative

Timber preservatives are divided into three groups:

- tar oils
- water-borne
- organic solvents.

Tar oils

Tar oils are derived from coal tar, are very effective and relatively cheap. However, they give off a very strong odour which may contaminate other materials. Creosote is the most common type of this preservative.

Water-borne

Water-borne preservatives are mainly solutions of copper, zinc, mercury or chrome. Water is used to carry the chemical into the timber and then allowed to evaporate, leaving the chemical in the timber. They are very effective against fungi and insects and are able to penetrate into the timber. They are also easily painted over and relatively inexpensive.

Organic solvents

Organic solvents are the most effective, but also the most costly of the preservatives. They have excellent penetrating qualities and dry out rapidly. Many of this type are proprietary brands such as those manufactured by Cuprinol™.

Remember

Contact a professional for advice where load bearing timbers have been attacked, as there may be a need to prop or shore up to prevent collapse of the structure

Methods of application

Preservative can be applied in two ways:

1. non-pressure methods – brushing, spraying, dipping or steeping

2. pressure methods – empty cell, full cell or double vacuum.

Non-pressure methods

Although satisfactory results can be achieved, there are disadvantages with using non-pressure methods. The depth of penetration is uneven and, with certain timbers, impregnation is insufficient to prevent leaking out (leaching). Table 9.4 lists non-pressure methods and how and where to employ them.

Method	How and where to employ it
Brushing	The most commonly used method of applying preservative, it is important to apply the preservative liberally and allow it to soak in
Spraying	Usually used where brushing is difficult to carry out, in areas such as roof spaces
Dipping	Timbers are submerged in a bath of preservative for up to 15 minutes
Steeping	Similar to dipping only the timber is left submerged for up to two weeks

Table 9.4 Non-pressure methods

Pressure methods

Pressure methods generally give better results with deeper penetration and less leaching. Table 9.5 lists pressure methods and how and where to employ them.

Method	How and where to employ it
Empty cell	Preservative is forced into the timber under pressure. When the pressure is released the air within the cells expands and blows out the surplus for re-use. This method is suitable for water-borne and organic solvent preservatives.
Full cell	Similar to empty cell, but prior to impregnation a vacuum is applied to the timber. The preservative is then introduced under pressure to fill the cells completely. Suitable for tar oils and water-borne preservatives.
Double vacuum	A vacuum is applied to remove air from the cells, the preservative introduced, the vacuum released and pressure applied. The pressure is released and a second vacuum applied to recover surplus preservative. This method is used for organic solvent preservatives.

Table 9.5 Pressure methods

> **Safety tip**
>
> All preservatives are toxic and care should be taken at all times. Protective clothing should always be worn when using preservatives

Manufactured boards

The most common types of wood-based manufactured boards are:

- plywood
- laminated board
- chipboard (particle board)
- fibre board.

Plywood

Plywood is made from thin layers of timber called **veneers**. The veneers are glued together to form boards. There is normally an odd number of veneers with the grain alternating across and along the sheet, which gives both strength and stability to the board. See Figure 9.26.

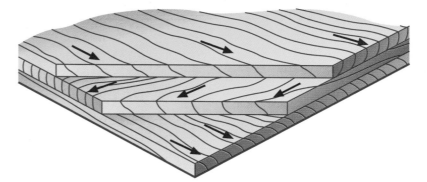

Figure 9.26 Plywood section showing grain direction

Plywood is graded according to the type of glue used in its manufacture and also according to the situation in which it will be used. The grade is usually stamped on the board by the manufacturer. A list of grades is shown in Table 9.6.

Stamp	Grade	Use
INT	Interior	Internal use only – has low resistance to humidity or dampness
MR	Moisture resistant	Has a fair resistance to humidity and dampness
BR	Boil resistant	Has a fairly high resistance in exposed conditions
WBP	Weather and boil proof	Can be used in extreme conditions under continuous exposure (boats, buildings etc.)

Table 9.6 Plywood grade table

Blockboard
Strips are up to 25 mm wide. Good quality hardwood veneers are sometimes used.

Laminboard
Strips are 7–8 mm wide. This produces a better quality board.

Battenboard
Strips are up to 75 mm wide, producing a poorer quality board.

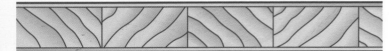

Figure 9.27 Laminated boards

Laminated boards

Laminated boards are made from strips of wood laminated together and sandwiched between two veneers. Three common varieties are shown in Figure 9.27.

Chipboard

Chipboard is made from compressed wood chips and wood flakes bonded with a synthetic resin glue. See Figure 9.28. There are various grades, which include those used to make floor panels etc.

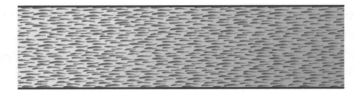

Figure 9.28 Chipboard

Fibreboard

Fibreboard is made from pulped wood that is mixed with an adhesive and pressed into sheets. It usually comes in three forms:

1. hardboard
2. insulation board (or softboard)
3. medium density fibreboard (**MDF**).

Hardboard

Hardboard is manufactured from sugar cane pulp and available in sheets 3–6 mm in thickness. Oil-tempered hardboard offers a reasonable resistance to moisture. See Figure 9.29.

Hardboard is also manufactured with various finishes, which include:

- plastic faced
- perforated
- reeded
- enamelled.

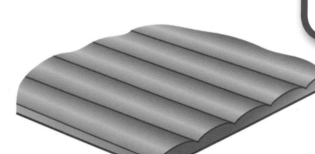

Figure 9.29 Hardboard

Safety tip

Dust extraction or a respirator should be worn when cutting or sanding any timber, but particularly some tropical hardwoods and MDF, which can be harmful

Insulation board (or softboard)

Also called softboard, insulation board is made from the same material as hardboard but not compressed. It is used as a wall and ceiling covering to give very good insulation. See Figure 9.30.

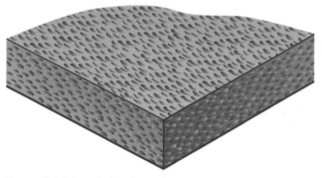

Figure 9.30 Insulation board

Did you know?

Architrave is a name for mouldings around windows and doors

Medium density fibreboard (MDF)

Medium density fibreboard (MDF) is now used to form skirting boards, and mouldings such as architraves etc. It is easy to work with both hand and powered tools. It is mainly used internally but moisture-resistant sheets are available.

FAQ

How do I choose between hardwood and softwood for a job?

The type of timber you should use for a job is usually detailed in the specifications. There are a number of reasons why one type of wood is chosen over another. Hardwood is usually more expensive and long-lasting than softwood. Often grown in the hot climates of equatorial countries (e.g. African and South American countries), hardwood is used for jobs where the wood will be visible (i.e. high-class joinery). Softwood is usually grown in countries with cooler climates. It is often cheaper than hardwood and used for jobs where the wood will be concealed (e.g. floorboards and rafters).

On the job: Twisted timber

Wayne goes to the store to get some timber for a job. When Wayne lifts the timber from the shelf, he notices that the wood is in wind (i.e. twisted). How could this have happened? What could have been done to prevent it from happening? Do you think Wayne can still use the timber? Give reasons for your answer.

Knowledge check

1. What is meant by timber conversion?

2. How can dry rot be eradicated?

3. Name four types of insects that attack timber.

4. How is plywood manufactured?

5. Name three non-pressure methods of applying preservatives.

6. Briefly describe the differences between hardwood and softwood.

7. Name four examples of seasoning defects.

8. Name four examples of natural defects.

9. What does MDF stand for?

10. What do the words 'deciduous' and 'evergreen' mean?

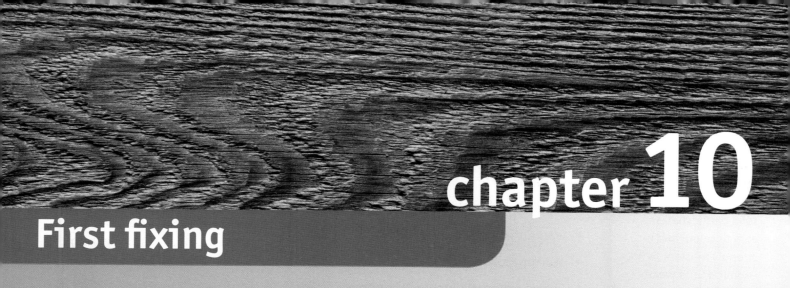

chapter 10

First fixing

OVERVIEW

Carpenters and joiners will undertake many different types of work on site. One of the main areas, and probably the most important, is called 'first fixing'.

This is work carried out before plastering takes place.

This chapter will cover the following topics:

- Windows
- Stairs
- Timber studwork
- Joist and stud coverings
- Door frames and linings
- Ground lats.

These topics can be found in the following modules:

CC 1001K   CC 2008K
CC 1001S   CC 2008S

Windows

Windows are generally designed to allow daylight and air into a room and give people an outside view, but they must also conserve heat and be able to withstand the weather.

There are many forms and designs, the most common being:

- traditional casement windows
- stormproof windows
- boxed frame sliding sash windows
- bay windows.

They are traditionally made from timber, softwood and hardwood but, with new materials and improved manufacturing methods, they are also available in metal, aluminium and pvc (sold under a variety of names including PVC, PVCu, uPVC, PVC-Upvc and others).

As windows are subjected to extreme weather conditions they must be constructed in such a way as to give maximum protection against the worst possible weather conditions.

A window should also be secure, durable, with easy opening operation and provide, where necessary, good sound and thermal insulation.

This section will look at:

- main window types
- assembly of frames and sashes
- installing windows
- window boards.

Window parts

Different types of windows have different parts, but the majority will have:

- **sill** – the bottom horizontal member of the frame
- **head** – the top horizontal member of the frame
- **jambs** – the outside vertical members of the frame
- **mullion** – intermediate vertical member between the head and sill
- **transom** – intermediate horizontal member between the jambs.

If a window has a sash, the sash parts will consist of:

- **top rail** – the top horizontal part of the sash frame

- **bottom rail** – the bottom horizontal part of the sash frame
- **stiles** – the outer vertical parts of the sash frame
- **glazing bars** – intermediate horizontal and vertical members of the sash frame.

Main window types
Traditional casement windows

Casement windows comprise a solid outer frame with one or more smaller and lighter frames within it, called sashes or casements.

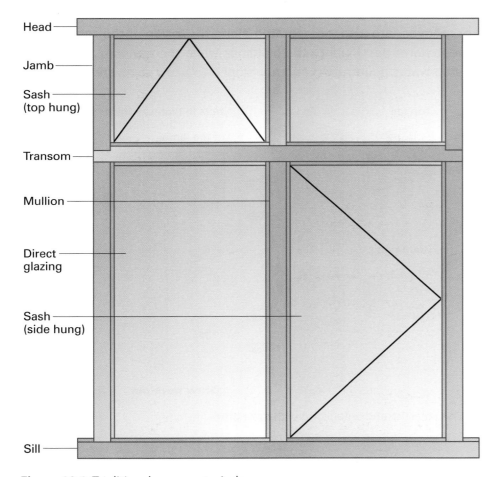

Figure 10.1 Traditional casement window

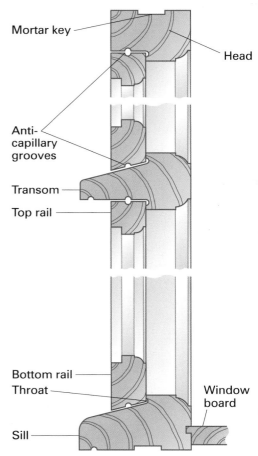

Figure 10.2 Section through traditional window incorporating top and side hung sashes

The main components are:

- **Frame.** The main frame consists of a head, sill and two vertical jambs. Intermediate members are incorporated to form openings for sashes or to alter the design. The vertical sections are known as mullions and horizontal members are transoms.

- **Opening casement.** Consists of top and bottom rails with two stiles. If a casement is divided up, these members are known as glazing bars. When glass is fixed into the main frame itself this is known as direct glazing.

- **Joints.** All joints used on a traditional casement frame are mortise and tenon, the mortises being formed in the head and sill, while the jambs are tenoned. Casements are haunched mortise and tenons and both are held together using wedges or draw pins and star dowels, draw pins being more suitable when the **horns** are to be cut off later. Joints are covered in more detail in Chapter 14 *Marking and setting out joinery products*.

- **Weather proofing.** During manufacture a groove is formed around the outside edges of the head, sill and jambs to provide a mortar key. Grooves, known as anti-capillary grooves, are also incorporated along the inside of the rebates in order to prevent water passing into the building; while the transom and sill have a 'throat' to stop water penetrating and are sloped to allow rainwater to run off.

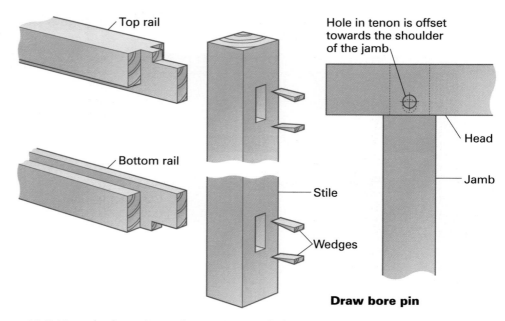

Top rail

Bottom rail

Stile

Wedges

Hole in tenon is offset towards the shoulder of the jamb

Head

Jamb

Draw bore pin

Figure 10.3 Haunched mortise and tenon joint with draw pin

Stormproof windows

Stormproof windows are a variation on the traditional casement windows and are designed to increase their performance against bad weather.

The outer frame is basically the same but the sashes are rebated to cover the gap between sash and frame. This reduces the possibility of driving rain entering the building.

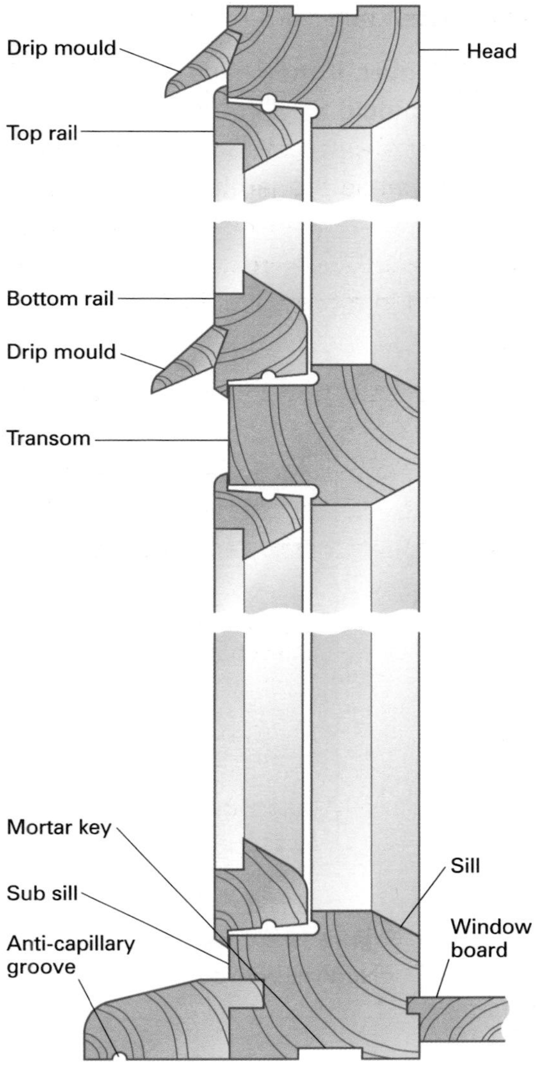

Drip mould
Head
Top rail
Bottom rail
Drip mould
Transom
Mortar key
Sub sill
Anti-capillary groove
Sill
Window board

Figure 10.4 Section through typical stormproof window

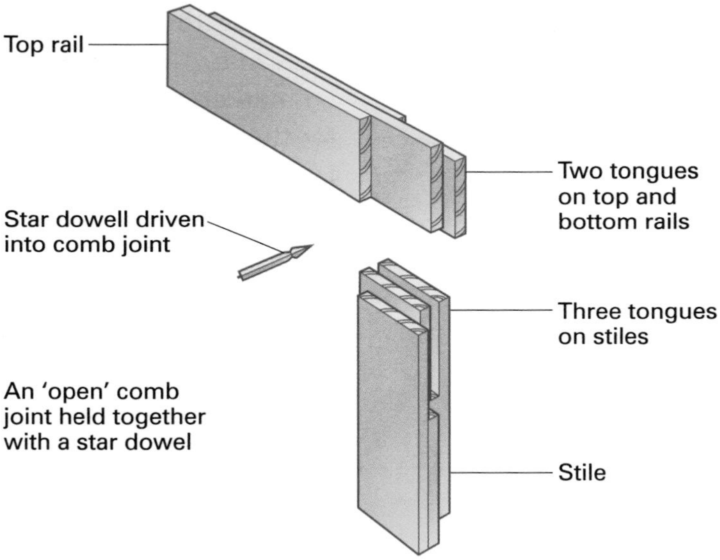

Top rail
Two tongues on top and bottom rails
Star dowell driven into comb joint
Three tongues on stiles
An 'open' comb joint held together with a star dowel
Stile

Figure 10.5 Joining of main frame

The main frame is joined together using mortise and tenon joints, but the sashes are held together using a comb joint and star dowels, although mortise and tenon can be used.

During manufacture anti-capillary grooves are incorporated along the inside of the rebates to prevent water passing into a building.

Assembly of frames and sashes

Windows may be delivered fully constructed. However, if they have to be assembled, the procedure below shows how this should be done:

Step 1 Prior to gluing, frames and sashes should be assembled dry to check the joints, sizes and that they are square.

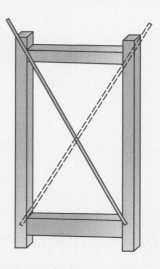

Step 2 Lay a **squaring rod** across the diagonals to check that each one measures the same. The length of a diagonal should be marked on the rod, then do the same for the other diagonal. If the pencil mark is in the same place the frame is square. New joints may have to be made if the frame is distorted.

Figure 10.6 Step 2 Check diagonals with a squaring rod

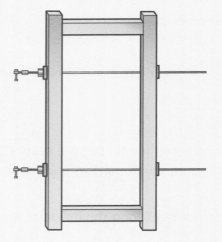

Step 3 Once a sash or frame has been assembled dry and checked, a waterproof adhesive can be applied to the faces of the tenons and shoulders and the items glued and cramped together to dry.

Figure 10.7 Step 3 Glue together and cramp

Definition

Squaring rod
– any piece of straight timber long enough to lie across the diagonal

Step 4 Check again that the sash is square and not in wind, as well as ensuring the overall size is correct. Then wedges should be lightly tapped into each tenon joint and, if everything is correct, they can be driven in fully, outside wedges first, and star dowels inserted if required.

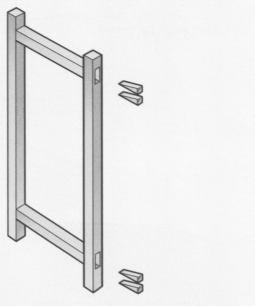

Figure 10.8 Step 4 Secure with wedges

Installing windows
As brickwork rises

Casement windows made of wood are usually built in as the brickwork proceeds. They are secured with separate fixing devices, traditionally referred to as frame clamps.

Essentially, according to the type used, they are either screwed or hammered into the wooden side-jambs as the brickwork rises, where they will be bedded into the mortar between the bricks. Two or three per side is usual, like built-in door frames covered later.

Initially the windows are put in position, checked that they are plumb and then supported at the head with one or two weighted scaffold boards.

If the windows have a separate sill of stone or pre-cast concrete these must be bedded first and protected with temporary boards on their outer face, sides and edges.

Frame clamp

Definition

Brick-on-edge – a way of laying bricks to form an attractive, projecting window or door sill

Projecting sills, formed with sloping **brick-on-edge**, are usually built at a later stage, the windows having been packed up with spare timber to leave space for these to be inserted.

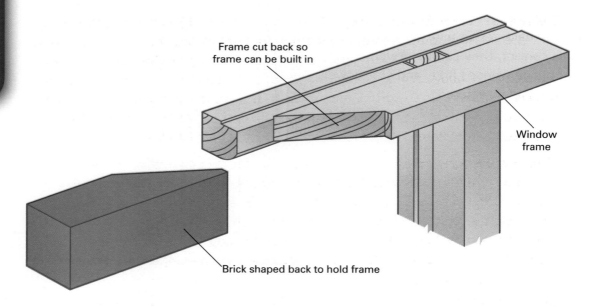

Frame cut back so frame can be built in

Window frame

Brick shaped back to hold frame

Figure 10.9 Built-in window frame

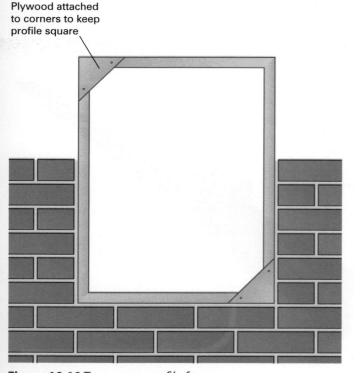

Plywood attached to corners to keep profile square

Figure 10.10 Temporary profile frame

Temporary profile frames

Sometimes it is not possible to build the windows in as the brickwork rises. If this is the case a temporary profile frame may be made on site or in the workshop to match the window size and put in place during building. It can be replaced by the window as soon as the brickwork is dry and stable, using suitable fixings.

Fixing components

There is a range of commercially available fixing components. Common ones are described below:

- **Galvanised steel frame cramps.** These provide good fixings and, because they are screwed to the frame, any cramps (or ties) already fixed and bedded in mortar are not disturbed by hammering. Also, by resting on the last laid bricks, the next brick above the tie is easily bedded. The disadvantages are their small screws, requiring a screwdriver and bradawl. If there is no groove in the frame, the upturned end of the cramp inhibits the next brick from touching the frame. There will be problems later, if rust-proofed screws are not used.

- **Zinc-plated screw ties.** These also provide good fixings and are screwed to the frame, avoiding vibration from hammering. They do not require screws, bradawl or screwdriver and can be offset or skewed to avoid the cavities in hollow blocks. On the negative side, the brickwork has to be stopped one course below the required fixing to allow rotation of the loop when screwing in, then the brick or block beneath the tie is bedded – with some difficulty and loss of normal bedding adhesion.

- **Sherardised holdfast.** Holdfasts are fixed quickly and are easily driven in by hammer. The spiked ends spread outwards when driven into the wood, forming a fishtail with good holding power. The main disadvantage is that the hammering disturbs the frame and permanently loosens any holdfasts already positioned in the **still green** mortar.

Definition

Still green – mortar that is unset and not at full strength

Figure 10.11 Galvanized steel frame cramp

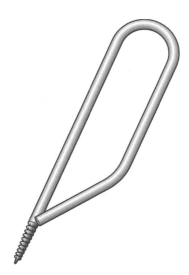

Figure 10.12 Zinc-plated screw tie

Box frame sliding sash window

Often referred to as a box sash window, this is rarely used nowadays – the casement window is preferred as it is easier to manufacture and maintain. Box sash windows are mainly fitted in listed buildings or in buildings where like-for-like replacements are to be used. Sometimes box sash windows are fitted because the client prefers them.

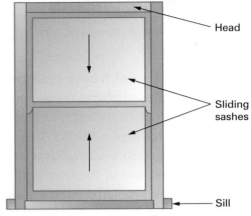

Figure 10.13 Box frame window

The box sash window is constructed with a frame and two sashes, with the top sash sliding down and the bottom sash sliding up. Traditionally the sashes work via pulleys, with lead weights attached to act as a counterbalance. The weights must be correctly balanced so that the sashes will slide with the minimum effort and stay in the required position: for the top sash, they must be only slightly heavier than the weight of the glazed sash; for the bottom sash, slightly lighter than the weight of the glazed sash.

Most pulley-controlled systems now use cast-iron weights instead of lead. Newer box sash windows use helical springs instead of the pulley system.

The parts of a box sash window are as follows:

- **head** – the top horizontal member housing the parting bead, to which an inner and outer lining are fixed

- **pulley stiles** – the vertical members of the frame housing the parting bead, to which inner and outer linings are fixed. The pulley stiles act as a guide for the sliding sashes

- **inner and outer linings** – the horizontal and vertical boards attached to the head and pulley stiles to form the inner and outer sides of the boxed frame

- **back lining** – a vertical board attached to the back of the inner and outer linings on the side and enclosing the weights, forming a box

- **parting bead** – a piece of timber housed into the pulley stiles and the head, separating the sliding sashes

- **sash weights** – cylindrical cast iron weights used to counterbalance the sashes. These are attached to the sashes via cords or chains running over pulleys housed into the top of the pulley stiles

- **sill** – the bottom horizontal part of the frame housing the pulley stiles and vertical inner and outer linings

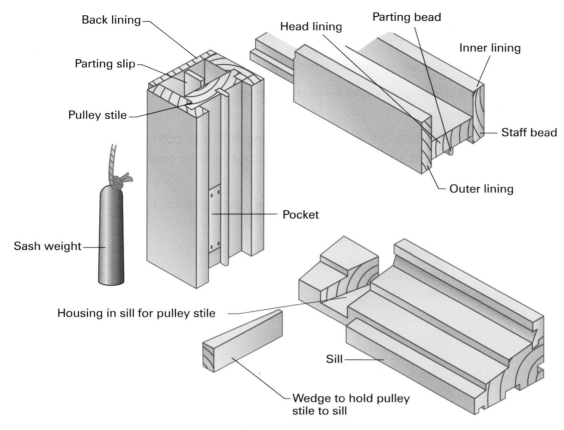

Figure 10.14 Parts of a box sash window frame

- **staff bead** – a piece of timber fixed to the inner lining, keeping the lower sash in place
- **parting slip** – a piece of timber fixed into the back of the pulley stiles, used to keep the sash weights apart
- **pockets** – openings in the pulley stile that give access to the sash cord and weights, allowing the window to be fitted and maintained.

The sash in a box frame is made slightly differently to normal sashes, and has the following parts:

- **bottom rail** – the bottom horizontal part of the bottom sash
- **meeting rails** – the top horizontal part of the bottom sash and the bottom horizontal part of the top sash
- **top rail** – the top horizontal part of the top sash
- **stiles** – the outer vertical members of the sashes.

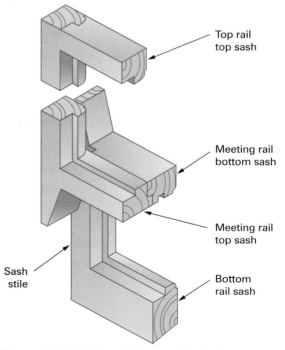

Figure 10.15 Sashes of a box sash window

Installing a box sash window

This can be done in one go with the sashes and weights already fixed, but for the purposes of this book we will show a more traditional method.

1. First strip the frame down and have the sashes, parting bead, staff bead and pockets removed.

2. Then fix the frame into the opening, ensuring that it is both plumb and level, and fit the sash cord and sashes (the fitting of the sash cord to the sashes will be covered in Chapter 13).

3. Next slide the top sash into place and fit the parting bead, to keep the top sash in place.

4. Now fit the bottom sash, then the staff bead, to keep the bottom sash in place.

5. Finally, seal the outside with a suitable sealer, fit the ironmongery and make good to the inside of the window.

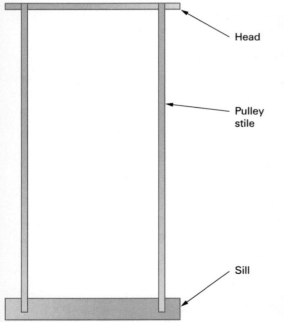

Figure 10.16 Stripped down frame

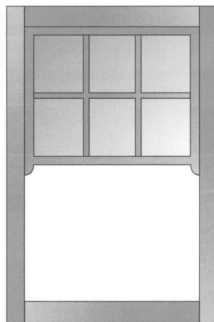

Figure 10.17 Top sash in place

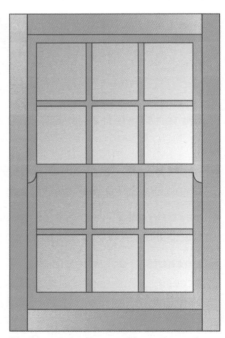

Figure 10.18 Top and bottom sashes in place

The marking out of windows will be covered in Chapter 14 and the maintenance of a box sash window in Chapter 13.

Bay window

Bay windows project out from the face of the main outer walls of the dwelling. There are several different styles, but only three ways of building or incorporating one into a house.

Method 1 Pre-built bay

This is the simplest method, as it is part of the original design of the house and is constructed when the base walls of the house are. The area in the bay contributes to the size of the interior of the room.

Method 2 Built bay

With this method a bay is added onto the original structure of the house. The builder may open up the existing wall and build the bay into it, giving a larger room, or leave the wall and have a large window board area that can be used for storage – the choice depends on both the bay size and the client's wishes.

Method 3 Supported bay

This is only used for small bays with a small projection. Here a support system carries the weight of the bay, but no wall is built. The room is no bigger, but there will be a larger window board area.

Remember

When replacing a bay, it is important to ensure that the roof is adequately supported *before* removing the old window

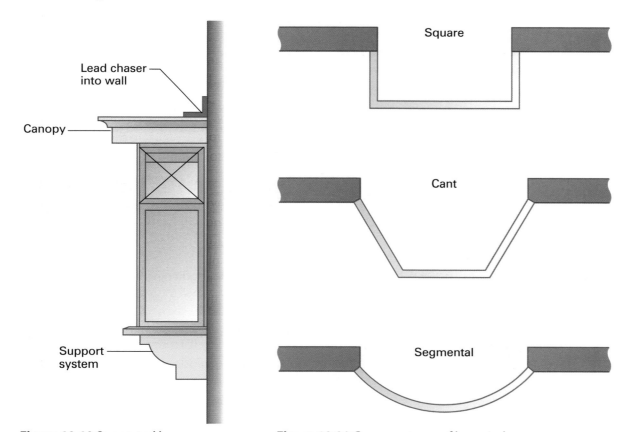

Figure 10.19 Supported bay

Figure 10.20 Common types of bay window

Types of bay window

Several types of bay window can be constructed. The choice of style will depend on the size and shape of the opening or existing bay.

Bay windows are usually made up as a series of windows joined together rather than a single window. Bay windows are joined together by either manufacturing the windows to allow them to be screwed together or by using pre-made posts.

If the window is a square bay, the windows can be joined together like this:

If the window is a splayed or cant bay, they can be joined together as shown in Figure 10.22.

There may need to be a joint along the front of a bay window (usually only on very large windows), in which case the windows are joined together as shown in Figure 10.23.

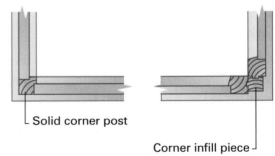

Solid corner post

Corner infill piece

Figure 10.21 Square bay joints

Installing a bay window

The first thing to do is to remove the old window. The old window will be load bearing, so the roof above it must be suitably supported before it is removed.

Next, fit and level the window sill. Then fit one of the end windows, level and plumb it, then fix it to the wall.

Now fit the corner block and the front window, screwing it to the corner block and ensuring it is level, plumb and square to the first window. Do not fix the window to the sill or roof/wall above yet as it may need to be moved slightly to allow the last window to be fitted.

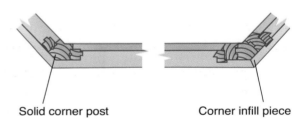

Solid corner post

Corner infill piece

Figure 10.22 Cant bay joints

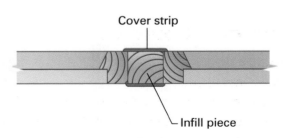

Cover strip

Infill piece

Figure 10.23 Lengthening bay joints

Next fit the final corner block and last window, screwing it to the wall, ensuring it is level, plumb and square to the front window. Now the windows can be fixed to the sill and roof/wall above and, if required, the cover strips at the joints can be fitted. Running a small bead of silicone along the back of the cover strip prior to fixing will help with waterproofing.

Remember

When removing an old window, the less damage you do, the less you will have to make good later

Remember

When fitting the first window, it is a good idea to run a small bead of silicone between the windows and corner posts to help with waterproofing

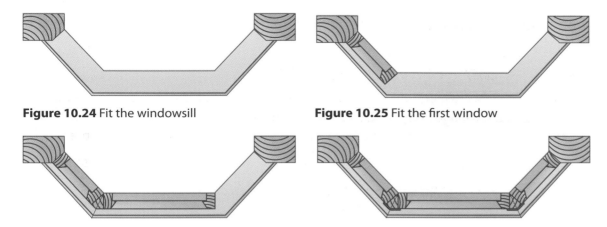

Figure 10.24 Fit the windowsill

Figure 10.25 Fit the first window

Figure 10.26 Fit the front window

Figure 10.27 Window fitted including cover strips

To finish, glaze the window, fit any ironmongery, seal the exterior with silicone, fit the window board and make good to the inside.

Bow window

Bow windows are similar to segmental bay windows. They are made as a single window if the radius of the arch is not too great; otherwise they are made as separate windows.

The segmental bow window is fitted in the same way as the bay window, with the sill fitted first and the windows fitted from one side to the other. A bow window made as a single window is fitted in the same way as a normal window, with extra care being taken when handling the large window.

Window boards

The window board forms the finish or trim to the top of the inner leaf of the cavity wall, where it finishes at the window opening. Window boards can be formed from solid timber, blockboard, MDF, plywood or plastic. They are used to give a neat finish on the inside of the window sill. They are usually fixed before plastering and, thus, are part of the first fix.

The window board shown in Figure 10.29 has been fixed to wooden plugs (similar to fixing a door lining, which is covered later). The back edge of the window board is fitted into a groove in the sill to allow for any movement in the timber. This will prevent a gap showing should shrinkage occur, which can often happen as radiators are usually situated below the windows.

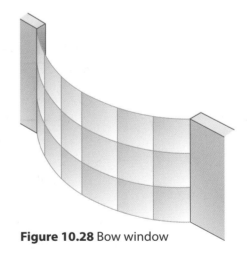

Figure 10.28 Bow window

Window board

Plug or pad

Plaster on inner skin

Figure 10.29 Fixed window board

The front end of the window board is usually nosed or rounded and the ends cut to run past the brick jambs.

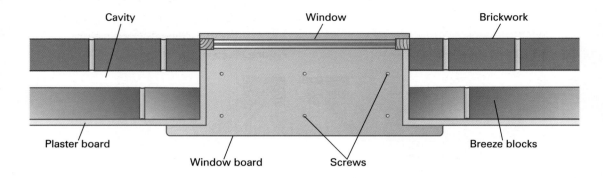

Figure 10.30 Plan view of window board fixed in position

The sequence for fixing window boards is as follows:

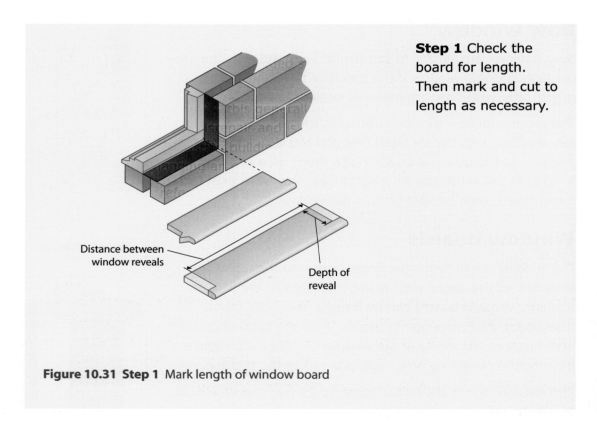

Step 1 Check the board for length. Then mark and cut to length as necessary.

Figure 10.31 Step 1 Mark length of window board

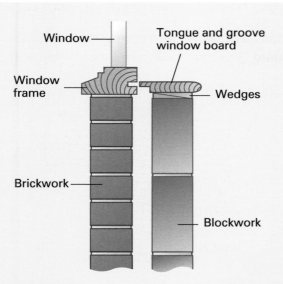

Window

Window frame

Brickwork

Tongue and groove window board

Wedges

Blockwork

Figure 10.32 Step 2 Fit into position

Step 2 Check for height and level, ensuring engagement of tongue and groove, then make and fix **packings** until height and level are achieved.

Definition

Packing – any material, generally waste wood, used to fill a gap

Step 3 Fix board with nails through packings, if this can be done into material suitable to receive nails; otherwise use plugs.

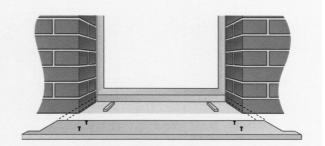

Figure 10.33 Step 3 Fix window board with nails

Did you know?

The back edge of a window board can be rebated to form a tongue to fit into a groove on the inside sill of a window

Stairs

Stairs are a means of providing access from one floor to another. They are made up of a number of steps and each continuous set of steps running in one direction is called a flight. Hence, a staircase may have several flights. They are often the cause of accidents in buildings and are governed very strictly by *Building Regulations*.

Stairs in this section will be looked at under the headings of:

- terminology
- building regulations
- setting and marking out
- how to cut and construct stairs
- how to install them.

Terminology

There are a number of terms and definitions associated with construction of stairs, which you will need to know and understand. The main ones are listed below and shown in Figure 10.34.

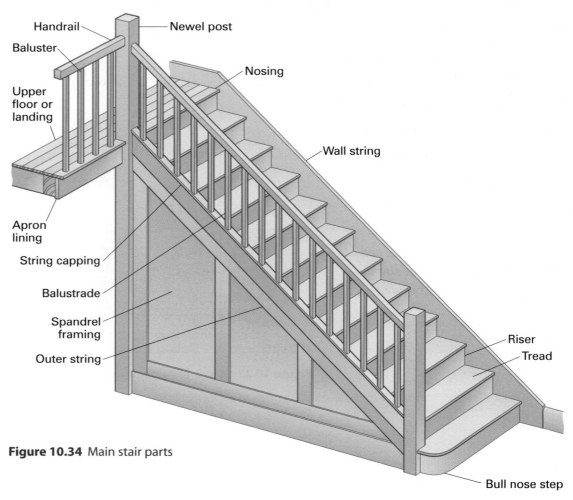

Figure 10.34 Main stair parts

landings – between floor levels to break up the overall length of a flight and can be used to change the direction of a flight of stairs

string – main board to which treads and risers are fixed, including: wall string, outer string, close string and cut string

tread – flat, horizontal part of the step

riser – vertical part of the step

rise – the height of the step (the measurement from the top of one step to the top of another)

step – combination of one tread and one riser

going – horizontal distance measured from the nosing of one step to the nosing of the step directly above or below

newel – heavy vertical member at each end of the stair to which the handrail is fixed

balustrade – unit comprising handrail, newels and the infill between it and the string, which provides a barrier for the open side of the stair

balusters – vertical members forming the infill between the string and the handrails

bull nose step – quarter-rounded step at the bottom of a stair

stair well – opening formed in the floor layout to accommodate a stair

nosing – front edge of a tread, or loose narrow top tread which sits on the trimmer joist at the top of the stairs

capping – fixed on the top edge of strings to take the fixing of a balustrade

cap – shaped top of a newel post, which can be fixed on or turned on the solid newel

spandrel framing – where the triangular area is formed under the stairs, which can be framed to form a cupboard.

Straight flight stairs

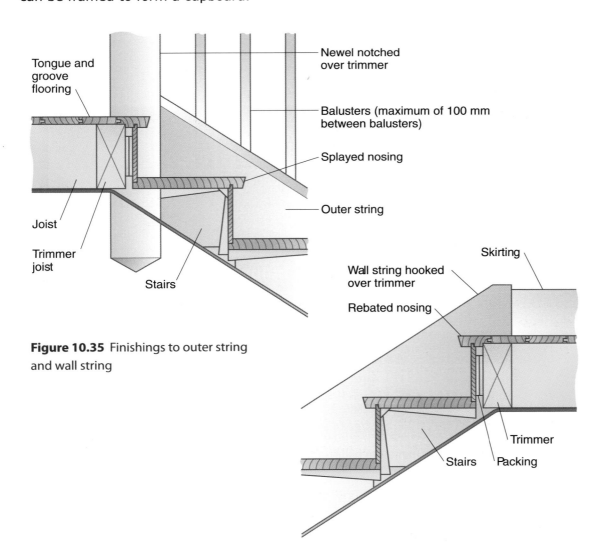

Figure 10.35 Finishings to outer string and wall string

Building Regulations

The construction and design of a stair is controlled under *Building Regulations 2000: Part K*. These specify the requirements for different types of stairs depending on the type of building and what it is to be used for.

Tables 10.1 to 10.3 and Figures 10.36–10.38 detail some of the requirements in the regulations.

Description of stair (to meet)	Max rise (mm)	Min going (mm)	Range (to meet pitch limitation) (mm)
Private stair	220	220	155–220 rise with 245–260 going or 165–200 rise with 220–350 going
Common stair	190	350	155–190 rise with 240–320 going
Stairway in institutional building (except stairs only used by staff)	180	280	
Stairway in assembly area (except areas under 100 m square)	180	250	
Any other stairway	190	250	

Table 10.1 Regulations for rise and going

Description of stair	Minimum balustrade height (mm)	
	Flight	Landing
Private stair	840	900
Common stairway	900	1000
Other stairway	900	1100

Table 10.2 Regulations for minimum balustrade heights

Description of stair	Minimum width (mm)
Private stair giving access to one room only (except kitchens and living room)	600
Other private stair	800
Common stair	900
Stairway in institutional building (except stairs used only by staff)	1000
Stairway in assembly area (except areas under 100 m square)	1000
Other stairway serving an area that can be used by more than 50 people	1000
Any other stairway	800

Table 10.3 Regulations for width of stairs

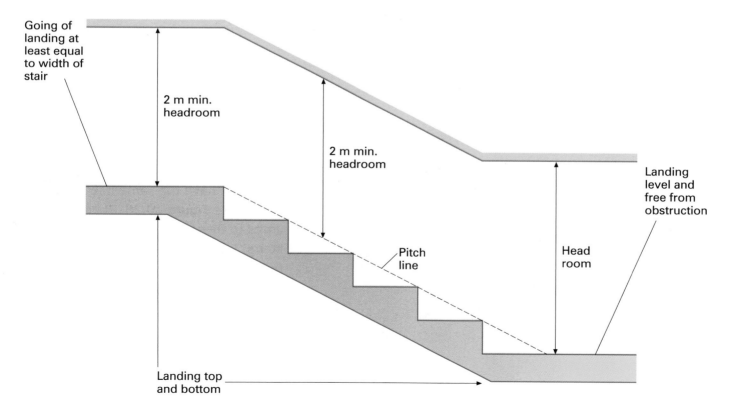

Figure 10.36 Regulations for minimum headroom height and landing requirements

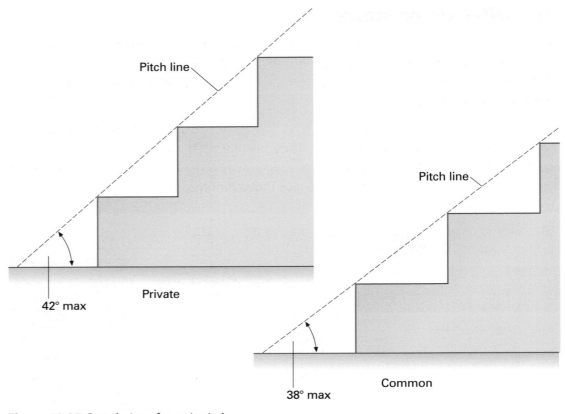

Figure 10.37 Regulations for stair pitch

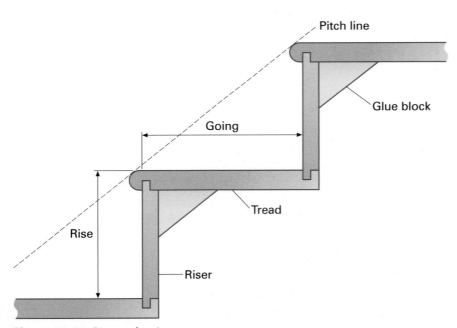

Figure 10.38 Rise and going

Figure 10.37 shows how the pitch or steepness is limited to a maximum of 42° for a private stair and a maximum of 38° for a common stair (i.e. where a stairway is used by more than one dwelling).

The steps for a straight flight of stairs should all have the same rise and going. See Figure 10.38. There are limits to these dimensions for different stairs. In all situations twice the rise plus the going should work out between 550 and 700 mm (2R+G).

Installation of stairs

Installing a staircase on a construction site is a major part in the construction of the building. The same basic procedures are followed, though there will be differences depending on what type of staircase is to be installed as well as variations due to the site.

A staircase is a large and fairly expensive item and should be handled with care when delivered to site and offloaded. They are normally delivered in completed form as far as possible, but components such as the handrail, balusters and newel posts will have to be fitted to suit the site requirements.

Step 1 Fix the wall string, cutting it off at floor level to suit the skirting height.

Step 2 Use a hand saw to cut the seat cut on the string at the foot of the stairs.

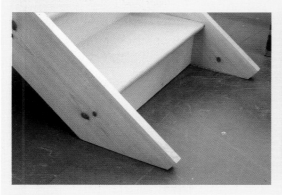

Step 3 The staircase is now level on the floor.

Step 4 Steps 1 and 2 can be repeated at the top with the underside cut out to sit on to the floor trimmer and the top tread is cut away so that it sits on the trimmer.

Step 5 Mortise the outer string into the newel posts at each end.

Step 6 Fix newel posts in place. The bottom newel can be held in position using various methods and this will depend on the composition of the floor. For rigidity and maximum strength the top newel should be notched over the trimmer joist and screwed or bolted to it.

Step 7 Fix the wall string to the wall in approximately four places below the steps, usually with 75 mm screws and plugs. Fixing should be below the steps unless access is difficult, in which case the resulting hole should be plugged. The balustrades and hand-rail can be fitted once the stairs are secure.

Step 8 Once the staircase has been fitted, it should be protected to prevent damage. Strips of hardboard should be pinned to the top of each tread with a lath to ensure the nosing is protected. Use the same method to protect the newel posts.

Timber studwork

Timber studwork covers partitions or walls, usually of light construction, used to divide a building or large area into compartments. The NVQ syllabus only covers non-load-bearing walls.

This section will cover:

- basic construction
- in-situ partitions
- pre-made partitions
- holes and notches
- insulation.

Basic construction

Timber partitions are made up of a head and sole plate with studs at regular centres fixed between them. These are stiffened up with noggings, which also provide extra fixings for sheet material, usually plasterboard, and fixing points for heavy components such as wash basins, toilets etc. See Figure 10.39.

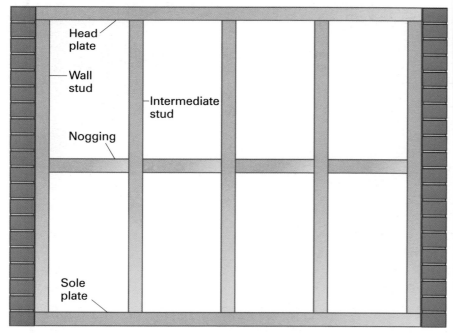

Head plate

Wall stud

Intermediate stud

Nogging

Sole plate

Figure 10.39 Components of a stud partition

The timber used is normally 75 mm × 50 mm for heights up to around 2.4 m and 100 mm × 50 mm for studwork above this height. Planed all round timber is preferred because its cross-section is uniform and is better to handle.

Partitions are either made in-situ or pre-made. In-situ components are cut, fixed together and fitted on site. Pre-made partitions are assembled on site or in a factory, for erection on site later.

The intermediate studs are measured and fixed to suit the sheet material to be fixed. For 9.5 mm plasterboard studs are spaced at 400 mm centre to centre and for 12 mm plasterboard at 600 mm. These sheets should cover 2.4 m × 1.2 m. A sheet size of 2.4 m × 0.9 m would have studs spaced at 450 mm.

The majority of stud partitions are fixed together by butt joints and skew-nailed. Another method is to use framing anchors, which are strong and quick to fit.

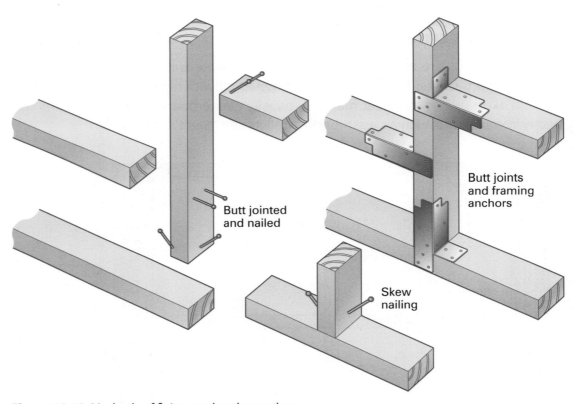

Butt jointed and nailed

Skew nailing

Butt joints and framing anchors

Figure 10.40 Methods of fixing studwork together

In-situ partitions
Fixing partitions in position

Step 1 Mark out positions for partitions on the floor. Then check the ceiling to see how to fix the head plate. Ideally it will run at a right angle to the joists but, if not, further noggings should be fixed between each joist to provide a fixing point. The head plate can now be fixed to the ceiling, usually with 100 mm wire nails or screws.

Step 2 Plumb down from the head plate with a plumb bob, or a level and straight edge, and mark the position on the floor.

Step 3 The sole plate can be fixed in the same way as the head plate. If on to a concrete floor it should be plugged and screwed.

Step 4 Cut the wall studs and fix to the wall by plugs and screws. They should be skew-nailed to the head and sole plate.

Step 5 Mark the position of intermediate studs on the side of the head and sole plates at the appropriate centres, that is 400, 450 or 600 mm depending on sheet material size and thickness.

Step 6 Measure each stud individually, as they may vary in length between head and sole plate, then skew-nail with 100 mm wire nails. Studs should be a tight fit and, with the bottom in place, the tip should be forced over into position.

Step 7 If overall partition height is greater than 2.4 m, position noggings centrally at the top edge of the boards. Through-nail one end and skew-nail the other.

Step 8 Fix additional noggings at various heights from 600 mm, 1200 mm to 1800 mm to give extra strength and rigidity to the frame, and provide extra fixing points for the sheet material and any components that may have to be fixed to the walls.

Forming openings

Where an opening is to be formed in a studwork partition, studs and noggings should provide fixing points for items such as door linings, windows, hatches etc. They also provide fixings for the edges of any sheet material to be used. Details are shown in Figures 10.41, 10.42 and photographs.

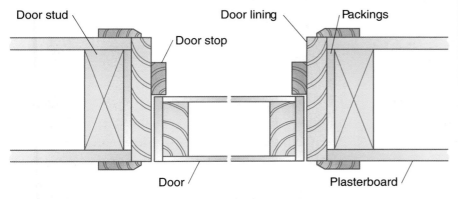

Figure 10.41 Section through a door opening

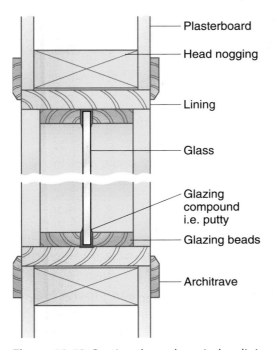

Figure 10.42 Section through a window lining

Noggings provide fixings for sheet materials

Opening formed to take a window lining

Corners and junctions

When a partition has to be returned at right angles or when a 'tee' junction is required, extra studs must be fixed to provide support and fixing for the covering material.

Corner detail with extra studs   T-junction detail with extra studs

Pre-made partitions

Pre-made partitions are the same specification as in-situ partitions but can be made up either in a workshop or on site. They are made slightly under size in height and width to allow for the frame to be offered into position.

Joints are usually butt-jointed and nailed but can also be housed-in or framing anchors used.

Folding wedges can be used to keep the frame in position while fixing takes place.

Holes and notches

It is common practice for pipes and cables to be concealed within timber partitions, though it is recommended that they are kept to a minimum so as not to affect partition strength.

Holes for pipework

Notches with wire

Positioning is important. Holes for carrying cables, and notches for water pipes, should be kept away from areas where there is a possibility they may be punctured by a nail or screw. This includes areas where, for example, kitchen units, cupboards, skirting boards, dado or picture rails may be fixed.

Notches can be protected by fixing a metal plate over them.

Insulation

Certain situations may require studwork to have insulation between the wall coverings to form a sound or thermal barrier. *Building Regulations 2000* control the methods of achieving this and should be referred to for technical information.

Extra fire resistance can be achieved by double boarding the studwork or by using fire-lined board.

Mineral wool or glass fibre insulation are most commonly used for sound and thermal insulation.

Fire-resistant insulation

Sound and thermal insulation

Joist and stud coverings

Lath and plaster

Joists are decked with a suitable flooring material, and the underside of the joists is usually covered with plasterboard. Stud partitions are also usually covered with plasterboard, with the occasional exception. For example, areas such as shower rooms are likely to be tiled, so a sturdier board such as WBP plywood is used.

Joist and stud coverings have come a long way in the past few years, and the traditional method of lath and plaster is now only really used in listed buildings.

There are two main ways of plasterboarding a room:

Method 1

The plasterboard is fixed to the stud or joist with the back face of the plasterboard showing. The plasterer covers the whole wall or ceiling with a thin skim of plaster, leaving a smooth finish.

Plastered wall

Taped wall

Method 2

The plasterboard is fixed with the front face showing, and the plasterer uses a special tape to cover any joints, then a ready mixed filler is applied over the tape and is used to fill in the nail and screw holes. Once the taped area is dry, the plasterer then gives the area a light sand to even it out.

Plasterboard comes with a choice of two different edges, and the right edge must be used:

- square edge – for use with Method 1, the whole wall/ceiling is plastered
- tapered edge – for use with Method 2, so that the plasterer can fix the jointing tape.

As well having the correct edging, the right type of plasterboard must be used.

Various types of plasterboard are available so a colour coding system is used (NB this may vary from manufacturer to manufacturer):

- fire resistant plasterboard (usually red or pink) – used to give fire resistance between rooms such as party walls
- moisture resistant plasterboard (usually green) – used in areas where it will be subjected to moisture such as bathrooms and kitchens
- vapour resistant plasterboard – with a thin layer of metal foil attached to the back face to act as a vapour barrier, usually used on outside walls or the underside of a flat roof
- sound resistant plasterboard – used to reduce the transfer of noise between rooms and can also help acoustics, used in places such as cinemas
- thermal check board – comes with a foam or polystyrene backing, so used to prevent heat loss and give good thermal insulation
- tough plasterboard – stronger than average plasterboard, it will stand up to more impact, used in public places such as schools and hospitals.

Sometimes plasterboard is doubled up – the walls are 'double sheeted' – to give fire resistance and sound insulation without using special plasterboard.

As well as a variety of types plasterboard comes in a variety of sizes, though the most standard size is 2400 × 1200 × 12.5 mm.

Remember

If you double sheet plasterboard, the joints in the two layers must be staggered

Cutting plasterboard

Plasterboard can be cut in two main ways: with a plasterboard saw or with a craft knife. The plasterboard saw can be used when accuracy is needed but most jobs will be touched up with a bit of plaster, so accuracy is not vital. Using a craft knife is certainly quicker.

This is how to cut plasterboard with a craft knife:

Step 1 Mark the line to be cut with a pencil.

Step 2 Score evenly along the line with the craft knife ensuring hands, legs, etc. are out of the way of the blade.

Step 3 Turn the board round and give the area behind the cut a slap, to split the board in two.

Step 4 Run the knife along the back of the cut to separate the two pieces of board.

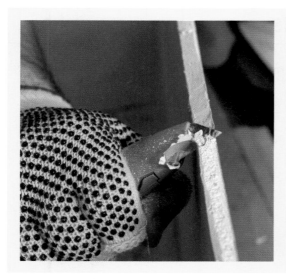

Step 5 Trim the cut with a slight back bevel to give a neater finish.

Fixing plasterboard

Plasterboard can be fixed by three different methods, which are:

- **Nail –** Plasterboard can be fixed using clout or special plasterboard nails. When nailing, make sure the nail is driven below the surface with the hammer, leaving a dimple or hollow that can be filled by plaster later. The nails used must be galvanised: if not, when the plaster is put on to fill the hollows, the nails will rust and will 'bleed' – the rust from the nails will stain the plaster and can even show through several coats of paint.

- **Screw –** Screws can be used but these too must be galvanised to prevent bleeding.

- **Dot and dab –** This is used where the plasterboard is fitted to a pre-plastered or flat surface such as block work. With dot and dab, there is no need for a stud or frame to fix the plasterboard, thus increasing the room size (it is often used in stairwells where space is limited). Dot and dab involves mixing up plaster and dabbing it onto the back of the board, then pushing the board directly onto the wall.

Remember

When fixing plasterboard using nails or screws, be aware of where the services are. Otherwise, you could hit an electricity cable or a water pipe

Remember

Dot and dab is a skilled technique, as with this method getting the boards plumb and flat can be difficult

Door frames and linings

The timber frames to which doors are hung may be divided into two main groups:

- linings (or casings)
- frames.

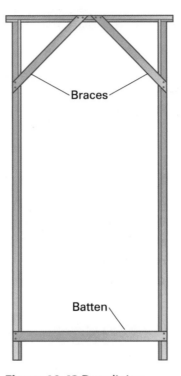

Figure 10.43 Door lining

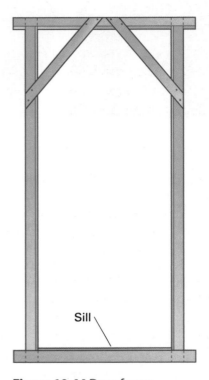

Figure 10.44 Door frame

Door frames are usually constructed from heavier material with solid rebates and generally used for external entrance doors. Linings are lighter with planted door stops and generally used for internal doors.

This section will look at doors under the following headings:

- door linings or casings
- door frames
- built-in frames
- storey frames.

Door linings or casings

Door linings or casings are lightweight internal frames that can be fixed into position by nailing through the jambs into fixing blocks, pads, plugs, or directly into a timber stud wall.

Blocks or pads are usually built into the door opening during construction by the bricklayer. Three or four will be needed on each jamb.

Where fixing blocks or pads have not been used, it may be necessary to plug the opening, using a plugging chisel to chase out four mortar joints on each side of the frame. Wooden plugs can be cut from scrap timber, driven into these slots and sawn off plumb. The sawn ends must be kept exactly in line and square. See Figure 10.45.

The door opening should have been built with a clearance of about 20 mm, so that the lining fits loosely and can easily be plumbed and aligned with the walls. Packing will be needed between lining and brickwork so that it is held securely in place once fitted.

The lining can now be offered into the opening and checked that it is plumb and level. It is normal practice to fix one jamb first, making sure that it is plumb and straight and will suit the line of the finished plasterwork. Once this is done the second jamb can be fixed and sighted through to make certain that it is exactly parallel.

It is good practice to make a temporary fixing at the top and bottom of each jamb, leaving the fixing protruding. The lining can then be checked finally for accuracy and position, and adjustment made before it is finally fixed. See Figure 10.46.

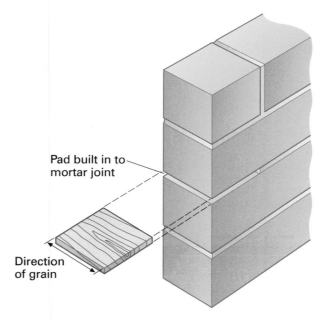

Pad built in to mortar joint

Direction of grain

Figure 10.45 Fixing blocks

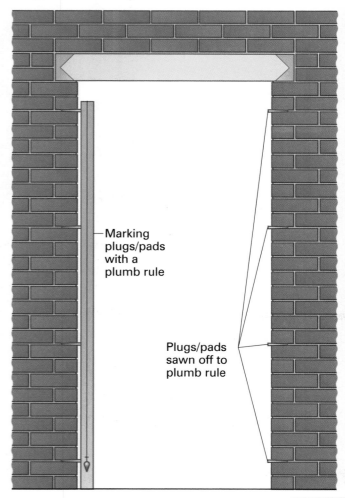

Marking plugs/pads with a plumb rule

Plugs/pads sawn off to plumb rule

Figure 10.46 Plugging the joints for fixing a door lining

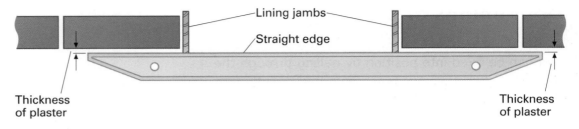

Thickness of plaster

Lining jambs

Straight edge

Thickness of plaster

Figure 10.47 Checking alignment of a door lining

The following sequence of operations is recommended:

Step 1 Firstly remove the door stops, which should have been nailed lightly, and put to one side. The door is unfinished so check the finished floor level (FFL) in relation to the bottom of the lining. The most accurate way of doing this is to measure down from a datum line (usually 1 m or 900 mm above FFL) and pack the lining to suit.

Step 2 Wedge the lining approximately in position by putting a wedge above each leg in the gap between the lintel and the head of the lining. The head should be checked with a spirit level to see that it is level. The lining can now be packed down the sides with either hardboard or plywood, at the top fixing positions. The top of the lining should be moved to ensure equal projection either side of the wall.

Step 3 Fix the lining at the top, through the packing, with two nails at each fixing point. An alternative to nailing is using screws, which is more common practice nowadays.

Step 4 Plumb the lining on the face side and edge using a level. Pack the bottom position as required and fix through one set of packing. Check that the fixed leg is square. Then complete the fixings on the same leg. The amount of fixing points should be about five (not less than four).

Step 5 Next remove the stretcher from the bottom of the lining, hold it at the head and mark the inside width to make a pinch rod.

Step 6 The pinch rod can now be fitted in the bottom position of the lining and the lining packed to suit.

Step 7 Check the leg to see that it is plumb and also the alignment of the frame. The bottom can be fixed, and then moving the pinch rod to ensure the frame is parallel, fix the intermediate positions. The head will only require fixing if the opening exceeds normal width. All nails should be punched to about 3 mm below the surface. The door stops can now be replaced and protection strips fixed if required.

Door frames

Frames can be fixed to timber stud partitions using 75 mm oval nails or alternatively they can be screwed to solid walls using plastic plugs.

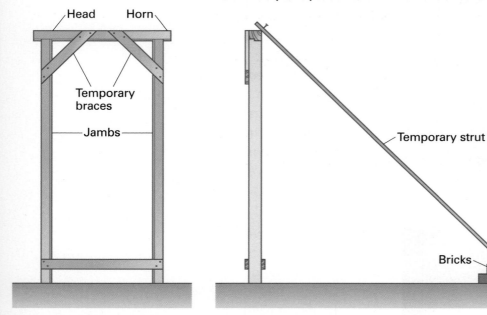

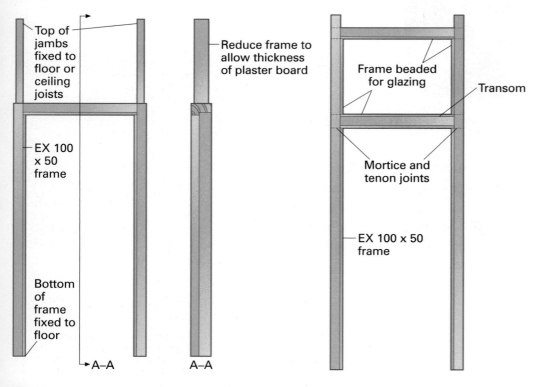

Figure 10.48 Temporary bracing and support

Built-in frames

The majority of frames are 'built-in' by the bricklayer as the brickwork proceeds. Prior to this the frame has to be accurately positioned, plumbed, levelled and struts temporarily inserted by the joiner.

Storey frames

Storey frames are designed to give stability when doorways, or fanlights, occur in thin non-loading walls. They are known as storey frames because their height is from floor to floor (one storey). They are used to borrow light from one room to another.

Figure 10.49 Plain storey frame

Figure 10.50 Fanlight storey frame

Ground lats

Ground lats (or just grounds) are timber battens which are fixed to the wall surface, particularly in high-class joinery work where hardwood skirting boards are to be used. They:

- provide a flat and level surface on which a covering (panelling, plasterboard, hardboard etc.) can be fixed to create a decorative finish
- provide a line for the plasterer to work to
- provide continuous fixing points for skirting boards
- enable these to be fixed during Second fixing (covered in Chapter 11) and avoid damage during plastering.

This section will look at:

- types of ground
- setting out of grounds
- fixing grounds.

Types of ground

Grounds are defined by their intended usage and include:

- grounds for skirting
- framed grounds
- counter battening
- separate grounds.

Setting out of grounds

Skirting grounds must be fixed carefully, keeping them straight and flush with the plaster line.

Grounds to which frames, sheets or panels are to be fixed need to be spaced to match their size.

The only requirements are a spirit level and straight edge, laser or snap chalk line, to mark their location.

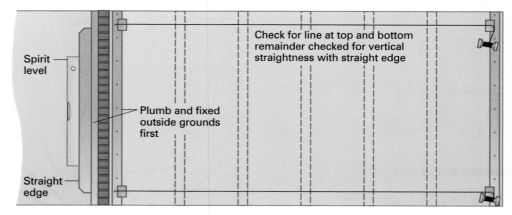

Spirit level

Check for line at top and bottom remainder checked for vertical straightness with straight edge

Plumb and fixed outside grounds first

Straight edge

Elevation

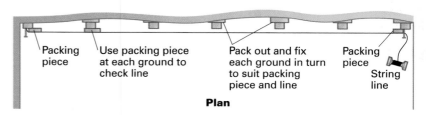

Packing piece

Use packing piece at each ground to check line

Pack out and fix each ground in turn to suit packing piece and line

Packing piece

String line

Plan

Figure 10.51 Setting out of grounds

Panelling or sheet material fixed to grounds

Fixing grounds

Grounds can be fixed in several ways, some of which are shown in the photographs. They include:

- screwing into timber plugs set into mortar joints
- screwing into plastic plugs set in masonry
- masonry nails
- adhesive.

Packing may be required at fixing points.

Twisted timber plug

Proprietary plastic plugs

 **Safety tip**

Always wear goggles when driving in masonry nails

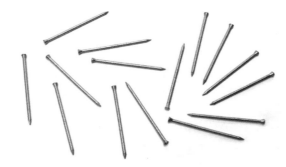

Masonry nails

FAQ

My maths isn't very good and so I often struggle with the geometry and trigonometry skills needed. What can I do?

Your training will include some basic numeracy skills lessons, which if you struggle with maths, you should take advantage of. Be patient and pay attention to any help and advice you are given regarding these sometimes complicated calculations. You will probably find that the more practical experience you get, the better your maths skills will become.

On the job: Finishing off

Scott and Nick are working in a large house plasterboarding the stud partition walls. When they come to the last wall, they find a note left by the plumber saying 'Don't plasterboard'. They are on price work and Nick says that they should just finish the work and get paid. Scott says that they should ask the plumber first. Who is right? What can the consequences be?

Knowledge check

1. Name the component parts of a window.

2. What is a mortar key used for in a window frame?

3. Why is it important to assemble windows flat?

4. How do you check a window frame is square as you assemble it?

5. Name two ways of fixing window board into position.

6. How do you level window board?

7. Name three materials used as window boards.

8. What types of weight are used in a box sash window?

9. Name three types of bay window.

10. List five components you will find on a flight of stairs.

11. What is the minimum headroom allowed as stated in the *Building Regulations* covering stairs?

12. Name the vertical member used to provide support to a handrail and infill on an open balustrade.

13. After stairs have been fitted what three measures can be taken to prevent damage during building work?

14. Name three ways of fixing grounds.

15. What PPE should be worn when using masonry nails?

chapter 11

Second fixing

OVERVIEW

Second fixing covers all joinery work after plaster work is completed.

This chapter is designed to help you to identify the main activities associated with second fixing work. It is also designed to provide you with the knowledge and understanding required to carry out the associated work activities.

This chapter will cover the following topics:

- Mouldings
- Internal doors
- External doors
- Double doors
- General door ironmongery
- Encasing services
- Wall and floor units.

These topics can be found in the following modules:

CC 1001K	CC 2009K
CC 1001S	CC 2009S

Mouldings

This section will provide you with the knowledge and understanding required to enable you to select and fit internal mouldings.

Typical timber mouldings found within a building

Moulding refers to the pattern put on a length of timber. This is normally done by a machine such as a spindle moulder. However, specialist hand planes could be used to match existing patterns.

Common types of moulding that you will meet are:

- architrave
- skirting board
- plinth block

- picture rail
- dado rail
- cornice.

Where you will find these within a room is shown on Figure 11.1.

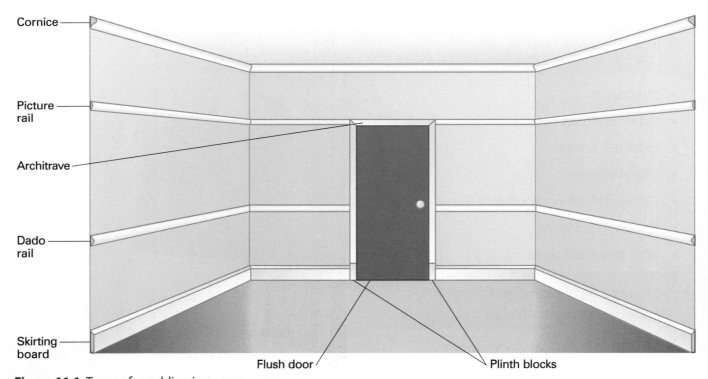

Figure 11.1 Types of moulding in a room

Mouldings are usually produced using premium grade timber or MDF and should only be fixed at the 'second fixing' stage, once a building is weather tight and plastered.

Architrave

Architraves provide a decorative finish around internal openings, especially doors, and cover the joint between frame and wall finish. They are available in lengths ranging from 2.1 m to 5.1 m, increasing in 300 mm multiples; or as sets consisting of 2 × 2.1 m legs (the sides of the opening) and 1 × 900 mm head (the top of the opening). Architraves are between 50 mm and 75 mm wide and are usually 19 mm to 25 mm thick before they are planed. There is a range of commercially available architrave mouldings.

Commercially available architrave mouldings

The steps below and on page 280 show how to fit an architrave to a smooth surface. If the surface is not smooth, the back of the architrave may have to be scribed to fit to it before finally fixing in place. This is described in the second section 'Scribing to walls'.

How to fit an architrave

Step 1 Architraves are kept back from the front edge of a frame by 6 mm –10 mm, which is known as the margin. This is the line we work to when fitting architraves. Hence, as Step 1 shows, draw the margin on the front edge of the legs and head of the frame until they meet.

Step 1 Drawing a working margin

Step 2 With the heel (narrow point or inner edge) of the architrave to the margin, place the leg on the floor and mark the architrave where the margin marks intersect each other on the frame.

Step 2 Marking the architrave

Step 3 Place the architrave into a mitre box and cut using a tenon saw. Alternatively, use a combination square to mark a 45° angle on the architrave and cut freehand.

Step 3 Mitring the architrave legs

Step 4 Position the cut leg against the margin marks and fix by nails. (When possible try and nail through the **quirks** in the moulding to help hide the nails.) If the surface is uneven you may need to shape the back of the architrave to fit to it before finally fixing, as described in the next section 'Scribing to walls'.

Step 4 Fixing the architrave legs

Step 5 Cut a 45° angle on the head piece and offer it to the angle of the leg that has just been fixed. If you have a good fit, repeat Step 2 with the head and then the remaining leg. Should the mitres not go together then the head piece will need planing until it does.

Step 5 Fixing the architrave head piece

Step 6 Punch all nails below the surface and remove the **arris** from the toe (wide part of the architrave – its outer edge).

Step 6 Finishing off the architrave

Definition

Quirk – the name used for the hollow in a moulding

Remember

Use a block plane when planing end grain

Definition

Arris – the sharp edge formed when two flat or curved surfaces meet

Scribing to walls

Scribing is to mark the profile of something onto the surface against which it is to be butted. In the case of architraves (or skirting) it is used so that the back can be shaped to fit against an uneven surface.

Step 1 Lightly nail the architrave 20 mm to 25 mm in front of the door frame or lining of the opening, and parallel to it.

Step 1 Temporarily fix forward of the door frame

Step 2 Measure the distance between the front edge of the architrave and the margin mark.

Step 2 Measure to the margin mark

Step 3 Cut a block, or set a pair of compasses, to the measurement you have just worked out. Take the architrave off the frame. Copy the shape of the uneven surface onto its face by running a compass, or block and pencil, down the uneven surface.

Step 3 Copy the surface shape onto the architrave

Step 4 Using a saw or a plane cut down to the line. Try in position, making sure the front of the architrave is level with the margin mark. Fix with nails.

Step 4 Shape the architrave

Skirting board

Skirting boards are used to provide a decorative finish between floor and walls. They also protect the wall finish from damage. Skirting boards are available in lengths ranging from 1.8 m up to 6.3 m, increasing by multiples of 300 mm. They are between 75 mm and 175 mm deep and 19 mm to 25 mm in thickness.

There are only ever three joints to cut when fitting skirting boards:

- internal joints, which should be scribed
- external joints, which should be mitred
- angle-lengthening joints.

How to fit skirting

Skirting boards can be fixed by:

- adhesive
- masonry nails
- screws and plugs
- oval/lost head nails (for internal stud walls).

Before starting, remove all obstructions from the floor, such as lumps of plaster. Then check lengths of timber and choose the most economical lengths.

The boards should be fixed on the top edge and at the bottom, and nails or screws made as inconspicuous as possible:

- Masonry nails should be **dovetailed**.
- Always plug the hole made by the screw or nail, using either a wooden plug or filler.

To get a board tight to the floor it is a good idea to use a kneeling board.

Dovetailed nails

Joiner using a kneeling board

Scribed joint

Skirting boards are shaped and frequently have mouldings to make them look attractive. Where they meet in a corner, the board butting up against the moulding needs to have its end shaped so that the boards fit neatly together. With one board fitted in place right into the corner, we now need to shape the board that will butt against it.

Method 1

Step 1 Place the square end of the board to be cut against the fixed board and copy the shape of the moulding onto the face of the board to be cut.

Step 1 Copy moulding shape to be scribed

Step 2 Back cut with a coping saw and/or tenon saw to the line. Keep trying in place until a good joint has been achieved and then fit in place.

Step 2 Back cut to the line

Method 2

Step 1 Using a mitre block, internally mitre across the width of the skirting board to be shaped. This will now show the end grain on the board.

Step 2 Using a coping saw and/or tenon saw back cut the mitre by letting the saw blade follow the line where the straight grain on the fixed board meets the end grain that is showing on the face of the board being shaped. Try in position and fit when a good joint has been achieved.

Step 1 Mitred skirting board with end grain to be shaped

Mitre joint

A completed mitre joint is shown in Figure 11.2. A step-by-step process for achieving this is as described on page 285.

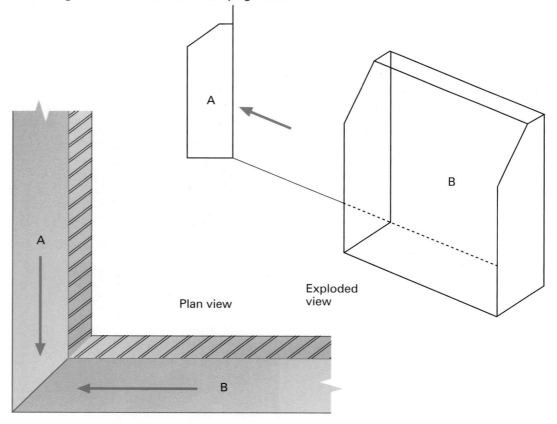

Plan view

Exploded view

Figure 11.2 Mitre joint on skirting board

Step 1 Lay the skirting board flat against the wall and draw a line on the floor where the board touches it, until the line reaches the adjoining wall. Repeat this with the board flat against the other wall. When the board is removed the lines on the floor should meet at a point near to the angle of the wall.

Step 1 Mark floor lines

Step 2 Place the stock of a sliding bevel right into the angle of the wall and move the blade around until it touches where the lines on the floor meet. You have just bisected the angle!

Step 2 Bisect the angle

Step 3 Place the bevel on the top edge of the skirting board and draw the line for the mitre. Cut accurately. Try in position and fit when a good joint has been achieved. Any adjustments can be made using a block plane.

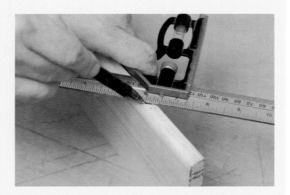

Step 3 Mark the mitre angle on the skirting board

Angle-lengthening joint

Angle-lengthening joints are used to join two pieces of board together in length.

This will be required if the skirting board available is too short for the wall.

Simply cut a 45° angle across the width of one board and do the opposite cut on the board to which it will be joined. Put them together and **skew-nail** through the joint to hold it in position. The finished joint is shown in Figure 11.3.

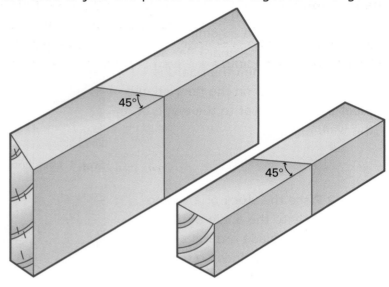

Figure 11.3 Angle-lengthening joint

Plinth block

Plinth blocks are also referred to as architrave blocks. They are fixed at the bottom of architraves and allow skirting to run up to them. See Figure 11.4.

They are used because:

- it is then possible to use skirting board that is thicker than the architrave

- skirting can have a moulded back edge

- they provide protection to the moulding on the architrave

- they look elegant.

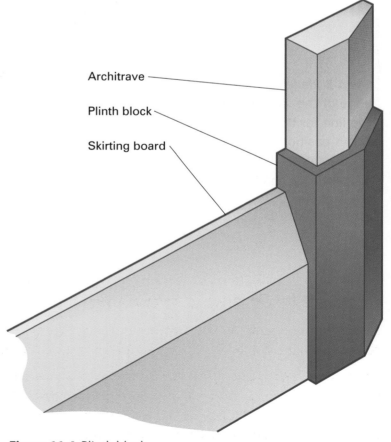

Architrave

Plinth block

Skirting board

Figure 11.4 Plinth block

Plinth blocks should be approximately 15 mm taller than the skirting board and follow the basic profile of the architrave with at least a 6 mm excess. They should be **butt jointed** to the architrave or preferably fixed using a barefaced tenon as shown in Figure 11.5. The skirting board can also be butt jointed to the plinth block, although a housing joint would be a better option, as shown in Figure 11.6. Mortise and tenon joints are covered in more detail in Chapter 14.

> ## Definition
>
> **Butt joint** – the simplest joint between two pieces of wood, with the end grain of one meeting the long grain of the other and glued, screwed or nailed together

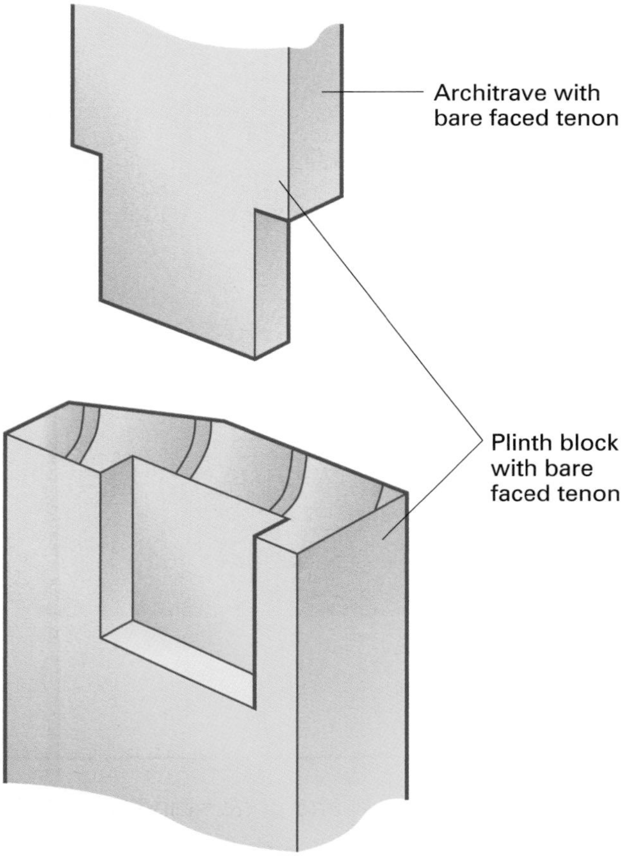

Architrave with bare faced tenon

Plinth block with bare faced tenon

Plinth block housed out to take skirting board

Skirting board

Figure 11.5 Plinth block joined by barefaced tenon

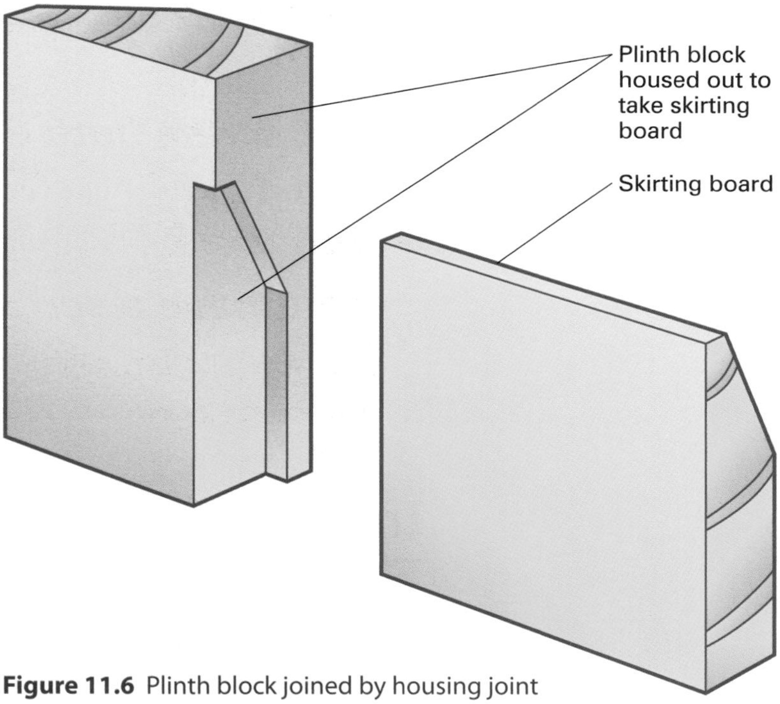

Figure 11.6 Plinth block joined by housing joint

Picture rails and dado rails

Pictures can be hung from a picture rail using special clips that are fastened to the picture frame, but the rails are often now just used for effect.

Dado is the name used for the lower part of a wall when it is visually different from the upper part. Dado rails run along the dividing line. They were originally designed to protect expensive hand-printed wallpapers from chair and sofa backs but, like picture rails, are generally now used for effect.

Both dado and picture rails help reduce the visual height of rooms with high ceilings.

The same sequence and methods employed to fix skirting boards are used to fix dado and picture rails.

FAQ

What is a scribe?

A scribe is a copy of the surface it fits over.

What does bisecting an angle mean?

Cutting an angle equally in two, as when cutting a mitre to create an internal angle.

Do I scribe both ends of a length of skirting?

No. It looks neater if one end butts to the wall and the other end is scribed to the butted end of the next skirting.

Can I have a scribe and a mitre?

No. It would be very difficult to get the scribed edges to meet correctly.

Internal doors

This section has been designed to provide the knowledge and understanding to select and hang internal doors. It will cover:

- purpose of internal doors
- types of internal door
- choosing a door
- fitting a door.

Purpose of internal doors

The purpose of an internal door is to provide a means of privacy, a thermal, sound or security barrier, a means of access and egress (i.e. a way out) and, in some instances, a fire-resisting barrier.

Types of internal door

Framed doors

Framed doors are doors made from hardwood or softwood and are constructed using mortise and tenon joints (described in Chapter 14 *Marking and setting out joinery products), or dowelled joints*. The frame is rebated or grooved, into which a board panel can be fitted; alternatively glass could be used and held in place with beading. Figure 11.7 shows an exploded view of a door including the types of joint used.

Figure 11.8 also shows types of sections used for door panels.

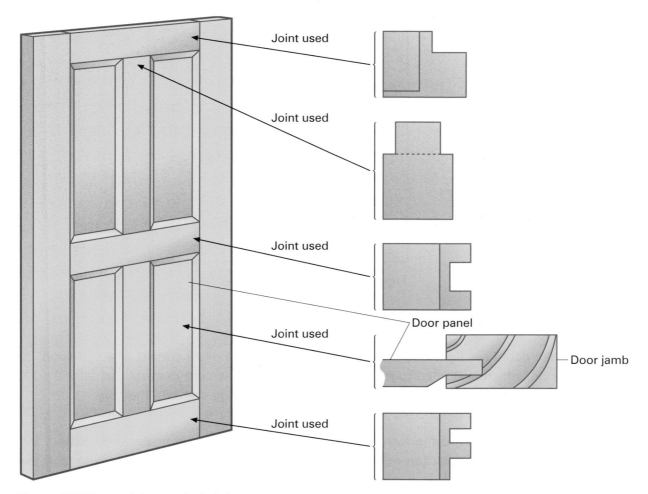

Joint used

Joint used

Joint used

Door panel

Joint used

Door jamb

Joint used

Figure 11.7 Framed door exploded view

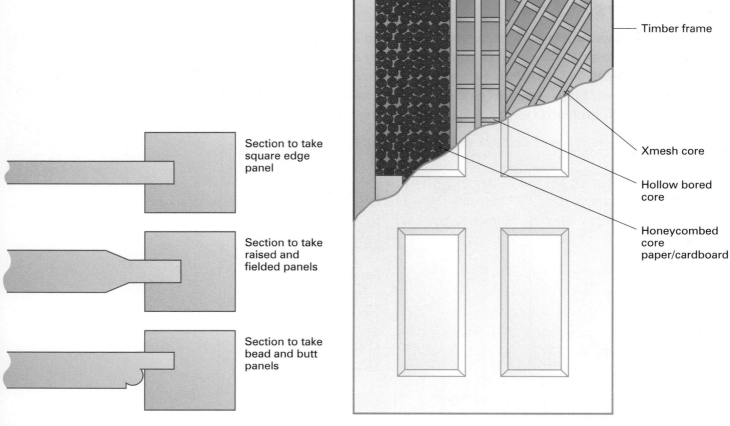

Figure 11.8 Sections used for door panels

Section to take square edge panel

Section to take raised and fielded panels

Section to take bead and butt panels

Timber frame

Xmesh core

Hollow bored core

Honeycombed core paper/cardboard

Figure 11.9 Flush door exploded view

Definition

Stile – the longest vertical timber in the frame of a door etc.

Did you know?

A material that swells when subjected to heat or damp is called intumescent

Flush doors

Flush doors are lightweight, cheap and simply made. Most consist of a softwood frame, which is stapled together and houses a hollow-core material (usually cardboard honeycomb that has the appearance of egg crates). They are then faced with hardboard or plywood. See Figure 11.9.

Flush doors often come with fitting instructions and these indicate which **stile** contains a lock block (an extra block of wood included in the door frame to take the door lock). Failing this, a symbol is normally printed on the head or bottom rail that indicates where the lock block can be found.

Fire-resisting doors

Fire-resisting doors having a core of solid, fire-retarding material and are frequently flush panelled. Their ratings are designated by their performance (resistance to penetration by flames or smoke through splits or gaps) with the prefix FD. For example, FD30 has a 30-minute rating, and an S suffix indicates the ability to resist smoke. The manufacturers also insert a coloured plastic plug into the door frame to indicate the rating.

Fire doors are available in standard sizes and in thicknesses of 44 mm for the FD30 and 54 mm for the FD60. Glazed fire doors must contain fire-rated plain or wired glass that is bedded in material that swells up when heated, hence increasing the resistance to smoke.

Choosing a door

On a newly built property the door schedule (see Figure 11.10) and plans of the building will determine what door to use and which direction it shall operate; also how it should be furnished with ironmongery.

Fire door plugs – 30- and 60-minute (note that colours used differ between manufacturers)

Description	D1	D2	D2	D2	D2	D2	D2	D2	D2	D10
Type										
External glazed A1					●					
External panel A2	●									
Internal flush B1									●	
Internal flush B2		●				●	●	●		●
Internal glazed B3			●	●						
Size										
813 mm x 2032 mm x 44 mm	●				●					
762 mm x 1981 x 35 mm		●	●	●		●	●	●		●
610 mm x 1981 x 35 mm									●	
Material										
Hardwood	●									
Softwood			●	●	●					
Plywood/polished		●								
Plywood/painted						●	●	●	●	●
Infill										
6 mm tempered safety glass										
Clear			●	●	●					
Obscured	●									

NOTES

BBS DESIGN

JOB TITLE

DRAWING TITLE
Door Schedule/doors

JOB NO.	DRAWING NO.

SCALE	DATE	DRAWN	CHECKED

Figure 11.10 A door schedule

Modern buildings tend to have standardised joinery throughout. Standard door heights are 2 m, 2.03 m and occasionally 2.17 m; widths range from 600 mm to 900 mm, usually in 75 mm increments. Thicknesses vary from 35 mm to 44 mm. Standard single door sizes that are available by type of door from a typical manufacturer are shown in Table 11.1. They include several that match the old imperial sizes (quoted in millimetres), as these are still required to replace existing doors. There are additional standard sizes available for double leaf, and single and double garage doors.

Height	1981	1981	1981	1981	2000	2032	2040	2040	2040
Width	610	686	762	838	807	813	626	726	826
External	*	*	*	*	*	*		*	*
Internal	*	*	*	*			*	*	*

Table 11.1 Standard single door sizes (mm)

Older properties often have relatively large doors to main rooms and may be of a non-standard size. In this case, the door will have to be made to measure; or the nearest available size, or larger, purchased and trimmed to fit by removing an equal amount from each stile to keep the frames symmetrical.

Fitting a door

Whenever possible doors should be stored flat for a few days prior to fitting in the location where they are to be hung. This is sometimes referred to as 'storing out of twist'. This gives the timber time to acclimatise to its new surroundings and any final shrinkage or reduction in moisture content can take place (which sometimes happens with modern central heating systems).

The following actions should be completed before starting to fit a door:

- Check building plans and door schedule to determine which type of door is to be used and in which direction it should operate. Failing this ask the supervisor or client.
- Check the frame is square and aligned properly.
- Measure the new door to make sure it will fit within the frame and check it is not twisted.
- Cut off horns, if fitted.
- Find where the lock block is located. The other stile is the hanging stile to which hinges will be attached.

Did you know?

Doors should normally conceal the largest area of the room when open to allow maximum privacy to those inside

Fitting is then carried out as follows:

Step 1 There must be a gap of 2 mm between the head of the door and the head of the frame and at least 6 mm clearance between the floor covering and the bottom of the door. If the door is too tall, always cut excess off the bottom rail.

Step 1 Make sure there is a gap between the door and the frame

Step 2 There must be a gap of 2 mm between both door styles and the frame. If necessary, plane evenly from both sides of the door. The door should now fit in the frame. With the hanging stile against the frame, wedge the door up so there is a 2 mm gap at the top and mark, if necessary, to shape the hanging style to match the frame. Plane off as necessary.

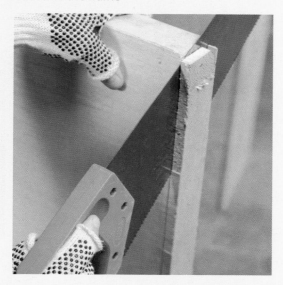

Step 2 Cut door to width and shape styles

Step 3 If necessary, mark and shape the opposite stile (containing the lock block), then slightly bevel the edge that will lead into the frame so that it will not catch when closing.

Step 3 Bevel the leading edge

Did you know?

Because the hinge knuckle goes on the outside edge of the door the arc followed by the inside edge is different; so a door requires a bevelled lead-in edge to close properly

Step 4 With the door fitting snugly in the frame and wedged up to give 2 mm clearance at the top, mark where the hinges are to go, on both door and frame at the same time. The top hinge normally sits 150 mm from the top of the door and the bottom hinge 225 mm from the bottom of the door.

Step 4 Mark hinges

Step 5 Remove the door from the frame and accurately mark where the hinges are to go, with the aid of a square and marking gauge. Do the same on the frame and then chop out recesses on both door and frame.

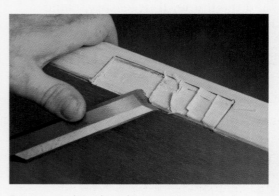

Step 5 Mark and cut hinge recesses

Step 6 Fit the hinges into the door recesses, putting in all the screws. Take the door to the frame and place the hinges into the frame recesses, securing the hinges using only one screw. Check the door swings without binding. Alter or adjust until there is an equal gap on both sides and the correct clearance at the top, and no resistance is encountered when closing the door. Put in all remaining screws on the stile hinges.

Step 6 Fit hinges

Step 7 Fit all remaining ironmongery, or furniture, as instructed by the schedule or client. (Ironmongery is covered in the next section.)

Step 7 Fit all remaining ironmongery

Remember

When fitting a fire-resisting door all tolerances, as specified in *Building Regulations*, must be adhered to, so check with the supervisor

External doors

This section aims to provide the knowledge and understanding needed to select, hang and fix the required ironmongery to an external door.

Types of external door

External doors are always solid or framed – a flush or hollow door will not provide the required strength or security.

In some cases external doors are made from UPVC. UPVC doors require little maintenance and the locking system normally locks the door at three or four different locations, making it more secure. High-quality external doors are usually made from hard-wearing hardwoods such as mahogany or oak, but can be made from softwoods such as pine.

External doors come in the same dimensions as internal doors except external doors are thicker (44 mm rather than 40 mm). On older properties external doors often have to be specially made as they tend not to be of standard size.

There are four main types of external door:

- framed, ledged and braced
- panelled doors
- half/full glazed doors
- stable doors.

Did you know?

A framed, ledged and braced door comes with the bracing unattached. The bracing can then be attached to suit the side on which the door is hung

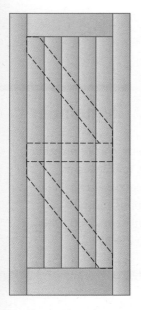
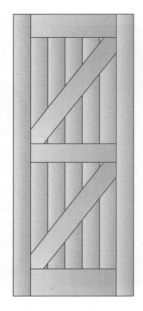

Front elevation   Back elevation

Figure 11.11 Front and back of a framed, ledged and braced door

Framed, ledged and braced doors

Framed, ledged and braced (FLB) doors consist of an outer frame clad on one side with tongued and grooved boarding, with a bracing on the back to support the door's weight.

FLB doors are usually used for gates and garages, and sometimes for back doors. When hanging an FLB door it is vital that the bracing is fitted in the correct way; if not, the door will start to sag and will not operate properly.

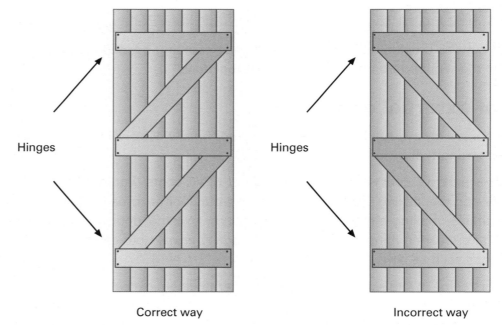

Hinges   Hinges

Correct way   Incorrect way

Figure 11.12 Bracing fitted correctly and incorrectly

Panelled doors

Panelled doors consist of a frame made up from stiles, rails, muntins and panels. Some panel doors are solid, but most front and back doors have a glazed section at the top to allow natural light into the room.

Half-/full-glazed doors

Half-glazed doors are panelled doors with the top half of the door glazed. These doors usually have diminished rails to give a larger glass area.

Full-glazed doors come either fully glazed or with glazed top and bottom panels, separated by a middle rail.

Full-glazed doors are used mainly for French doors or back doors where there is no need for a letterbox.

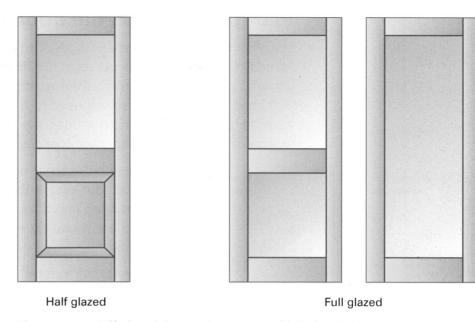

Half glazed   Full glazed

Figure 11.13 Half-glazed door and two types of full-glazed door

Stable doors

As their name implies, stable doors are modelled on the doors for horses' stables and they are now most commonly used in country or farm properties. A stable door consists of two doors hung on the same frame, with the top part opening independently of the bottom. The make-up of a stable door is similar to the framed, ledged and braced door, but the middle rail will be split and rebated as shown in Figure 11.14.

To hang a stable door, first secure the two leaves together with temporary fixings, then hang just like any other door, remembering that four hinges are used instead of three.

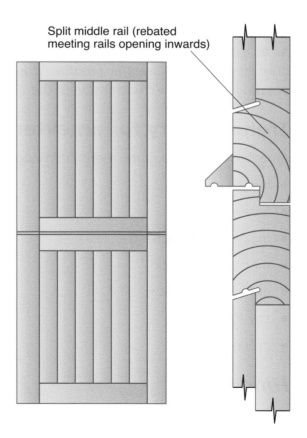

Split middle rail (rebated meeting rails opening inwards)

Figure 11.14 Stable door with section showing the rebate on the middle rails

Hanging an external door

External doors usually open inwards, into the building. Where a building opens directly onto a street, this prevents the door knocking into unsuspecting passers-by, but, even where there is a front garden, having a door that opens outwards is not good practice as callers will need to move out of the way to allow the door to be opened. Externally opening front doors are usually only used where there is limited space, or where there is another door nearby which affects the front door's usage.

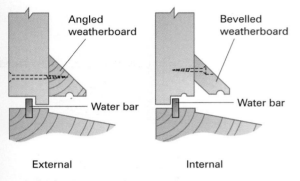

Figure 11.15 Internally and externally opening doors

Hanging an exterior door is largely the same as hanging an interior door except that the weight of the door requires three hinges sited into a frame rather than a lining. Because of this, there is usually a threshold or sill at the bottom of the doorframe.

If a water bar is fitted into the threshold to prevent water entering the dwelling, you will need to rebate the bottom of the door to allow it to open over the water bar. The way the door is then hung will depend on which side of the door is rebated. A weatherboard must also be fitted to the bottom of an external door, to stop driving rain entering the premises.

As an exterior door is exposed to the elements it is important that the opening is draught proofed: a draught-proofing strip can be fitted to the frame, or draught proofing can be fitted to the side of the door.

Ironmongery for an external door

An external door requires more ironmongery than an internal door, and may need:

- hinges
- letter plate
- mortise lock/latch
- mortise dead lock
- cylinder night latch
- spy hole
- security chains.

Hinges

External door hinges are usually butt hinges, though framed, ledged and braced doors often use T hinges. Three 4-inch butt hinges are usually sufficient, though it is advisable to use security hinges (hinges with a small steel rod fixed to one leaf, with a hole on the other leaf) to prevent the door being forced at the hinge side.

Letter plate

The position of a letter plate depends on the type of door. Letter plates are usually fitted into the middle rail, but could be fitted in the bottom rail (for full-glazed doors) or even the stile (with the letter plate fitted vertically).

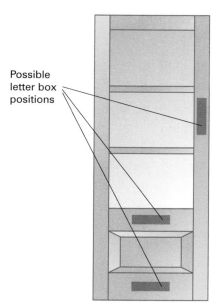

Possible letter box positions

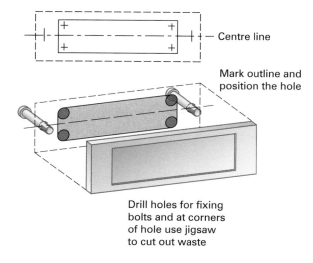

Centre line

Mark outline and position the hole

Drill holes for fixing bolts and at corners of hole use jigsaw to cut out waste

Figure 11.16 Different options for fitting a letter plate

A letter plate can be fitted either by drilling a series of holes and cutting out the shape with a jigsaw, or by using a router with a guide.

Once the opening has been formed, the letter plate can be screwed into place.

Mortise lock/latch

Mortise locks are locks housed into the stile and are fitted as follows:

Step 1 Mark out the position for the lock (usually 900 mm from the floor to the centre of the spindle) and mark the width and thickness of the lock on the stile.

Step 2 Using the correct sized auger/flat bit, drill a series of holes to the correct depth.

Step 3 Using a sharp chisel remove the excess timber, leaving a neat opening.

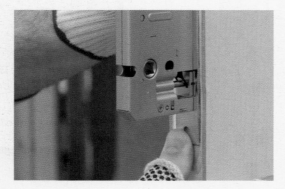

Step 4 Mark around the faceplate and with a sharp chisel, remove the timber so that the faceplate sits flush with the stile.

Step 5 Mark where the spindle and keyhole are, then drill out to allow them both to be fitted and operate properly.

Step 6 Fix the lock and handles in place.

Step 7 Mark the position of the striking plate on the doorframe, then house it into the frame.

Step 8 Check that the lock operates freely.

A mortise lock on its own does not usually provide sufficient security for an exterior door, so most doors will also have one of the following:

- **Mortise deadlock** – fitted like a mortise lock except that it has no latch or handles, so an escutcheon is used to cover the keyhole opening. It is usually fitted three-quarters of the way up the door.
- **Cylinder night latch** – preferred to a mortise deadlock, as it does not weaken the door or frame as much. It is also usually fitted three-quarters of the way up the door. The manufacturer will provide fitting instructions with the lock.

Spy hole

A spy hole is usually fitted to a door that is solid or has no glass, and is used as a security measure so that the occupier can see who is at the door without having to open the door.

To fit a spy hole, simply drill a hole of the correct size, unscrew the two pieces, place the outer part in the hole, then re-screw the inner part to the outer part.

Basic security ironmongery

Security chain

A security chain allows the door to be opened a little without allowing the person outside into the dwelling. The chain slides into a receiver, then as the door opens the chain tightens, stopping the door from opening too far.

Double doors

Here two doors are fitted within one single larger frame/lining, with 'meeting stiles', that is two stiles that meet in the middle, rebated so that one fits over the other.

Double doors are usually used where a number of people will be walking in the same direction, while double swing doors are used where people will be walking in different directions. Double doors allow traffic to flow without a 'bottleneck' effect, and let large items such as trolleys pass from one room to another. Double doors are used in places such as large offices, or public buildings such as schools and hospitals.

Did you know?

Spy holes work by having a small curved lens, which creates a wider field of vision than normal. This means that, even though the hole is small, you can see a full-height image of the person standing outside the door. The lens bends the image so that it appears curved, like the reflection in a Christmas bauble

Hanging double doors is the same as for any other door, though extra care must be taken to ensure that the stiles meet evenly, and that there is a suitable gap around the doors.

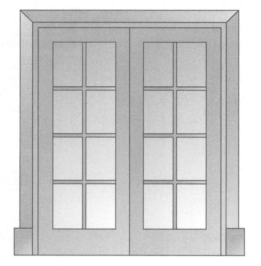

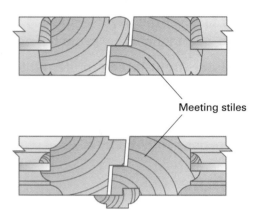

Meeting stiles

Figure 11.17 Meeting stiles

Extra ironmongery is required on double doors as follows:

- **parliament hinges** – these project from the face of the door and allow the door to open 180 degrees

Parliament hinge

- **rebated mortise lock** – similar to a standard mortise lock but the lock and striking plate are rebated to allow for the rebates in the meeting stiles

Rebated mortise lock

- **push/kick plates** – fixed to the meeting stiles and the bottom rails to stop the doors getting damaged

Pull handles

- **door pull handles** – fixed to the meeting stiles on the opposite side from the push plates, these allow the door to be opened

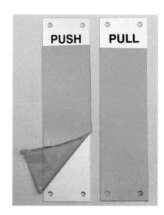

Push or kick plates

- **barrel bolts** – usually fixed to one of the doors at the top or the bottom to secure the door when not in use.

Most double doors also have door closers to ensure that they will close on their own, to prevent the spread of fire or draughts throughout the building.

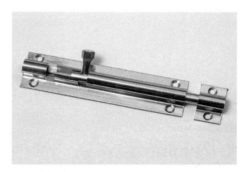

Barrel bolt

There are four main types of door closer: overhead, concealed, floor springs and helical spring hinges. Floor springs and **helical hinges** will be looked at on page 305 when we deal with double swing doors, so for the moment we will look at concealed and overhead door closers.

Concealed door closers

Concealed door closers work through a spring and chain mechanism housed into a tube.

To fit a concealed door closer, you must house the tube into the edge of the hinge side of the door, with the tension-retaining plate fitted into the frame.

Concealed door spring closer

Overhead door closers

As the name implies, overhead door closers are fitted to either the top of the door or the frame above. They work through either a spring or a hydraulic system fitted inside a casing, with an arm to pull the door closed.

There are different strengths of overhead door closer to choose from depending on the size and weight of the door. Table 11.2 below shows the strengths available.

Power no.	Recommended door width	Door weight
1	750 mm	20 kg
2	850 mm	40 kg
3	950 mm	60 kg
4	1100 mm	80 kg
5	1250 mm	100 kg
6	1400 mm	120 kg
7	1600 mm	160 kg

Table 11.2 Strengths of overhead door closures

Remember

When choosing an overhead door closer, you must take into account any air pressure from the wind. If the pressure is strong, you may require a more powerful closer

There are a number of different ways to fit an overhead door closer depending on where the door is situated. The two main ways are:

- fitting the main body of the closer to the door, with the arm attached to the frame
- inverting the closer so that the main body is fitted to the frame and the arm is fitted to the door.

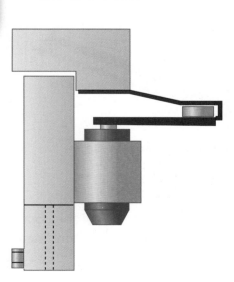

Figure 11.18 Closer fitted to door

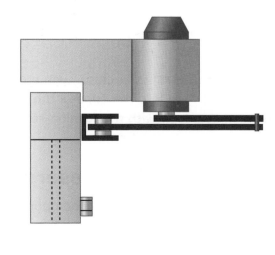

Figure 11.19 Closer fitted to frame

Door closers come with instructions and a template, to make fitting easier.

Double swing doors need to open both ways to accommodate traffic going in both directions. The main difference work-wise between standard double doors and double swing doors is in the ironmongery. Double swing doors need special hinges, and must have some form of door closer fitted – and it is usually best to combine these, in one of two ways.

- **Helical spring hinges** – Helical spring hinges are three-leaf hinges with springs integrated into the barrels. The way the leaves are positioned allows the doors to open both ways. The hinges are fitted just like normal hinges and the tension on the springs are adjusted via a bar inserted into the hinge collar.

- **Floor springs** – Doors using floor springs are hung via the floor spring at the bottom and a pivot plate at the top. The floor spring is housed into the floor, and the bottom of the door is recessed to accept the shoe attached to the floor spring.

The pivot plate at the top is attached to the frame, and a socket is fixed to the top of the door.

Helical or double action hinge

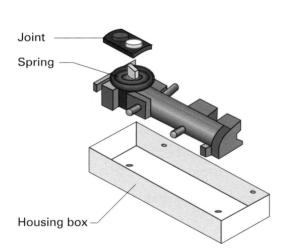

Figure 11.20 Helical floor springs

Joint
Spring
Housing box

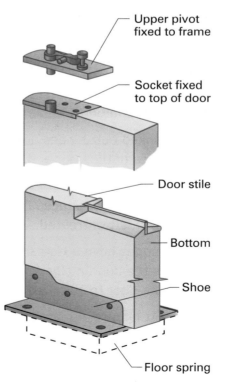

Upper pivot fixed to frame
Socket fixed to top of door
Door stile
Bottom
Shoe
Floor spring

Figure 11.21 Floor spring and top of door fixing

General door ironmongery

This section has been designed to enable you to choose the correct fixtures and fittings for the work you are involved in.

It offers just a small guide to what is available and the range of ironmongery is continually increasing. As a carpenter it will be your responsibility to choose and supply the correct items to your client. Therefore you would be well advised to look around your local DIY superstore to see what is available. Also look at completed work to see what other people have used.

Functions of ironmongery

Ironmongery is also referred to as hardware and relates to the components used to fix or decorate work.

To fix a door into a frame a carpenter would require a minimum of hinges, screws and some form of lock or latch, unless the client specified otherwise. All are ironmongery and can be classified as:

- fixings that allow movement
- fixings that provide security
- fixings that penetrate timber.

Fixings that allow movement

These are principally hinges and made of metal, the main ones of which are described, with photographs here.

Standard cranked butt hinges are fitted so that the knuckle of the hinge protrudes beyond the face of the door and the edge of the door jamb. These hinges are available in sizes from 25 mm to 100 mm, and produced in brass or steel.

Cast butt hinges are produced from cast iron and are classified as heavy duty. They will rust if left untreated.

Rising butt hinges have a helical knuckle that allows the door to rise when opened, which assists the door in clearing carpets and uneven floors. The helical knuckle also allows the door to close under its own weight.

Loose pin butt hinges enable the door to be removed from the frame by removing the pin from the hinge knuckle.

Storm-proof hinges are used in the manufacture of casement timber windows and have a cranked appearance.

Standard cranked butt hinges

Cast butt hinges

Rising butt hinges

Loose pin butt hinge

Friction hinge

Tee hinges

Bands and gudgeons

Friction hinges are used on uPVC window systems and act as both a pivot and a stay.

Tee hinges are frequently used for hanging external doors on buildings like sheds. Light-duty versions are generally made from thin gauged steel and either galvanised or painted in a black epoxy lacquer. Heavy duty versions are made of stronger steel with a steel or brass pin.

Bands or straps of strong steel may be used instead of tee hinges for heavy-duty tasks, often doubled over and leaving a **gudgeon** to fit over a hinge pin welded to a plate on the frame; thus enabling the door or gate to be lifted off.

> **Definition**
>
> **Gudgeons** – tubes at the end of hinges to take the pin around which the hinge rotates

Fixings that provide security

Locks

Mortise dead locks have no latch and therefore they do not require standard handle furniture. Both the sash lock and dead lock can be obtained with hook bolts for use on sliding doors.

Mortise locks are available in 3-lever and 5-lever versions. 5-lever versions are stronger than their 3-lever counterparts and therefore more suitable for external door security. They also come in two depths: 75 mm for standard doors and 65 mm for narrow stile doors. Good quality locks also have a

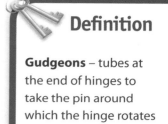

Mortise dead lock

3- and 5-lever mortise locks

Cylinder lock

reversible latch which means they can be fitted to LH (left hand) and RH (right hand) opening doors. Fitting instructions will be supplied with the locks and should be read thoroughly.

Cylinder locks consist of a body, a cylinder containing the key mechanism and a staple into which the latch engages.

Bolts and other fastenings

A range of bolts and other fastenings is available on the market. A selection of those most frequently used in second fixing is listed below.

A hasp and staple comprise a hinged strap (the hasp) that closes over a strong wire staple; a padlock is then put through the staple. It is normally used to secure garages and sheds. However, it is versatile, being used for many applications.

Hasp and staple

Tower bolts consist of a long bar that is fixed to the door and is pushed to locate it in a keeper on the frame.

Tower bolt

Pad bolts work on the same principle but when the bolt is closed, the handle locates over a staple to which a padlock can be fixed.

Bow handle bolts are much longer and more suited to heavy-duty applications, such as garage doors. The long bow handle means that the bolt can be operated even when the locking point is out of normal reach. They are made from square section metal and are spring loaded to prevent the bolt slipping out, especially under its own weight, if fitted vertically.

Pad bolt

Monkey tail bolts are very similar to bow handle bolts but have a ball on the end of the bolt, rather than a bow handle.

Monkey tail bolt

Panic bar bolts are available for single and double doors. They incorporate a bolt that runs the length of the door, passing through a latch that is fixed at approximately waist height. When the latch is pulled the bolts locate in keepers positioned in the frame; when the latch is pushed the bolts disengage from the keepers. They are fitted on doors that are used as a means of escape, for example fire exits.

Panic bar bolt

Fixings that penetrate timber

Fixings are items of ironmongery that enable a carpenter to connect components together. When choosing a fixing we must consider certain factors, which include:

- What strength must the fixing have?
- Where will the fixing be used?
- Will the fixing need to be removed at a later date?
- Cost.

The best sources for finding out what fixings are available are trade catalogues, local builders' merchants or DIY superstores. New types of fixing are regularly added to an already extensive range.

Although there are many specialist fixings available the most common are:

- nails
- screws
- wall fixings for solid walls
- wall fixings for hollow walls
- adhesives.

Did you know?

Some hardwoods are acidic and when unprotected ferrous metals are inserted the process of oxidisation (rusting) is accelerated and they stain the timber

Nails

Nails consist of a head and shank and are inserted by a hammer or mechanical tool. There are several types, made from either ferrous or non-ferrous metal. Ferrous metal contains iron and will therefore rust unless protected. The carpenter must decide the most appropriate nail for the required application.

Round wire nails are available in sizes from 25 mm to 150 mm. They should not be driven below the surface of the timber and are relatively easy to remove. They are used for low-quality work where they will not be seen, such as roofing, studwork etc.

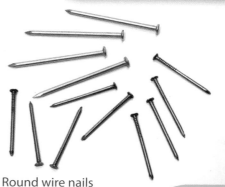

Round wire nails

Annular ring nails are available in sizes from 20 mm to 75 mm and also **sherardised** to prevent rusting. These nails are similar to the round wire nail but feature a series of rings along the shank that makes them much harder to remove, and also provides a stronger hold.

Annular ring nails

Oval wire nails are available in sizes from 25 mm to 100 mm. They are manufactured from ferrous metal and can be punched below the surface of the timber. They are less likely to split the grain of the timber and are usually used for higher-quality work than the round wire nail.

Oval wire nails

Lost head nails are available in sizes from 40 mm to 75 mm. The head can be punched below the surface of the timber for concealment.

Panel pins are available in sizes from 20 mm to 40 mm. They are easy to punch below the surface, causing little damage to the face of the work. They are used for fine applications. Variations include sherardised and brass versions that resist rust, and veneer pins for extra fine work.

Lost head nails

Other nails include plasterboard, felt and clout nails, also plastic-headed nails for use with uPVC systems and double-headed nails for shuttering work. All are designed for specific applications.

Panel pins

Other nails

Screws

Most modern screws are computer designed. Like a nail, screws consist of a head and a shank. However, the shank is threaded and designed to pull the fixing into the material into which it is being inserted.

Screws are manufactured from both ferrous and non-ferrous materials and are defined by:

- head type
- length, measured from the tip to the part of the head that will be flush with the work surface, ranging from 12 mm to 150 mm

- gauge, the diameter of the shank, ranging from 2 mm to 6.5 mm.

Once again, it is the carpenter's responsibility to choose the correct screw for the application in which it is being used.

Head types

Screws with countersunk heads are used when the screw has to be flush with the work or below it.

Raised head screws are usually used for attaching metal components, such as door handles. Round heads are usually used for attaching sheet material to timber that is too thin to countersink.

Mirror screws have a thread within the head to which a decorative dome can be attached. As the name suggests, these are used mainly for fixing mirrors.

Pan head and flange heads are commonly found on self-tapping screws where the fixing of sheet metal is involved.

| Countersunk | Raised head screws | Mirror screw head | Pan head | Flange head |

Screwdriver types

There are screwdrivers available, designed to fit each type of screw head and size of screw. A selection is shown in the following photographs:

Posidrive screwdriver

Phillips screwdriver

Slotted screwdriver

Security screwdriver

Wall fixings for solid walls

Plug fixings

Nails and screws can be used to fix components to masonry. However, the carpenter must first plug the masonry, which can either be done using a plugging chisel or an electric drill.

If a plugging chisel is used carpenters can make their own plugs, but this is time-consuming and not commonly done now. When using an electric hammer drill to plug a wall several plug types are available, which include:

● moulded plastic plugs

● hammer plugs

● frame fixings.

Moulded plastic plug

Hammer plug

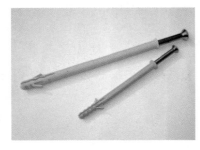

Frame fixings

All work on the same principle. A plastic segmented sleeve fits snugly into a hole that has been pre-drilled to the plug manufacturer's stated dimensions. A screw is then inserted into the plastic sleeve that pushes the segments apart to grip the side of the hole.

Anchor bolts

These are used for giving an extra strong fixing in concrete or masonry. They consist of a segmented metal sleeve that encases a conical plug on the end of a bolt. As the bolt is turned clockwise the conical plug rises up the thread of the bolt, expanding the metal sleeve.

Anchor bolt

Wall fixings for hollow walls

Hollow wall fixings work on the principle of the fixing opening out behind the wall panel and gripping it in some way. These include:

- nylon anchors
- plastic collapsible anchors
- metal collapsible anchors
- rubber sleeve anchors
- gravity toggles
- spring toggles.

Nylon anchors

Plastic collapsible anchors

Metal collapsible anchors

Rubber sleeve anchors

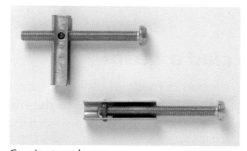

Gravity toggles

Spring toggles

Other fixings are available called EASI drivers™ but these are generally for use only in plasterboard, though they will support heavy items such as kitchen wall units or radiators. It is worth the effort to experiment with different types of fixing before deciding which one to use.

Adhesives

Adhesives, able to provide long-term fixing, are a relatively recent development within the building industry but already there are numerous products available. These range from dry lining adhesive through to expanding foam and trade name products such as No Nails™.

A selection of adhesives

Safety tip

With adhesive, or any other chemical, carefully read all instructions before use

Encasing services

This section has been designed to provide you with the knowledge and understanding to correctly encase a variety of services and structures.

The phrase 'encasing services' usually refers to the carcase, framework and trim that cover BSBs, service pipes, cables, steel and concrete columns or, in some instances, unsightly spaces such as where a bath is situated or where the bulkhead for a staircase cuts into a room.

Beams and columns

In some buildings BSBs have to be cladded to protect them from the effects of fire. The amount of protection required will depend upon the function and location of the beam.

How to clad a beam

If a beam to be clad is made of timber, or even concrete, it can have a frame fixed to it and then facing material put on to the frame, or the facing material can be fixed directly to it. The bigger problem is to clad a steel beam, often an BSB used as a load-bearing support.

Step 1 Fix **noggings** between the rolled edges of the beam. This can be done by accurately cutting lengths of 50 mm x 50 mm timber to match the vertical gap in the side of the BSB. Then drive them vertically into the gap on both sides of the beam so they wedge in place. They will provide attachment points for a cradle so should be no more than 600 mm apart.

Step 1 Fix noggings

Alternatively, timber supports can be fitted in the gap along the whole length of the beam and then bolted into place through holes in the beam (some come with them or they can be drilled). These similarly provide attachment points for a cradle.

Step 2 Create a cradle using 50 mm × 25 mm treated softwood and use a simple **halving joint** (or housing or lap joint) to fix the corners together. Screw the cradle to the bearers. (If the noggings or timber supports come flush with the sides of the metal beam it is possible to dispense with a cradle by fixing additional timber supports along the beam near the top and bottom as a support for facing material.)

Step 2 Create a cradle

Step 3 Run soffits along the length of the joist, fastening them into the cradle. The facing material can now be fixed to the soffits to conceal the sides of the beam.

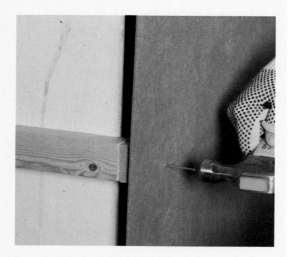

Step 3 Fitting soffits and facing material

Definition

Halving joint – the same amount is removed from each piece of timber so that when fixed together the joint is the same thickness as the uncut timber

How to clad a column

Concrete or steel columns are cladded by constructing a set of framed grounds (either two or four) slightly larger than the column itself. These are generally shaped like a ladder and made from rough sawn timber. They are assembled around the column and adjusted until plumb.

Any slackness is taken up, using wedges or packing pieces with a screw or nail driven into them to stop them slipping.

Facing material is then fixed to the framed grounds.

Did you know?

Plumb means vertical; hence, a plumb bob is used to check something is vertical

Service pipes and cables

Service pipes and cables are hidden from view whenever possible. However, there are occasions when it is not possible to do so; when the circuit of pipe work has to go from one storey to another, for instance. When this occurs they should be encased behind a timber stud frame. The method is very similar to cladding a column.

Encasing pipes and wires

Step 1 Measure how far the services protrude. The framework that is about to be fitted should not stand excessively from the object to be encased.

Step 2 By using halving joints or by butt nailing, construct two sets of framed grounds.

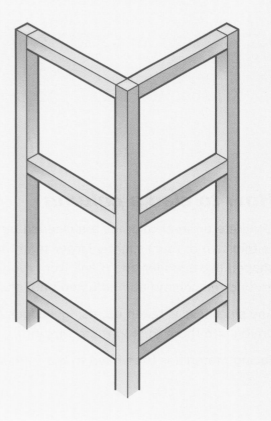

Step 3 Fix battens to the wall with plugs and screws or masonry nails.

Step 4 Fix the framed grounds to the battens.

Step 5 Clad with 6 mm ply or similar, scribing to the wall if necessary.

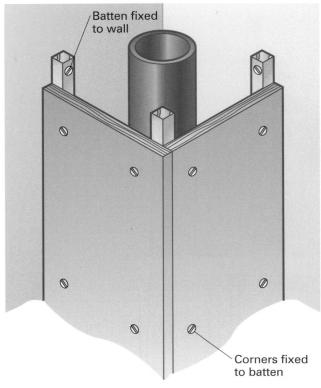

Figure 11.22 Corner casing for a large pipe

Batten fixed to wall

Corners fixed to batten

A corner casing for a large pipe is shown in Figures 11.22 and 11.23.

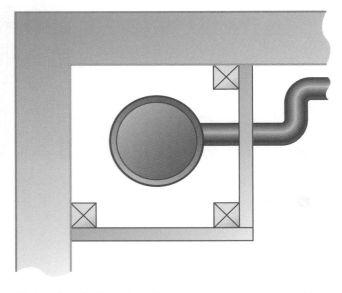

Figure 11.23 Plan view of casing

Alternative ways of encasing services are shown in Figures 11.24–11.27.

Remember

Where there is a service valve or stop cock a removable panel must be fitted for maintenance and emergencies

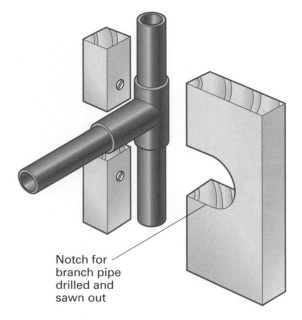

Notch for branch pipe drilled and sawn out

Figure 11.24 Branch pipe

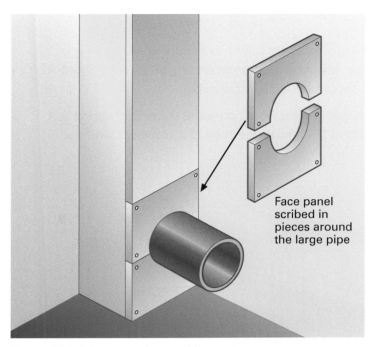

Figure 11.25 Face panel around large pipes

Face panel scribed in pieces around the large pipe

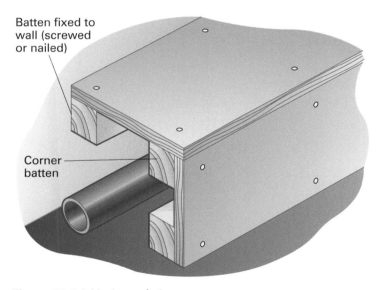

Batten fixed to wall (screwed or nailed)

Corner batten

Figure 11.26 Horizontal pipes

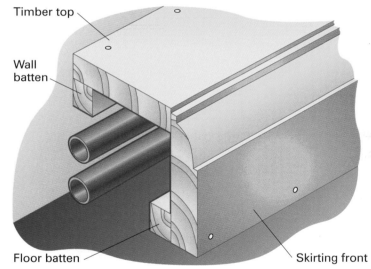

Timber top

Wall batten

Floor batten

Skirting front

Figure 11.27 Horizontal pipes with skirting board

Baths

Baths are often panelled, but care should be taken to ensure that suitable materials are used that can cope with the heat and condensation associated with bathrooms. Also the plumbing should be easily accessible.

Panelling a bath is similar to encasing pipe work, only the framed grounds will probably be wider. Figures 11.28–11.30 illustrate typical framework and panel fixing.

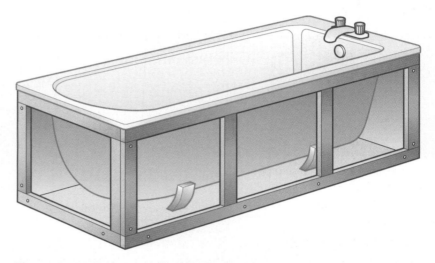

Figure 11.28 Framework for a bath panel

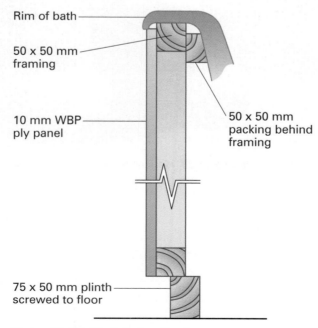

Rim of bath

50 x 50 mm framing

10 mm WBP ply panel

50 x 50 mm packing behind framing

75 x 50 mm plinth screwed to floor

Figure 11.29 Plinth below the frame as a toe space

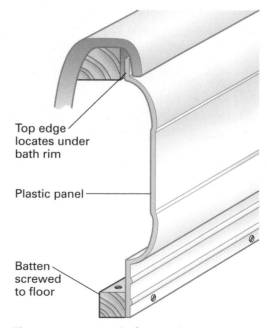

Top edge locates under bath rim

Plastic panel

Batten screwed to floor

Figure 11.30 Detail of a standard panel

Wall and floor units

This section is designed to provide you with the knowledge and understanding to assemble and fix both wall and floor units. Normally you will be fitting these in kitchens and bedrooms. Kitchens are the most difficult and are described below. If you learn to follow these instructions they will enable you to fit wall and floor units in a bedroom or anywhere else.

Kitchen safety

Most household accidents happen in the kitchen. Thus, when planning a kitchen, safety is the number one priority. Always:

- create a working triangle consisting of the sink, cooker and fridge, which avoids unnecessary movement and improves efficiency

- avoid a design that encourages people to use the working area of the kitchen as a cut through, to garden or living room for example

- try not to place a hob or sink at the end of a run of units; also make sure that children cannot easily reach either hob or sink

- try to design work surfaces either side of the hob and avoid placing the hob in front of a window.

A plan of a well laid out kitchen is shown in Figure 11.31.

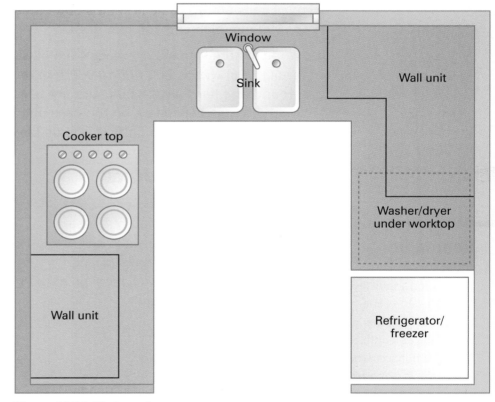

Figure 11.31 Kitchen plan

Kitchen units

The majority of kitchen units available today are constructed from melamine-faced particle board, and supplied as either base units or wall units. Base units are floor standing, whereas wall units are mounted on the wall.

Units are supplied ready-built or flat-pack. When supplied as flat-packs the assembly instructions must be followed with great care. Check whether shelves have to be fitted during assembly, rather than after installation, particularly in corner units.

Base units are available in widths 300 mm to 1200 mm, and normally 600 mm deep and 900 mm high. Corner units are available to match up with them. Most manufacturers incorporate adjustable height legs in their designs to allow for uneven floors.

Wall units are available in widths 300 mm to 1000 mm; typically 600 mm, 720 mm or 900 mm high and 300 mm deep.

Worktops are also mainly made from plastic-laminated particle board and available in widths 600 mm to 900 mm, but granite, stainless steel and proprietary, hard-wearing plastics are used in more expensive kitchens. The width fitted is dependent on the base unit used. Depth is usually 40 mm but can vary. Plastic-laminated particle boards have a square- or post-formed front edge.

Remember

When using a spirit level to mark a datum line always alternate the level along its length; this counteracts inaccuracies and prevents cumulative error

Fitting base units

Step 1 Carefully mark a horizontal datum line, approximately 1 metre from the floor, on the walls that are to have base or wall units placed against them. We now have a point to work up or down from.

Step 1 Mark a horizontal datum line

Step 2 Determine a user-friendly height for the units minus the worktop. Then, working down from the 1 metre datum, fix a back rail to the wall. We now have a datum that allows us to level across and along the units.

Step 2 Fixing a back rail

Step 3 Preferably working from a corner, place units where they should go, taking into consideration service pipes and cables. Adjust the legs so that all units are level and in line with the back rail.

Step 3 Adjust the unit's legs

Step 4 Once all units are in place they can be fixed together using connecting bolts and fixed to the wall or floor using screws.

Step 4 Base units fixed in position

Fitting worktops

Specialist tools and equipment are needed to cut and fix worktops. These include:

- jigsaw with downward cutting blades
- plunge router (minimum 1600 watt) capable of plunging more than the depth of the worktop, with 12.7 mm diameter collet and template guide
- 12.7 mm diameter straight-cutting router bits, minimum length equal to the depth of the worktop
- worktop jig
- biscuit jointer
- clamps (G clamp or similar)
- contact adhesive
- varnish
- panel connectors.

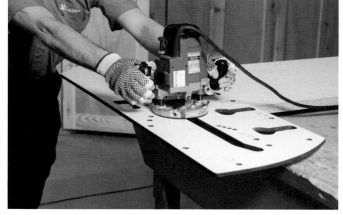

Joiner using router and jig

Step 1 Cut to length and form an internal corner using a purpose-made jig. These are produced by a variety of manufacturers and, when used in conjunction with a powerful router, a clean accurate cut is achievable.

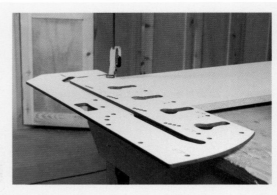

Step 1 Cut lengths and internal corners

Step 2 Strengthen joints using a biscuit jointer.

Step 2 Strengthen joints with biscuit jointer

Step 3 Cut worktops to house sink units or appliances using a jigsaw with a downward cutting blade. This prevents the plastic-laminated surface from being chipped. The chipboard that is exposed by the cut should be coated in varnish to prevent any moisture penetration.

Step 3 Cut worktops to house sink units etc.

Step 4 Fix worktops to the base units with screws and connectors.

Step 4 Fix worktops to base units

Fitting wall units

Most wall units are fastened to the wall by an adjustable bracket that hooks itself onto a steel hanger plate. The plans should show the clearance required between the worktop and base of the wall units. If not, check with a supervisor or the customer. It is normal to have the tops of all units at the same height.

Step 1 Measure the height of units and add this to the clearance from the work surface, which gives the height where the tops of the wall units should go. Measure and mark these, working upwards from our datum line.

Step 1 Mark the tops of wall units

Step 2 Mark the locations for the hanger plates and fix in place. Often the manufacturer of the wall units provides a template to help position them.

Step 2 Mark and fix hanger plates

Step 3 Hang the wall units on the hanger plates and adjust for height by turning a screw housed within the adjustable bracket. Once done, another screw enables the bracket to be tightened on to the steel plate.

Step 3 Hang and adjust wall units

Various plinths, cornices, doors and shelves

As with base units, wall units can be connected together by using connecting bolts.

Finishing off

Complete the job by fixing plinth boards, cornices, pelmets, doors, shelves and drawers. Details of how to do this should be provided with the kitchen units and often vary between manufacturers.

In general:

● Plinth boards are fixed to the legs of base units using clips supplied by the manufacturer. However, the boards may need scribing to the floor to avoid having any unsightly gaps.

● Cornices are often fitted on to wall units. This requires accurate marking out and either a chop saw or mitre saw used to give a clean, accurate cut. Care must be taken to ensure the cornice does not move or slip while it is being cut.

- Finally doors and shelves can be fitted to the units. This should be done last to avoid damage to the face of the work. The runners that hold the shelves, and hinges that hold the doors, both allow for adjustment. This ensures that an equal gap is seen between doors and shelves when the work has been completed.

Completed kitchen

On the job: Fitting mouldings

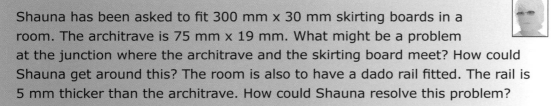

Shauna has been asked to fit 300 mm x 30 mm skirting boards in a room. The architrave is 75 mm x 19 mm. What might be a problem at the junction where the architrave and the skirting board meet? How could Shauna get around this? The room is also to have a dado rail fitted. The rail is 5 mm thicker than the architrave. How could Shauna resolve this problem?

Knowledge check

1. Name three examples of mouldings.

2. How can architrave be fitted to a surface that is not smooth?

3. What are the different ways in which skirting can be fitted?

4. Briefly describe the two different methods of shaping skirting where it meets in a corner.

5. Why are plinth blocks used?

6. Describe in your own words the following types of door: a framed door; a flush door; a fire-resisting door.

7. Why should a door be stored flat for a few days prior to fitting?

8. If a door is too tall for an opening, where should you cut excess wood from?

9. State the size, type and number of hinges usually fitted to an external door.

10. What is the purpose of a spyhole?

11. What are push plates for?

12. Why would a door closer be fitted to a door?

13. What is the special feature of a helical or double action hinge?

14. Why might a British standard beam (BSB) need to be cladded?

15. What should you always consider when encasing service pipes or cables?

16. What is the number one priority when planning a kitchen?

17. When fitting kitchen wall units, how do you know how much clearance is required between the worktop and the base of the unit?

18. Why are the doors and shelves of kitchen units fitted last?

chapter 12

Structural carcassing

OVERVIEW

Structural carcassing covers all carpentry work associated with the structural elements of a building such as floors and roofs. This chapter is designed to help you identify the main activities associated with structural carcassing and to provide you with the knowledge and understanding required to carry them out.

This chapter will cover the following topics:

- Basic terms
- Traditional pitched roofs
- Flat roofs
- Ground and upper floors.

These topics can be found in the following modules:

CC 1001K	CC 2010K
CC 1001S	CC 2010S

Basic terms

Roofing terminology

Roofs are made up of a number of different parts called 'elements'. These in turn are made up of 'members' or 'components'.

Elements

The main elements, shown in Figure 12.1, are:

- **gable** – the triangular part of the end wall of a building that has a pitched roof

- **hip** – where two external sloping surfaces meet

- **valley** – where two internal sloping surfaces meet

- **verge** – where the roof overhangs at the gable

- **eaves** – the lowest part of the roof surface where it meets the outside walls.

Members or components

The main members or components, shown in Figure 12.1, are:

- **ridge board** – a horizontal board at the apex acting as a spine, against which most of the rafters are fixed

- **wall plate** – a length of timber placed on top of the brickwork to spread the load of the roof through the outside walls and give a fixing point for the bottom of the rafters

- **rafter** – a piece of timber that forms the roof, of which there are several types

- **common rafters** – the main load-bearing timbers of the roof

- **hip rafters** – used where two sloping surfaces meet at an external angle, this provides a fixing for the jack rafters and transfers their load to the wall

- **crown rafter** – the centre rafter in a hip end that transfers the load to the wall

- **jack rafters** – these span from the wall plate to the hip rafter, enclosing the gaps between common and hip rafters, and crown and hip rafters

- **valley rafters** – like hip rafters but forming an internal angle, acting as a spine for fixing cripple rafters

- **cripple rafters** – similar to a jack rafter, these enclose the gap between the common and valley rafters

- **purlins** – horizontal beams that support the rafters mid-way between the ridge and wall plate.

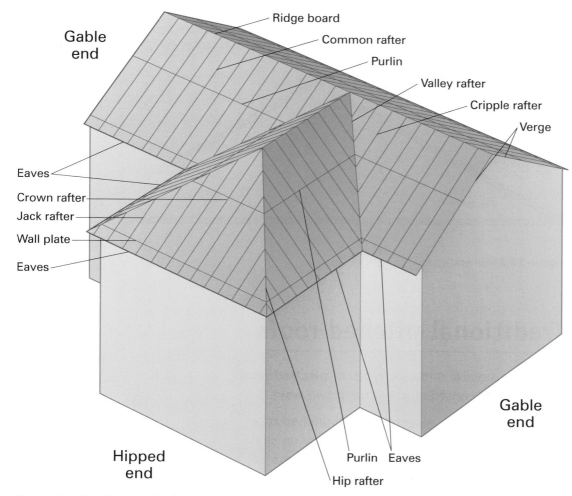

Figure 12.1 Roofing terminology

Basic terms for setting out

Setting out covers all the action necessary before commencing construction. Here are the most common terms you will need:

- **span** – the distance measured in the direction of the ceiling joists, from the outside of the wall plate to wall plate, known as the overall span

- **run** – equal to half the span

- **apex** – the peak or highest part of the roof

- **rise** – the distance from the outside of the wall plates at wall-plate level to the apex

- **pitch** – the angle or slope of the roof, calculated from the rise and the run

- **pitch line** – a line that is marked up from the underside of the rafter, one third of its depth to the top of the birdsmouth cut

- **plumb cut** – the angle of cut at the top of the rafter

- **seat cut** – the angle of cut at the bottom of the rafter
- **birdsmouth** – notch cut out at the bottom of the rafter to allow the rafter to sit on the wall plate.

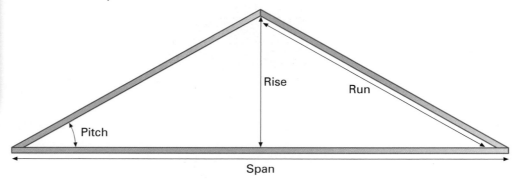

Figure 12.2 Rise and span

Traditional pitched roofs

There are several different types of **pitched roof** but most are constructed in one of two ways.

1. Trussed roof – A prefabricated pitched roof specially manufactured prior to delivery on site, saving timber as well as making the process easier and quicker. Trussed roofs can also span greater distances without the need for support from intermediate walls.

2. Traditional roof – A roof entirely constructed on site from loose timber sections using simple jointing methods.

Figure 12.3 Single roof

Roof types

A pitched roof can be constructed either as a single roof, where the rafters do not require any intermediate support, or a double roof where the rafters are supported. Single roofs are used over a short **span** such as a garage; double roofs are used to span a longer distance such as a house or factory.

There are many different types of pitched roof including:

- **mono pitch** with a single pitch
- **lean-to** with a single pitch, which butts up to an existing building

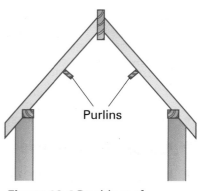

Figure 12.4 Double roof

Figure 12.5 Mono pitch roof

Figure 12.6 Lean-to roof

Figure 12.7 Duo pitch roof with gable ends

Figure 12.9 Over hip roof

Figure 12.10 Mansard roof

Figure 12.8 Hipped roof

- **duo pitch** with **gable ends**
- **hipped roof** with hip ends incorporating crown, hip and jack rafters
- **over hip** with gable ends, hips and valleys incorporating valley and cripple rafters
- **mansard** with gable ends and two different pitches used mainly when the roof space is to be used as a room
- **gable hip** or **gambrel** – double-pitched roof with a small gable (gablet) at the ridge and the lower part a half-hip
- **jerkin-head** or **barn hip** – double-pitched roof hipped from the ridge part-way to the eaves, with the remainder gabled.

The type of roof used will be selected by the client and architect.

Trussed rafters

Most roofing on domestic dwellings now comprises factory-made trussed rafters. These are made of stress graded, **p.a.r.** timber to a wide variety of designs, depending on requirements. All joints are butt jointed and held together with fixing plates, face fixed on either side. These plates are usually made of galvanised steel and either nailed or factory pressed. They may also be **gang-nailed** gusset plates made of 12 mm resin bonded plywood.

Definition

p.a.r. – a term used for timber that has been 'planed all round'

Gang-nailed – galvanised plate with spikes used to secure butt joints

One of the main advantages of this type of roof is the clear span achieved, as there is no need for intermediate, load-bearing partition walls. Standard trusses are strong enough to resist the eventual load of the roofing materials. However, they are not able to withstand pressures applied by lateral bending. Hence, damage is most likely to occur during delivery, movement across site, site storage or lifting into position.

Wall plates are bedded as described above. Following this, the positions of the trusses can be marked at a maximum of 600 mm between centres along each wall plate. The sequence of operations then varies between gable and hipped roofs.

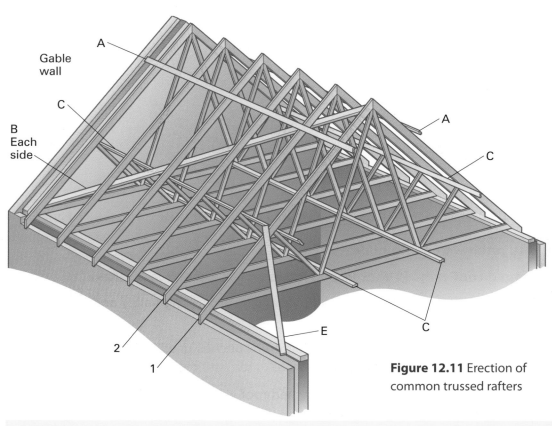

Figure 12.11 Erection of common trussed rafters

Step 1 Fix first truss using framing anchors 1.

Step 2 Stabilise and plumb first truss with temporary braces E.

Step 3 Fix temporary battens on each side of ridge A.

Step 4 Position next truss 2.

Step 5 Fix the wall plate and temporary battens A. Continue until last truss is positioned.

Step 6 Fix braces B.

Step 7 Fix braces C.

Step 8 Fix horizontal restraint straps at max 2.0 m centres across trusses on to the inner leaf of the gable walls.

Remember

Never alter a trussed rafter without the structural designer's approval

Gable ends

Setting out for a gable end

First set out and fix the wall plate. The wall plate is set on the brick or block work and either bedded in by the bricklayer or temporarily fixed by nailing through the joints. Once secured it is held in place with restraint straps (see Figure 12.13). If the wall plate is to be joined in length, a **halving joint** is used (see Figure 12.14). It is vital that the wall plate is fixed level to avoid serious problems later.

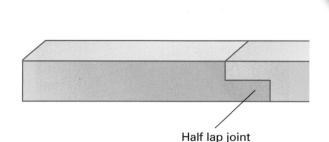

Wall plate

Restraint strap

Figure 12.12 Restraint straps

Half lap joint

Figure 12.13 Plate with halving joint

Once the wall plate is in place, you need to measure the span and the rise. You can use these measurements to work out the rafter length in different ways, using a roofing square, a **roofing ready reckoner**, **geometry** or **scale drawings**. Ready reckoners and geometry are covered later in this chapter, so we will start with scale drawings.

For this example we will use a span of 5 m and a rise of 2.3 m.

Using a scrap piece of plywood or hardboard we first draw the roof to a scale that will fit the scrap piece of plywood/hardboard (usually a scale of 1:20).

From this drawing we can measure at scale and find the true length of the rafter. Then by using a **sliding bevel** we can work out the plumb cut and seat cut.

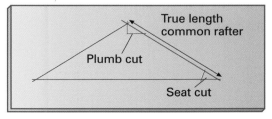

True length common rafter

Plumb cut

Seat cut

Figure 12.14 Sketch on a piece of scrap ply showing the true length, plumb and seat cut for a common rafter

Definition

Halving joint – the same amount is removed from each piece of timber so that when fixed together the joint is the same thickness as the uncut timber

Definition

Roofing reading reckoner – a set of mathematical tables giving a quick way to work out rafter lengths

Geometry – a form of mathematics using formulas, arithmetic and angles

Scale drawing – a drawing of a building or component in the right proportions, but scaled down to fit on a piece of paper. On a drawing at a scale of 1:50, a line 10 mm long would represent 500 mm on the actual object

Sliding bevel – a tool that can be set so that the user can mark out any angle

Definition

Pattern rafter – a
precisely cut rafter
used as a template to
mark out other rafters

Making and using a pattern rafter

From our scale drawing we can mark out one rafter, which we will then use as a
pattern rafter.

There are five easy steps to follow when marking out a pattern rafter.

Step 1 Mark the pitch line one-
third of the way up the width of
the rafter.

Step 2 Set the sliding bevel to the
plumb cut and mark the angle onto
the top of the rafter.

Step 3 Mark the true length on the
rafter, measuring along the pitch
line.

Step 4 Use the sliding bevel to
mark out the seat cut, then with a
combination square mark out the
birdsmouth at 90 degrees to the seat
cut.

Step 5 Re-mark the plumb cut to allow for half the thickness of the ridge.

Once it has all been marked out, this can be cut and used as a pattern rafter.

The pattern rafter can be used to mark out all the remaining common rafters, although it is advisable to mark out and cut only four, then place two at each end of the roof to check whether the roof is going to be level.

Once all the rafters are cut, mark out the wall plate and fix the rafters. Rafters are normally placed at 400 mm centres, with the first and last rafter 50 mm away from the gable wall. The rafters are usually fixed at the foot by skew-nailing into the wall plate and at the head by nailing through the ridge board.

Did you know?

The first and last rafters are placed 50 mm away from the wall to prevent moisture that penetrates the outside wall coming into contact with the rafters, thus preventing rot

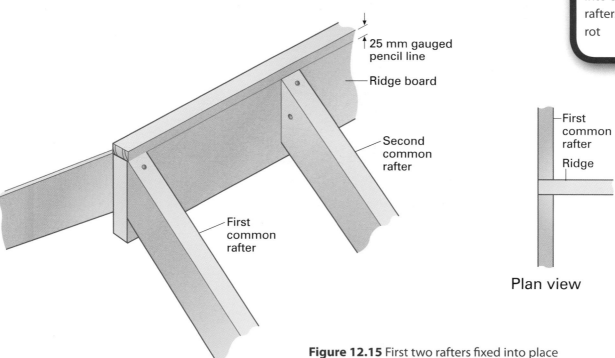

Figure 12.15 First two rafters fixed into place

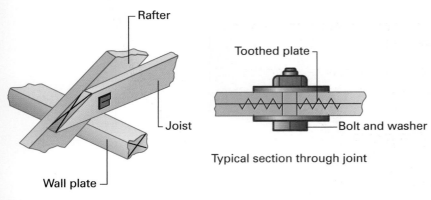

Figure 12.16 Rafter legs bolted to joists

If the roof requires a loft space, joists can be put in place and bolted to the rafters; if additional support is required, struts can be used.

For roofs with a large span, purlins provide adequate support. Purlins are usually built into the brickwork: either the gable wall will not have been finished being built yet or the bricklayer will have left a **sand course**. The same is true of roof ladders (see below).

Definition

Sand course – where the bricklayer beds in certain bricks with sand instead of cement so that they can be easily removed to accommodate things like purlins

Finishing a gable end

There are two types of finish for a gable end:

- a flush finish, where the bargeboard is fixed directly onto the gable wall
- a roof ladder – a frame built to give an overhang and to which the bargeboard and soffit are fixed.

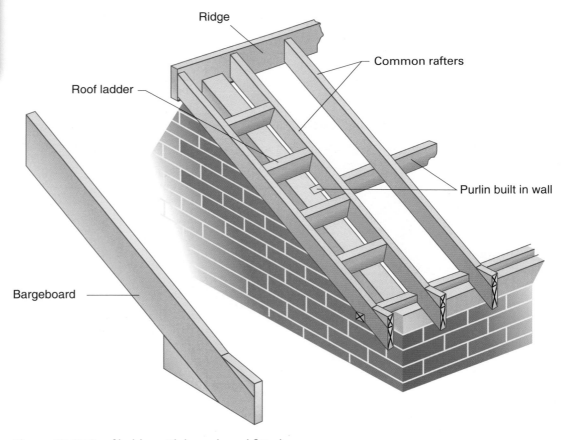

Figure 12.17 Roof ladder with bargeboard fitted

The most common way is to use a roof ladder, which when creating an overhang, stops rainwater running down the face of the gable wall.

The continuation of the fascia board around the verge of the roof is called the bargeboard. Usually the bargeboard is fixed to the roof ladder and has a built-up section at the bottom to encase the wall plate.

The simplest way of marking out the bargeboard is to temporarily fix it in place and use a level to mark the plumb and seat cut.

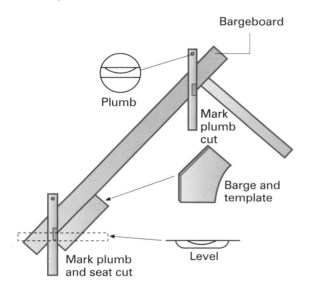

Bargeboard

Plumb

Mark plumb cut

Barge and template

Mark plumb and seat cut

Level

Figure 12.18 Marking out a bargeboard using a level

When fixing a bargeboard, the foot of the board may be mitred to the fascia, butted and finished flush with the fascia, or butted and extended slightly in front of the fascia to break the joint.

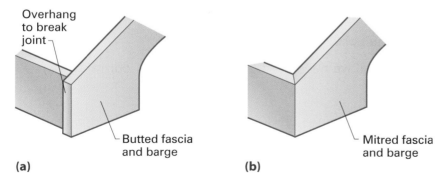

Overhang to break joint

Butted fascia and barge

(a)

Mitred fascia and barge

(b)

Figure 12.19 Fascia joined to bargeboard

The bargeboard should be fixed using oval nails or lost heads at least 2.5 times the thickness of the board so that a strong fixing is obtained. If there is to be a joint along the length of the bargeboard, the joint *must* be a mitre.

Hipped roofs

In a fully hipped roof there are no gables and the eaves run around the perimeter, so there is no roof ladder or bargeboard.

Marking out for a hipped roof

All bevels or angles cut on a hipped roof are based on the right-angled triangle and the roof members can be set out using the following two methods:

- **roofing ready reckoner** – a book that lists in table form all the angles and lengths of the various rafters for any span or rise of roof

- **geometry** – working with scale drawings and basic mathematic principles to give you the lengths and angles of all rafters.

The ready reckoner will be looked at later in this chapter, so for now we will concentrate on geometry.

Pythagoras' theorem

When setting out a hipped roof, you need to know Pythagoras' theorem. Pythagoras states that 'the square on the hypotenuse of a right-angled triangle is equal to the sum of the squares on the other two sides'. For the carpenter, the 'hypotenuse' is the rafter length, while the 'other two sides' are the run and the rise.

From Pythagoras' theorem, we get this calculation:

$A = \sqrt{B^2 + C^2}$ ($\sqrt{}$ means the square root and ^{2} means squared)

If we again look at our right-angled triangle we can break it down to:

A (the rafter length – the distance we want to know)

B^2 (the rise, multiplied by itself)

C^2 (the run, multiplied by itself).

Therefore we have all we need to find out the length of our rafter (A):

$A = \sqrt{4^2 + 3^2}$

$A = \sqrt{4 \times 4 + 3 \times 3}$

$A = \sqrt{16 + 9}$

$A = \sqrt{25}$

$A = 5$

so our rafter would be 5m long.

Did you know?

The three angles in a triangle always add up to 180 degrees

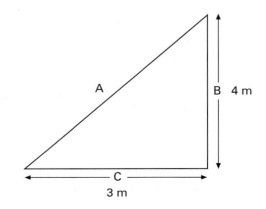

Finding true lengths

The next task is to find the hip rafter true length, plumb and seat cuts. This is done in two stages (the first will be familiar to you, as it is the same as for a common rafter).

The next step is to lengthen the common rafter true length by the amount of the rise, then join this line up to the base of the roof.

From this point, to make the geometrical drawings as clear as possible, abbreviated labels will be used. See Table 12.1.

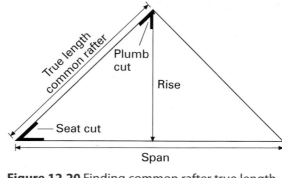

Figure 12.20 Finding common rafter true length

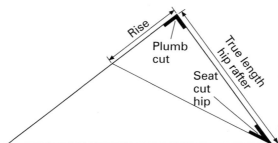

Figure 12.21 Finding hip rafter true length

Did you know?

Most drawings use abbreviations or symbols to avoid cluttering the drawings and make them easier to read

Abbreviation	Definition	Abbreviation	Definition
TL	true length	PCHR	plumb cut hip rafter
TLCR	true length common rafter	SCHR	seat cut hip rafter
TLHR	true length hip rafter	PCVR	plumb cut valley rafter
TLJR	true length jack rafter	SCVR	seat cut valley rafter
TLCrR	true length cripple rafter	EC	edge cut
PC	plumb cut	ECHR	edge cut hip rafter
SC	seat cut	ECJR	edge cut jack rafter
PCCR	plumb cut common rafter	ECVR	edge cut valley rafter
SCCR	seat cut common rafter	ECCrR	edge cut cripple rafter

Table 12.1 Abbreviations

There are two other angles that are concerned with hip rafters: the dihedral angle (or backing bevel for the hip) and the edge cut to the hip.

Finding the dihedral or backing bevel angle

The backing bevel angle is the angle between the two sloping roof surfaces. It provides a level surface so that the tile battens or roof boards can lie flat over the hip rafters. The backing bevel angle is rarely used in roofing today as the edge of the hip is usually worked square, but you should still know how to work it out.

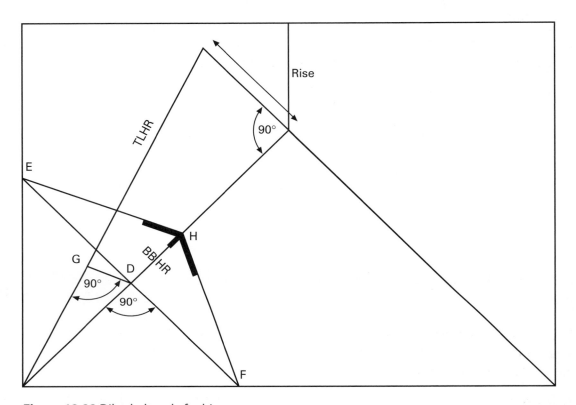

Figure 12.22 Dihedral angle for hip

1. Draw a plan of the roof and mark on the TLHR as before.

2. Draw a line at right angles to the hip on the plan at D, to touch the wall plates at E and F.

3. Draw a line at right angles to the TLHR at G, to touch point D.

4. With centre D and radius DG, draw an arc to touch the hip at H.

5. Join E to H and H to F. This gives the required backing bevel (BBHR).

Finding the edge cut

The edge cut is applied to both sides of the hip rafter at the plumb cut. It enables the hip to fit up to the ridge board between the crown and the common rafters.

1. Draw a plan of the roof and mark on the TLHR as before.

2. With centre I and radius IB, swing the TLHR down to J, making IJ, TLHR.

3. Draw lines at right angles from the ends of the hips and extend the ridge line. All three lines will intersect at K.

4. Join K to J. Angle IJK is the required edge cut (ECHR).

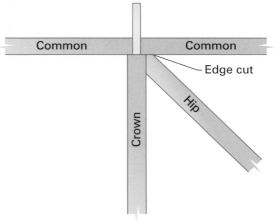

Figure 12.23 Edge cut on hip joined to ridge

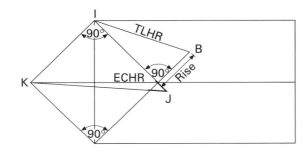

Figure 12.24 Edge cut on hip

The jack rafter's plumb and seat cuts are the same as those used in the common rafters so all you need to work out is the true length and edge cut.

- Draw the plan and section of the roof. Mark on the plan the jack rafters. Develop roof surfaces by swinging TLCR down to L and projecting down to M¹.

- With centre N and radius NM¹, draw arc M¹O. Join points M¹ and O to ends of hips as shown.

- Continue jack rafters on to development.

- Mark the true length of jack rafter (TLJR) and edge cut for jack rafter (ECJR).

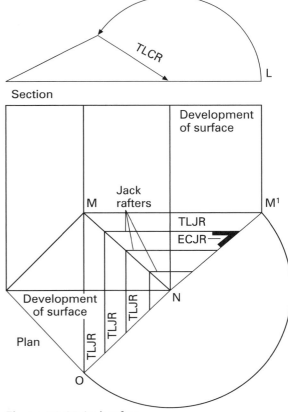

Figure 12.25 Jack rafter length and cuts

Setting out a hipped end

First we need to fit the wall plate, then the ridge and common rafters. To know where to place the common rafters you need to work out the true length of the rafter, then begin to mark out the wall plate.

The wall plate is joined at the corners and marked out as shown in Figure 12.28.

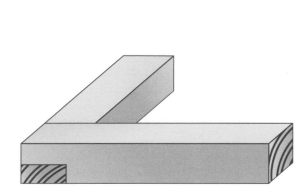

Figure 12.26 Corner halving

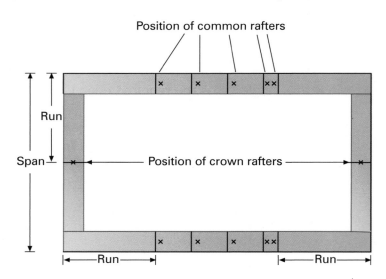

Figure 12.27 Wall plate marked out

Marking out for a hipped roof

To mark out for a hipped roof, follow these steps:

1. Measure the span and divide it by two to get the run, then mark this on the hipped ends – the centre of your **crown rafter** will line up with this mark.

2. Mark the run along the two longer wall plates – these marks will give you the position of your first and last common rafter.

3. The common rafter will sit to the side of this line, so a cross or other mark should be made to let you know which side of the line the rafter will sit.

4. Mark positions for the rest of the rafters on the wall plate at the required centres, again using a cross or other mark to show you which side of the line the rafter will sit. Note: the last two rafters may be closer together than the required centres but must not be wider apart than the required centres.

5. Cut and fit the common rafters using the same method as used for a gable roof.

6. Fit the crown rafter, which has the same plumb and seat cuts as the common rafter and is almost the same length – but here you should *not* remove half the ridge thickness.

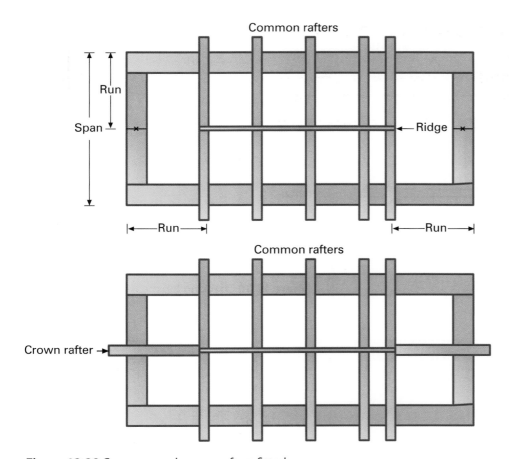

Figure 12.28 Common and crown rafters fitted

7. For the hip rafters, work out the true length, all the angles and bevels and mark out one hip as shown below, then cut the hip and try it in the four corners. If the hip fits in all four corners, you can use it as a template to mark out the rest of the hips; if not, the roof is out of square or level, but you can still use this hip to help mark out the remaining three corners.

Figure 12.29 Pitch line marked on hip

With a hip rafter it is important to remember that the pitch line is marked out differently. It is marked from the top of the rafter and is set at two-thirds of the depth of the common rafter.

The best way to check your rafter before cutting is to measure from the point of the ridge down to the corner of the wall plate. This distance should be the same as marked on your rafter.

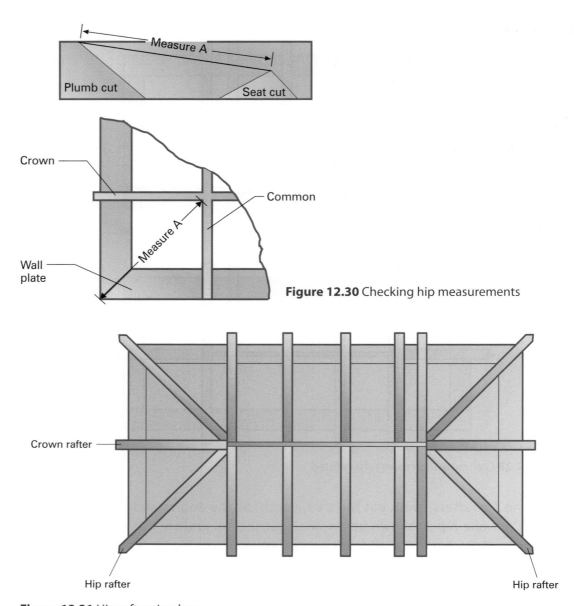

Figure 12.30 Checking hip measurements

Figure 12.31 Hip rafters in place

8. Cut in the jack rafters. Find out the true length and edge cut of all the jack rafters, then mark them out and cut them. As the jack rafters are of different sizes it is better to cut them individually to fit. They can still be used as template rafters on the opposite side of the hip.

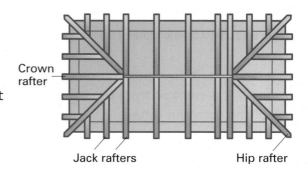

Figure 12.32 Jacks fitted

Valleys

The next section deals with valleys, which are formed when two sloping parts of a pitched roof meet at an internal corner.

Marking out for valleys

Valleys can be worked out in the same way as hips, using either a ready reckoner or geometry. Here we will look at geometry.

Working out the angles for valleys is similar to doing so for hips except that the key drawing is not a triangle but a plan drawing of the roof.

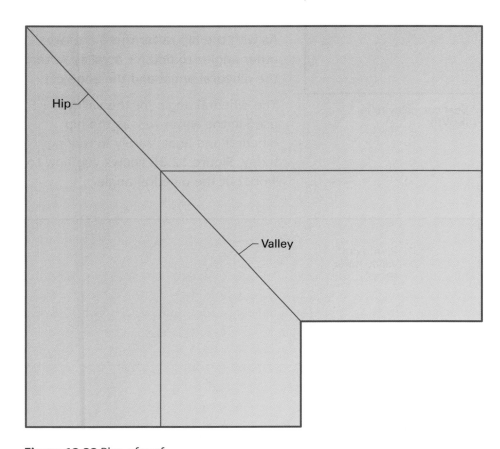

Figure 12.33 Plan of roof

First you need to find out the valley rafter true length, plumb and seat cut. Start by finding the rise of the roof and drawing a line this length at a right angle to the valley where it meets the ridge. Join this line to the point where the valley meets the wall plate. This will give you the true length of the valley rafter as well as the plumb and seat cuts.

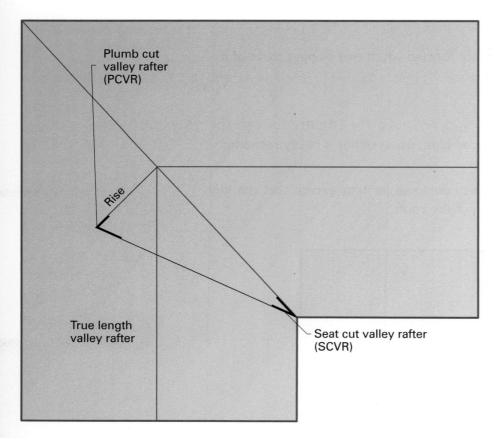

Figure 12.34 Valley true length

As with the hip rafter there are two other angles to find for a valley rafter: the dihedral angle and the edge cut.

The dihedral angle for the valley is used in the same way as the hip dihedral and again rarely in roofing today. Figure 12.36 shows you how to work out the dihedral angle.

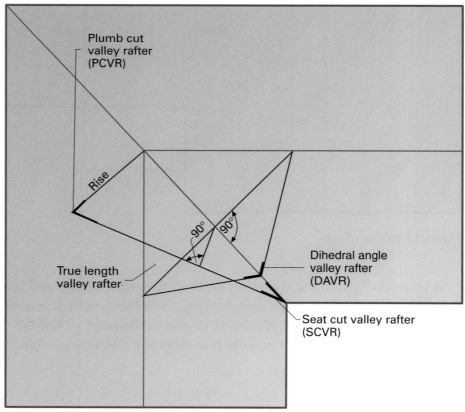

Figure 12.35 Dihedral angle hip

The final angle to find is the edge cut for the valley rafter, as follows.

1. Mark on the rise and true length of the valley rafter.

2. Draw a line at right angles to the valley where it meets the wall plate and extend this line to touch the ridge at A.

3. Set your compass to the true length of the valley and **swing an arc** towards the ridge at B.

4. Join up the line A–B to give you the edge cut.

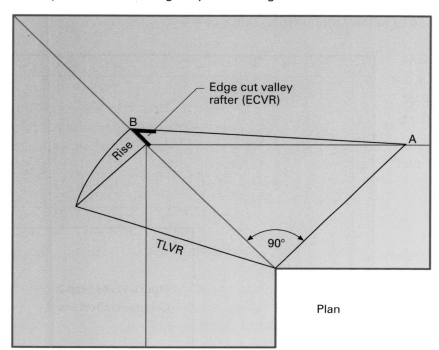

Plan

Figure 12.36 Edge cut

> ### Definition
>
> **Swing an arc** – set a compass to the required radius and lightly draw a circle or arc

The final part of valley geometry is to find the true length and edge cut for the cripple rafters, as follows.

1. Draw out the roof plan as usual, then to the side of your plan draw out a section of the roof.

2. Set your compass to the rafter length and swing an arc downwards.

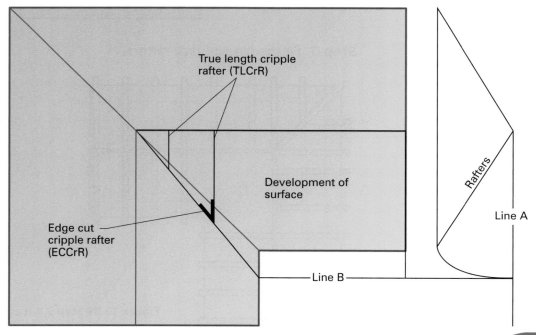

Figure 12.37 Cripple length and angle

3. Draw line A downwards until it meets the arc, then draw a line at right angles to line A until it hits the wall plate, creating line B.

4. Draw a line from where line B hits the wall plate up to where the valley meets the ridge. This will give you the appropriate true length (TLCrR) and edge cut (ECCrR).

Setting out a valley

There are four steps to follow when setting out a valley.

Step 1 Fit the wall plate and mark it out with the position of the common rafters.

Step 2 Fit the common rafters and ridge.

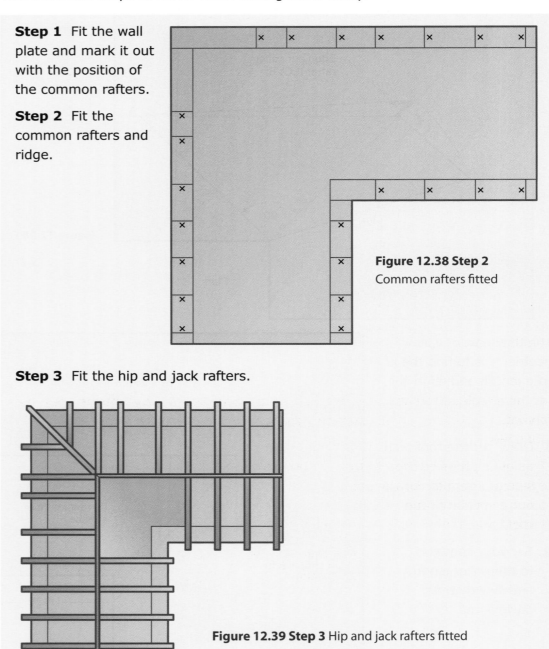

Figure 12.38 Step 2 Common rafters fitted

Step 3 Fit the hip and jack rafters.

Figure 12.39 Step 3 Hip and jack rafters fitted

Step 4 Fit the valley and cripple rafters, taking the true lengths and bevels from the drawings.

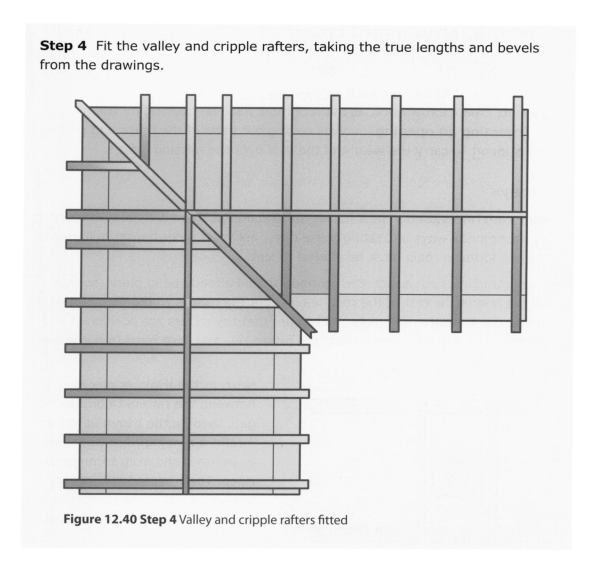

Figure 12.40 Step 4 Valley and cripple rafters fitted

An alternative to using valley rafters is to use a lay board. **Lay boards** are most commonly used with extensions to existing roofs or where there are dormer windows.

The lay board is fitted onto the existing rafters at the correct pitch, then the cripple rafters are cut and fixed to it.

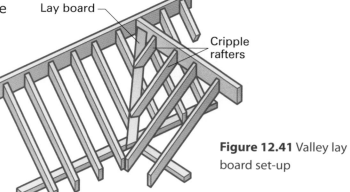

Lay board

Cripple rafters

Figure 12.41 Valley lay board set-up

Definition

Lay board – a piece of timber fitted to the common rafters of an existing roof to allow the cripple rafters to be fixed

Definition

Trimming an opening – removing structural timbers to allow a component such as a chimney or staircase to be fitted, and adding extra load-bearing timbers to spread the additional load

Trimming, eaves and covering

Trimming roof openings

Roofs often have components such as chimneys or roof windows. These components create extra work, as the roof must have an opening for them to be fitted. **Trimming an opening** involves cutting out parts of the rafter and putting in extra support to carry the weight of the roof over the missing rafters.

Chimneys

Chimneys are rarely used in new house construction as there are more efficient and environmental ways of heating these days, but most older houses will have chimneys and these roofs must be altered to suit.

When constructing such a roof, the chimney should already be in place, so you should cut and fit the rest of the roof, leaving out the rafters where the chimney is. When you mark out the wall plate, make sure that the rafters are positioned with a 50 mm gap between the chimney and the rafter. You may also need to put in extra rafters.

Next, fit the trimmer pieces between the rafters to bridge the gap, then fix the trimmed rafters – rafters running from wall plate to trimmer and from trimmer to ridge. The trimmed rafters are birdsmouthed at the bottom to sit over the wall plate, and the plumb and seat cut is the same as for common rafters.

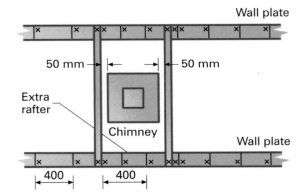

Figure 12.42 Wall plate marked out to allow for chimney

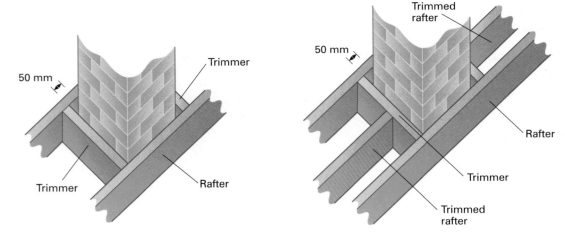

Figure 12.43 Opening trimmed around chimney with trimmers fitted

If the chimney is at **mid-pitch** rather than at the ridge, you will need to fit a chimney gutter: this ensures the roof remains watertight by preventing the water gathering at one point. The chimney gutter should be fixed at the back of the stack.

Definition

Mid-pitch – in the middle of the pitch rather than at the apex or eaves

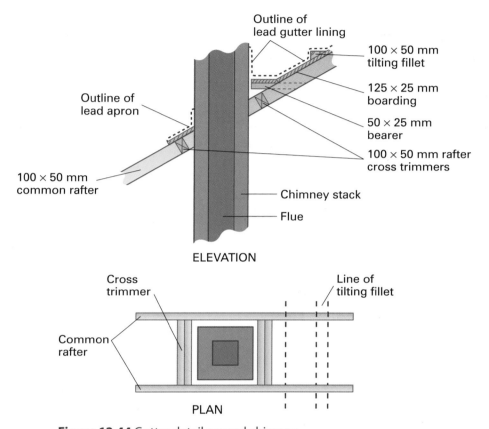

Figure 12.44 Gutter detail around chimney

Roof windows and skylights

You need to trim openings for roof windows and skylights too.

If the roof is new, you can plan it in the same way as for a chimney, remembering to make sure the area you leave for trimming is the same size as the window.

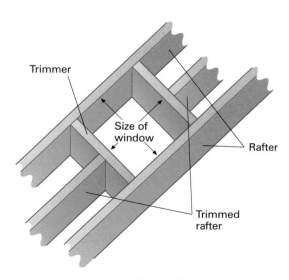

Figure 12.45 Roof trimmed for roof window or skylight

If you need to fit a roof window in an existing roof the procedure is different.

1. Strip all the tiles or slates from the area where you want to fit the window.

2. Remove the tile battens, felt and roof cladding.

3. Mark out the position of the window and cut the rafters to suit.

4. Now fit the trimmer and trimmed rafters – you may need to double up the rafters for additional strength – and fit the window.

5. Re-fit the cladding and felt, and fix any flashings.

6. Cut and fit new tile battens and finally re-tile or re-slate the roof.

Dormers

Dormer windows are different to roof windows: a roof window lays flat against the roof while a dormer window projects up from the sloping surface. A dormer window is often preferred to a roof window when there is limited headroom.

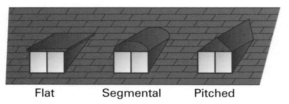

Flat   Segmental   Pitched

Figure 12.46 Dormer types

The construction of a dormer is similar to that of a roof window, except once the opening is trimmed, you need to add a framework to give the dormer shape. For a dormer you usually double up the rafters on either side to give support for the extra load the dormer puts on the roof.

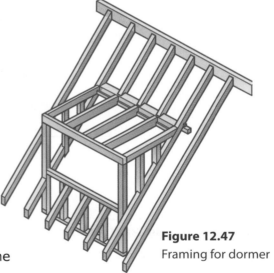

Figure 12.47
Framing for dormer

Eaves details

The eaves are how the lower part of the roof is finished where it meets the wall, and incorporates fascia and soffit. The fascia is the vertical board fixed to the ends of the rafters. It is used to close the eaves and allow fixing for rainwater pipes. The soffit is the horizontal board fixed to the bottom of the rafters and the wall. It is used to close the roof space to prevent birds or insects from nesting there, and usually incorporates ventilation to help prevent rot.

There are various ways of finishing a roof at the eaves; we will look at the four most common.

Flush eaves

Here the eaves are finished as close to the wall as possible. There is no soffit, but a small gap is left for ventilation.

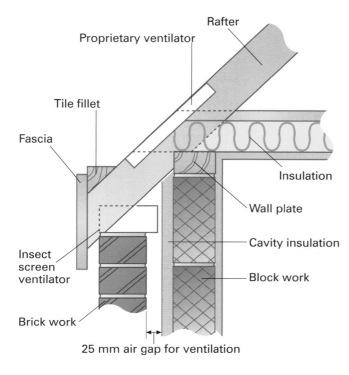

Figure labels: Rafter, Proprietary ventilator, Tile fillet, Fascia, Insulation, Wall plate, Cavity insulation, Insect screen ventilator, Block work, Brick work, 25 mm air gap for ventilation

Figure 12.48 Flush eaves

Open eaves

An open eaves is where the bottom of the rafter feet are planed as they are exposed. The rafter feet project beyond the outer wall and eaves boards are fitted to the top of the rafters to hide the underside of the roof cladding. The rainwater pipes are fitted via brackets fixed to the rafter ends.

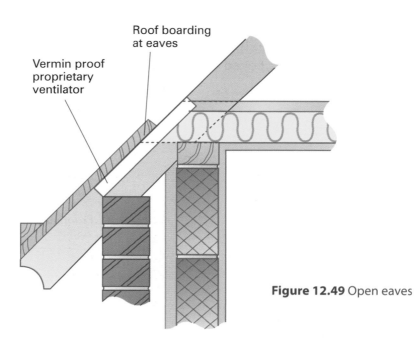

Figure labels: Roof boarding at eaves, Vermin proof proprietary ventilator

Figure 12.49 Open eaves

Closed eaves

Closed eaves are completely closed or boxed in. The ends of the rafters are cut to allow the fascia and soffit to be fitted. The roof is ventilated either by ventilation strips incorporated into the soffit or by holes drilled into the soffit with insect-proof netting over them. If closed eaves are to be re-clad due to rot you must ensure that the ventilation areas are not covered up.

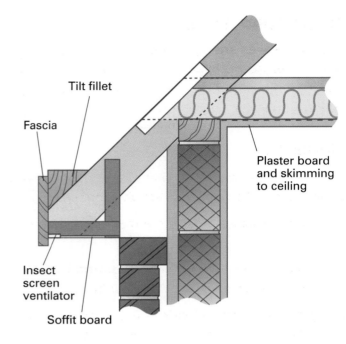

Figure 12.50 Closed eaves

Definition

Sprocket – a piece of timber bolted to the side of the rafter to reduce the pitch at the eaves

Sprocketed eaves

Sprocketed eaves are used where the roof has a sharp pitch. The **sprocket** reduces the pitch at the eaves, slowing down the flow of rainwater and stopping it overshooting the guttering. Sprockets can either be fixed to the top edge of the rafter or bolted onto the side.

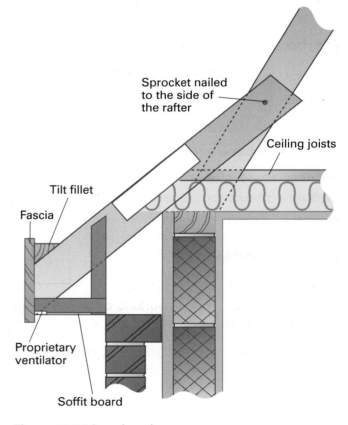

Figure 12.51 Sprocketed eaves

Roof coverings

Once all the rafters are on the roof, the final thing is to cover it. There are two main methods of covering a roof, each using different components. Factors affecting the choice of roof covering include what the local weather is like and what load the roof will have to take.

Method 1

This method is usually used in the north of the country where the roof may be expected to take additional weight from snow.

1. Clad the roof surface with a man-made board such as OSB or exterior grade plywood.

2. Cover the roof with roofing felt starting at the bottom and ensuring the felt is overlapped to stop water getting in.

3. Fit the felt battens (battens fixed vertically and placed to keep the felt down while allowing ventilation) and the tile battens (battens fixed horizontally and accurately spaced to allow the tiles to be fitted with the correct overlap).

4. Finally, fit the tiles and cement on the ridge.

Method 2

This is the most common way of covering a roof.

1. Fit felt directly onto the rafters.

2. Fit the tile battens at the correct spacing.

3. Fit the tiles and cement on the ridge.

Another way to cover a roof involves using slate instead of tiles. Slate-covered roofing is a specialised job as the slates often have to be cut to fit, so roofers usually carry this out.

Ready reckoner

A ready reckoner is a book used as an alternative to the geometry method and is often the simplest way of working out lengths and bevels. The book consists of a series of tables that are easy to follow once you understand the basics.

To use the ready reckoner you must know the span and the pitch of the roof.

Did you know?

Where you live may have an affect on your choice of roof: in areas more prone to bad weather, the roof will need to be stronger

Example

Take a hipped roof with a 36 degree pitch and a span of 8.46 m. First you halve the span, getting a run of 4.23 m. Referring to the tables in the ready reckoner, you can work out the lengths of the common rafter as follows:

RISE OF COMMON RAFTER 0.727 m PER METRE RUN   PITCH 36 degrees

BEVELS: COMMON RAFTER   SEAT 36
             COMMON RAFTER   PLUMB 54
             HIP OR VALLEY   SEAT 27
             HIP OR VALLEY   RIDGE 63
             JACK RAFTER   EDGE 39

JACK RAFTERS 333 mm CENTRES DECREASE   412 mm
                 400 mm CENTRES DECREASE   494 mm
                 500 mm CENTRES DECREASE   618 mm
                 600 mm CENTRES DECREASE   742 mm

Run of Rafter	0.1	0.2	0.3	0.4	0.5	0.6	0.7	0.8	0.9	1.0
Length of Rafter	0.124	0.247	0.371	0.494	0.618	0.742	0.865	0.989	1.112	1.236
Length of Hip	0.159	0.318	0.477	0.636	0.795	0.954	1.113	1.272	1.431	1.590

You can see that the length of a rafter for a run of 0.1 m is 0.124 m, therefore for a run of 1 m, the rafter length will be 1.24 m – but you need to find the length of a rafter for a run of 4.23 m. This is how:

       1.00 m = 1.24 m

So: 4.00 m = 4.994 m

       0.20 m = 0.247 m

       0.03 m = 0.037 m

       total   = **5.228 m**

So the length of the common rafter is 5.228 m. However there are a few adjustments you must make before finding the finished size.

You need to allow for an overhang and for the ridge, which can both easily be measured. For the purposes of this example, we will use an overhang of 556 mm and a ridge of 50 mm. The final calculation is:

Basic rafter	5.228 m
+ overhang	0.556 m
– half ridge	0.025 m
total	**5.759 m**

So the rafter length is 5.759 m.

Now you need to refer back to the table, which tells you that, for the common rafter, the seat cut is 36 degrees and the plumb cut is 54 degrees.

You can now mark out and cut your pattern rafter as before, remembering to mark on the pitch line and plumb cut. Measure the original size (5.228 m for our example) along the plumb cut, mark out the seat cut and finally cut.

The hip and valley rafters are worked out in the same way but jack and cripple rafters are different.

Jack and cripple rafters are marked on the table and are easy to work out. Continuing with the example, where the rafter length is 5.228 m, you look at the table and, if you are working at spacing the rafters at 400 mm centres, you reduce the length of the common rafter by 494 mm. The first jack/cripple rafter will be 5.228 m – 0.494 m = 4.734 m; the next jack/cripple rafter will be 4.734 m – 0.494 m = 4.240 m and so on. The angles for all the rest of the cuts are all shown on the table.

```
BEVELS:  COMMON RAFTER    SEAT 36
         COMMON RAFTER    PLUMB 54
         HIP OR VALLEY    SEAT 27
         HIP OR VALLEY    RIDGE 63
         JACK RAFTER      EDGE 39
```

```
JACK RAFTERS 333 mm CENTRES DECREASE    412 mm
             400 mm CENTRES DECREASE    494 mm
             500 mm CENTRES DECREASE    618 mm
             600 mm CENTRES DECREASE    742 mm
```

Remember

The plumb and seat cuts for the jack rafter are the same as for the common rafter

Flat roofs

A flat roof is any roof which has its upper surface inclined at an angle (also known as the fall, slope or pitch) not exceeding 10 degrees.

A flat roof has a fall to allow rainwater to run off, preventing puddles forming as they can put extra weight on the roof and cause leaks. Flat roofs will eventually leak, so most are guaranteed for only 10 years (every 10 years or so the roof will have to be stripped back and re-covered). Today **fibreglass** flat roofs are available that last much longer, so some companies will give a 25-year guarantee on their roof. Installing a fibreglass roof is a job for specialist roofers.

The amount of fall should be sufficient to clear water away to the outlet pipe(s) or guttering as quickly as possible across the whole roof surface. This may involve a single direction of fall or several directional changes of fall such as:

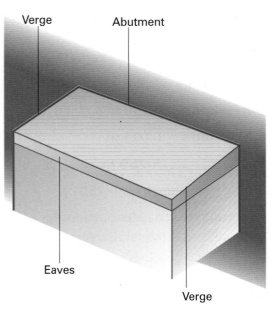

Figure 12.52 Flat roof terminology

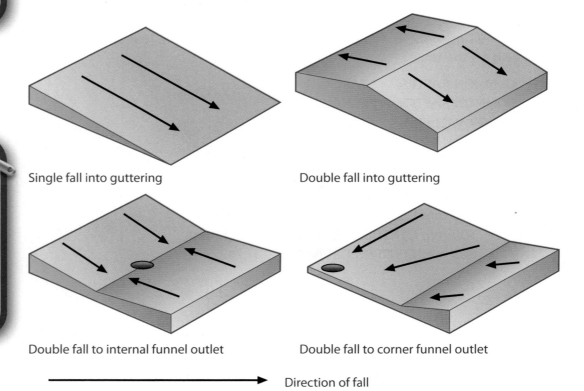

Single fall into guttering

Double fall into guttering

Double fall to internal funnel outlet

Double fall to corner funnel outlet

Direction of fall

Figure 12.53 Falls on a flat roof

Construction of a flat roof

Flat roof joists are similar in construction to floor joists (discussed later in this chapter) but unless they are to be accessible, they are not so heavily loaded. Joists are, therefore, of a smaller dimension than those used in flooring.

There are many ways to provide a fall on a flat roof. The method you choose depends on what the direction of fall is and where on a building the roof is situated.

Laying joists to a fall

This method is by far the easiest: all you have to do is ensure that the wall plate fixed to the wall is higher than the wall plate on the opposite wall or vice versa. The problem is that this method will also give the interior of the roof a sloping ceiling. This may be fine for a room such as a garage, but for a room such as a kitchen extension the client might not want the ceiling to be sloped and another method would have to be used.

Joists with firring pieces

Firring pieces provide a fall without disrupting the interior of the room, but involve more work. Firrings can be laid in two different ways, depending on the layout of the joists and the fall.

The layout of joists is explained in more detail in the flooring section of this chapter.

> ### Definition
>
> **Firring piece** – long wedge tapered at one end and fixed on top of joists to create the fall on a flat roof

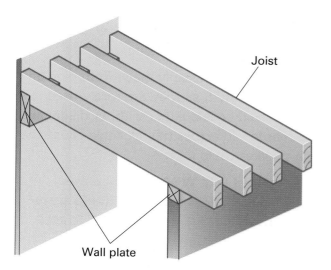

Figure 12.54 Joists laid to a fall

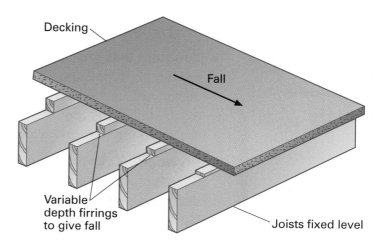

Figure 12.55 Joists with firring pieces

Using firring pieces

The basic construction of a flat roof with firrings begins with the building of the exterior walls. Once the walls are in place at the correct height and level, the carpenter fits the wall plate on the eaves wall (there is no need for the wall plate to be fitted to the verge walls). This can be bedded down with cement, or nailed through the joints in the brick or block work with restraining straps fitted for extra strength. The carpenter then fixes the **header** to the existing wall.

The header can either be the same depth as the joists or have a smaller depth. If it is the same depth, the whole of the joist butts up to the header and the joists are fixed using joist hangers; if the header has a smaller depth, the joists can be notched to sit on top of the header as well as using framing anchors.

Once the wall plate and header are fixed, they are marked out for the joists at the specified centres (300 mm, 400 mm or 600 mm). The joists are cut to length, checked for **camber (crown)** and fixed in place using joist hangers. Strutting or **noggings** are then fitted to help strengthen the joists.

Once the joists are fixed in place they must have restraining straps fitted. A strap must be fitted to a minimum of one joist per 2 m run, then firmly anchored to the wall to prevent movement in the roof under pressure from high winds.

The next step is to fix the firring pieces, which are either nailed or screwed down onto the top of the joists. Insulation is fitted between the joists, along with a vapour barrier to prevent the movement of moisture caused by condensation.

Decking

Once the insulation and vapour barrier are fitted, it is time to fit the decking. Decking a flat roof can be done with a range of materials including:

- **Tongued and grooved board** – These boards are usually made from pine and are not very moisture-resistant, even when treated, so they are rarely used for decking these days. If used, the boards should be laid either with, or diagonal to, the fall of the roof. Cupping of the boards laid across the fall could cause the roof covering to form hollows in which puddles could form.

- **Plywood** – Only roofing grade boards stamped with **WBP** should be used. Boards must be supported and securely fixed on all edges in case there is any distortion, which could rip or tear the felt covering. A 1mm gap must be left between each board along all edges in case there is any movement caused by moisture, which again could cause damage to the felt.

- **Chipboard** – Only the types with the required water resistance classified for this purpose must be used. Boards are available that have a covering of bituminous felt bonded to one surface, giving temporary protection against wet weather. Once laid, the edges and joints can be sealed. Edge support, laying and fixing are similar to floors (covered later in this chapter). Moisture movement will be greater than with plywood as chipboard is more **porous**, so a 2 mm gap should be allowed along all joints, with at least a 10 mm gap around the roof edges. Tongued and grooved chipboard sheets should be fitted as per the manufacturer's instructions.

- **Oriented strand board (OSB)** – Generally more stable than chipboard, but again only roofing grades must be used. Provision for moisture movement should be made as with chipboard.

- **Cement-bonded chipboard** – Strong and durable with high density (much heavier and greater moisture resistance than standard chipboard). Provisions for moisture movement should be the same as chipboard.

- **Metal decking** – Profiled sheets of aluminium or galvanised steel with a variety of factory-applied colour coatings and co-ordinated fixings are available. Metal decking is more usually associated with large steel sub-structures and fixed by specialist installers, but it can be used on small roof spans to some effect. Sheets can be rolled to different profiles and cut to any reasonable length to suit individual requirements.

- **Translucent sheeting** – This might be corrugated or flat (e.g. polycarbonate twin wall), and must be installed as per the manufacturer's instructions.

Metal decking and translucent sheeting are supplied as finished products but timber-based decking needs additional work to it to make it watertight, as explained in the next section.

Weatherproofing a flat roof

Once a flat roof has been constructed and decked the next step is to make it watertight. The roof decking material can be covered using different methods and with different materials. One basic way of covering the decking is as follows.

The roof decking is covered in a layer of hot **bitumen** to seal any gaps in the joints. Then another layer of bitumen is poured over the top and felt is rolled onto it, sticking fast when the bitumen sets. A second roll of felt is stuck down with bitumen, but is laid at 90 degrees to the first. Some people add more felt at this stage – sometimes up to five more layers.

Definition

Porous – full of pores or holes: if a board is porous, it may soak up moisture or water and swell

Translucent – allowing some light through but not clear like glass

Bitumen – also known as pitch or tar, bitumen is a black sticky substance that turns into a liquid when heated, and is used on flat roofs to provide a waterproof seal

Did you know?

Bitumen is bought as a solid material. It is then broken up, placed in a cauldron and heated until it becomes a liquid, which can be spread over the flat roof. The bitumen will only stay in liquid form for a few minutes before it starts to set. Once set, it forms a waterproof barrier

Definition

Tilt or angle fillet
– a triangular-shaped piece of wood used in flat roof construction

Did you know?

Flat roofs have stones or chippings placed on top to reflect the heat from the sun, which can melt the bitumen, causing problems with leaks

Did you know?

Dressing lead means moulding the lead so that it covers or fits into something. Lead dressing is usually done by plumbers and roofers

The final step can also be done in a number of ways. Some people put stones or chippings down on top of the final layer; some use felt that has stones or chippings imbedded and a layer of dried bitumen on the back. The felt with stones imbedded into it is laid by rolling the felt out and using a gas blowtorch to heat the back, which softens the bitumen allowing it to stick.

Finishing a flat roof at the abutment and verge
Abutment

The abutment finish needs to take into consideration the existing wall as well as the flat roof. The abutment is finished by cutting a slot into the brick or block work and fixing lead to give a waterproof seal, which prevents water running down the face of the wall and into the room. **Tilt** or **angle fillets** are used to help with the run of the water and to give a less severe angle for the lead to be dressed to.

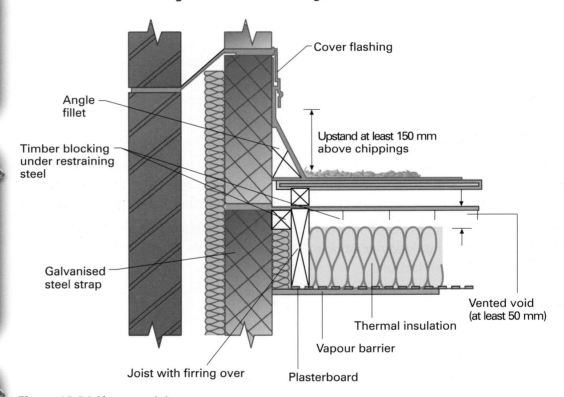

Figure 12.56 Abutment joint

Verge

Since a flat roof has such a shallow pitch, the verge needs some form of upstand to stop the water flowing over the sides at the verge instead of into the guttering at the eaves. This is done using tilt or angle fillets nailed to the decking down the full length of the verge prior to the roof being felted or finished.

Eaves details

The eaves details can be finished in the same way as the eaves on a pitched roof, with the soffit fitted to the underside of the joists and the fascia fitted to the ends.

Ground and upper floors

Several types of flooring are used in the construction of buildings, ranging from timber floors to large pre-cast concrete floors, which are used in large buildings such as residential flats. The main type of flooring a carpenter will deal with is suspended timber and that is what we will concentrate on in this section. A suspended timber floor can be fitted at any level from top floor to ground floor. In the next few pages we will look at:

- basic structure
- joists
- construction methods
- floor coverings.

Basic structure

Suspended timber floors are constructed with timbers known as joists, which are spaced parallel to each other spanning the distance of the building. Suspended timber floors are similar to traditional roofs in that they can be single or double, a single floor being supported at the two ends only and a double floor supported at the two ends and in the middle by way of a sleeper/dwarf wall, steel beam or load-bearing partition.

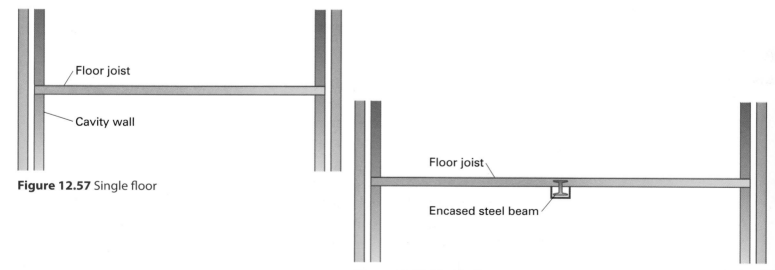

Floor joist

Cavity wall

Figure 12.57 Single floor

Floor joist

Encased steel beam

Figure 12.58 Double floor

All floors must be constructed to comply with the *Building Regulations*, in particular Part C, which is concerned with damp. The bricklayer must insert a **damp proof course (DPC)** between the brick or block work when building the walls, situated no less than 150 mm above ground level. This prevents moisture moving from the ground to the upper side of the floor. No timbers are allowed below the DPC. Air bricks, which are built into the external walls of the building, allow air to circulate round the underfloor area, keeping the moisture content below the dry rot limit of 20 per cent, thus preventing dry rot.

Joists

In domestic dwellings suspended upper floors are usually single floors, with the joist supported at each end by the structural walls but, if support is required, a load-bearing partition is used. The joists that span from one side of the building to the other are called bridging joists, but any joists that are affected by an opening in the floor such as a stairwell or chimney are called trimmer, trimming and trimmed joists.

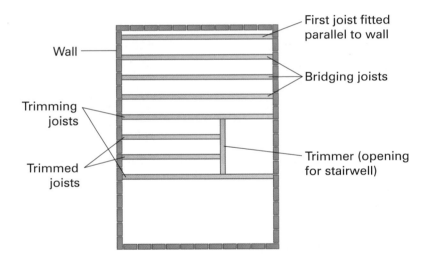

Figure 12.59 Joists and trimmers

Bridging joists are usually sawn timber 50 mm thick. The trimmer that carries the trimmed joists and transfers this load to the trimming must be thicker – usually 75 mm sawn timber, or in some instances two 50 mm bridging joists bolted together. The depth of the joist is easily worked out by using the calculation:

Depth of joist = span / 20 + 20

So, for example, if you have a span of 4 m

Depth = 4000/20 + 20

Depth = 200 + 20

Depth = 220

the depth of the joist required would be 220 mm.

If the span was 8 m, the depth would double to 440 mm. A depth of 440 mm is too great, so you would need to look at putting in a support to create a double floor.

Types of joist

As well as the traditional method of using solid timber joists, there are now alternatives available. These are the most common.

Laminated joists

These were originally used for spanning large distances, as a laminated beam could be made to any size, but now they are more commonly used as an environmental alternative to solid timber – recycled timber can be used in the laminating process. They are more expensive than solid timber, as the joists have to be manufactured.

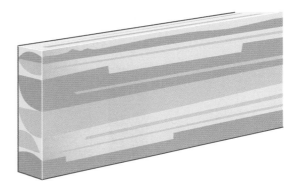

Figure 12.60 Laminated joist

I type joists

These are now some of the most commonly used joists in the construction industry: they are particularly popular in new build and are the only joists used in timber kit house construction. I type beam joists are lighter and more environmentally friendly as they use a composite panel in the centre, usually made from oriented strand board, which can be made from recycled timber.

The following construction method shows how to fit solid timber joists but whichever joists you use, the methods are the same.

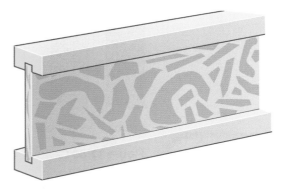

Figure 12.61 I type joist

Construction methods

A suspended timber floor must be supported either end. Figures 12.63 and 12.64 show ways of doing this.

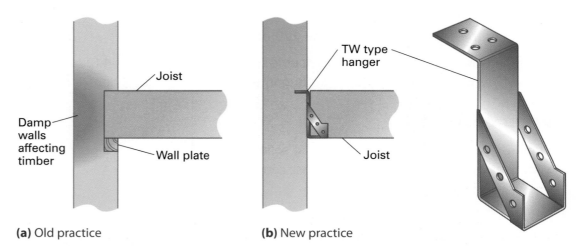

(a) Old practice **(b)** New practice

Figure 12.62 Solid floor bearings

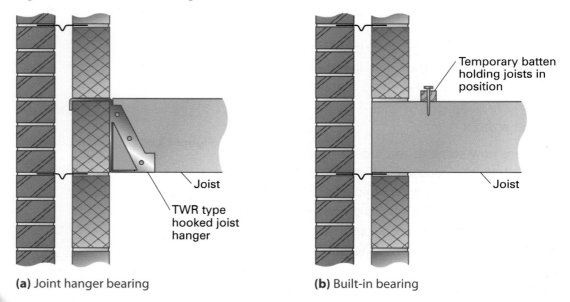

(a) Joint hanger bearing **(b)** Built-in bearing

Figure 12.63 Cavity wall bearings

If a timber floor has to trim an opening, there must be a joint between the trimming and the trimmer joists. Traditionally, a **tusk tenon joint** was used (even now, this is sometimes preferred) between the trimming and the trimmer joist. If the joint is formed correctly a tusk tenon is extremely strong, but making one is time-consuming. A more modern method is to use a metal framing anchor or timber-to-timber joist hanger.

Traditional tusk tenon joint

Joint hanger

Fitting floor joists

Before the carpenter can begin constructing the floor, the bricklayer needs to build the honeycomb sleeper walls. This type of walling has gaps in each course to allow the free flow of air through the underfloor area. It is on these sleeper walls that the carpenter lays his timber wall plate, which will provide a fixing for the floor joists.

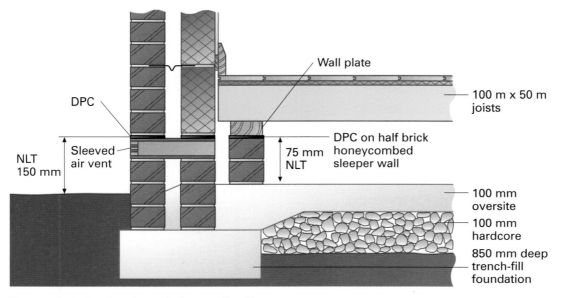

Figure 12.64 Section through floor and wall

The following pages describe the steps in fitting floor joists.

Step 1 Bed and level the wall plate onto the sleeper wall with the DPC under it.

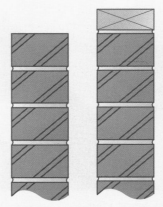

Figure 12.65 Step 1 Bed in the wall plate

Step 2 Cut joists to length and seal the ends with a coloured preservative. Mark out the wall plate with the required centres, space the joists out and fix temporary battens near each end to hold the joists in position. Ends should be kept away from walls by approximately 12 mm. It is important to ensure that the camber is turned upwards.

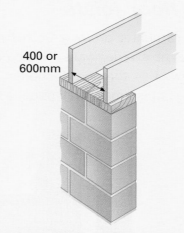

400 or 600mm

Figure 12.66 Step 2 Space out joists

Step 3 Fix the first joist parallel to the wall with a gap of 50 mm. Fix trimming and trimmer joists next to maintain the accuracy of the opening.

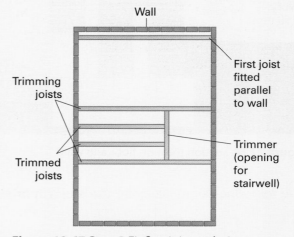

Wall

Trimming joists

Trimmed joists

First joist fitted parallel to wall

Trimmer (opening for stairwell)

Figure 12.67 Step 3 Fit first joist and trimmers

Step 4 Fix subsequent joists at the required spacing as far as the opposite wall. Spacing will depend on the size of joist and/or floor covering, but usually 400 mm to 600 mm centres are used.

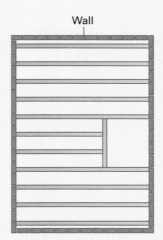

Wall

Figure 12.68 Step 4 Fit remaining joists

Step 5 Fit folding wedges to keep the end joists parallel to the wall. Overtightening is to be avoided in case the wall is strained.

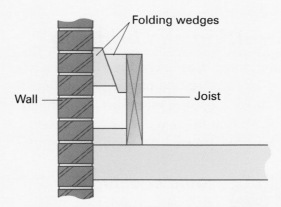

Folding wedges

Wall

Joist

Figure 12.69 Step 5 Fit folding wedges

Step 6 Check that the joists are level with a straight edge or line and, if necessary, pack with slate or DPC.

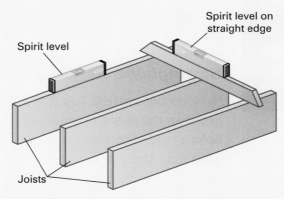

Spirit level on straight edge

Spirit level

Joists

Figure 12.70 Step 6 Ensure joists are level

Step 7 Fit restraining straps and, if the joists span more than 3.5 m, fit strutting and bridging, described in more detail next.

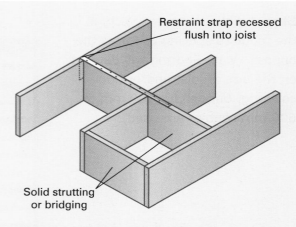

Figure 12.71 Step 7 Fix restraining straps, struts and bridges

Strutting and bridging

When joists span more than 3.5 m, a row of struts must be fixed midway between each joist. Strutting or bridging stiffens the floor in the same way that noggins stiffen timber stud partitions, preventing movement and twisting, which is useful when fitting flooring and ceiling covering. A number of methods are used, but the main ones are solid bridging, herringbone strutting and steel strutting.

Solid bridging

For solid bridging, timber struts the same depth as the joists are cut to fit tightly between each joist and **skew-nailed** in place. A disadvantage of solid bridging is that it tends to loosen when the joists shrink.

Solid bridging

Herringbone strutting

Here timber battens (usually 50 × 25 mm) are cut to fit diagonally between the joists. A small saw cut is put into the ends of the battens before nailing to avoid the battens splitting. This will remain tight even after joist shrinkage. The following steps describe the fitting of timber herringbone strutting.

Step 1 Nail a temporary batten near the line of strutting to keep the joists spaced at the correct centres.

Space joists

Step 2 Mark the depth of a joist across the edge of the two joists, then measure 12 mm inside one of the lines and remark the joists. The 12 mm less than the depth of the joist ensures that the struts will finish just below the floor and ceiling level (as shown in step 5).

Mark joist depths

Step 3 Lay the strut across two joists at a diagonal to the lines drawn in Step 2.

Lay struts across two joists at a diagonal

Step 4 Draw a pencil line underneath as shown in Step 3 and cut to the mark. This will provide the correct angle for nailing.

Cut to the mark

Step 5 Fix the strut between the two joists. The struts should finish just below the floor and ceiling level. This prevents the struts from interfering with the floor and ceiling if movement occurs.

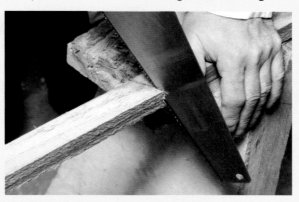

Fix the strut

Steel strutting

There are two types of galvanised steel herringbone struts available.

The first has angled lugs for fixing with the minimum 38 mm round head wire nails.

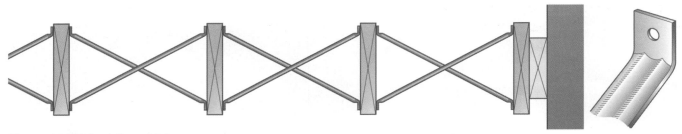

Figure 12.72 Catric® steel joist struts

The second has pointed ends, which bed themselves into joists when forced in at the bottom and pulled down at the top. Unlike other types of strutting, this type is best fixed from below.

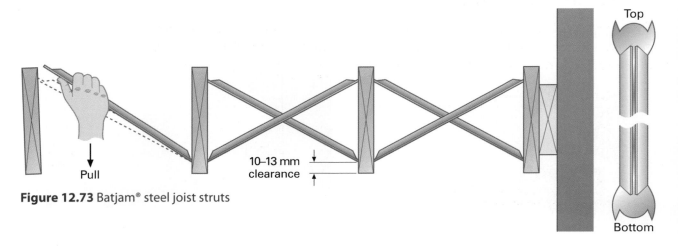

Figure 12.73 Batjam® steel joist struts

The disadvantage of steel strutting is that it only comes in set sizes, to fit centres of 400, 450 and 600 mm. This is a disadvantage as there will always be a space in the construction of a floor that is smaller than the required centres.

Restraint straps

Anchoring straps, normally referred to as restraint straps, are needed to restrict any possible movement of the floor and walls due to wind pressure. They are made from galvanised steel, 5 mm thick for horizontal restraints and 2.5 mm for vertical restraints, 30 mm wide and up to 1.2 m in length. Holes are punched along the length to provide fixing points.

When the joists run parallel to the walls, the straps will need to be housed into the joist to allow the strap to sit flush with the top of the joist, keeping the floor even. The anchors should be fixed at a maximum of 2 m centre to centre. More information can be found in schedule 7 of the *Building Regulations*.

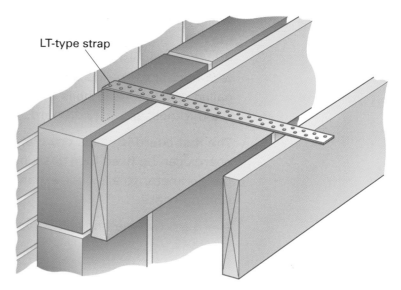

Figure 12.74 Restraint straps for joists parallel or at right angles to a wall

Floor coverings
Softwood flooring

Softwood flooring can be used at either ground or upper floor levels. It usually consists of 25 × 150 mm tongued and grooved (T&G) boards. The tongue is slightly off centre to provide extra wear on the surface that will be walked upon.

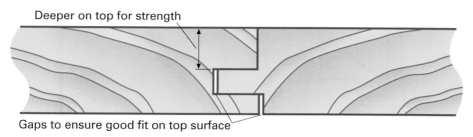

Figure 12.75 Section through softwood covering

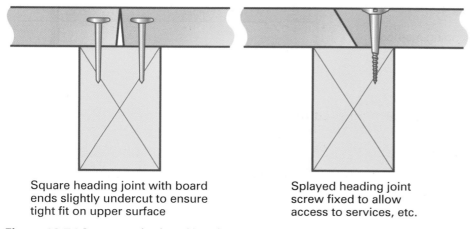

Square heading joint with board ends slightly undercut to ensure tight fit on upper surface

Splayed heading joint screw fixed to allow access to services, etc.

Figure 12.76 Square and splayed heading

375

When boards are joined together, the joints should be staggered evenly throughout the floor to give it strength. They should never be placed next to each other, as this prevents the joists from being tied together properly. The boards are either fixed with floor brads nailed through the surface and punched below flush, or secret nailed with lost head nails through the tongue. The nails used should be 2½ times the thickness of the floorboard.

The first board is nailed down about 10–12 mm from the wall. The remaining boards can be fixed four to six boards at a time, leaving a 10–12 mm gap around the perimeter to allow for expansion. This gap will eventually be covered by the skirting board.

There are two methods of clamping the boards before fixing:

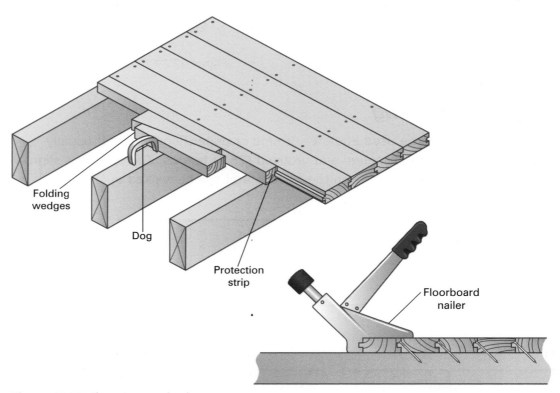

Figure 12.77 Clamping methods

Chipboard flooring

Flooring-grade chipboard is increasingly being used for domestic floors. It is available in sheets sizes of 2440 × 600 × 18 mm and can be square edged or tongued and grooved on all edges, the latter being preferred. If square-edged chipboard is used it must be supported along every joint.

Tongued and grooved boards are laid end to end, at right angles to the joists. Cross-joints should be staggered and, as with softwood flooring, expansion gaps of 10–12 mm left around the perimeter. The ends must be supported.

When setting out the floor joists, the spacing should be set to avoid any unnecessary wastage. The boards should be glued along all joints and fixed using either 50–65 mm annular ring shank nails or 50–65 mm screws. Access traps must be created in the flooring to allow access to services such as gas and water.

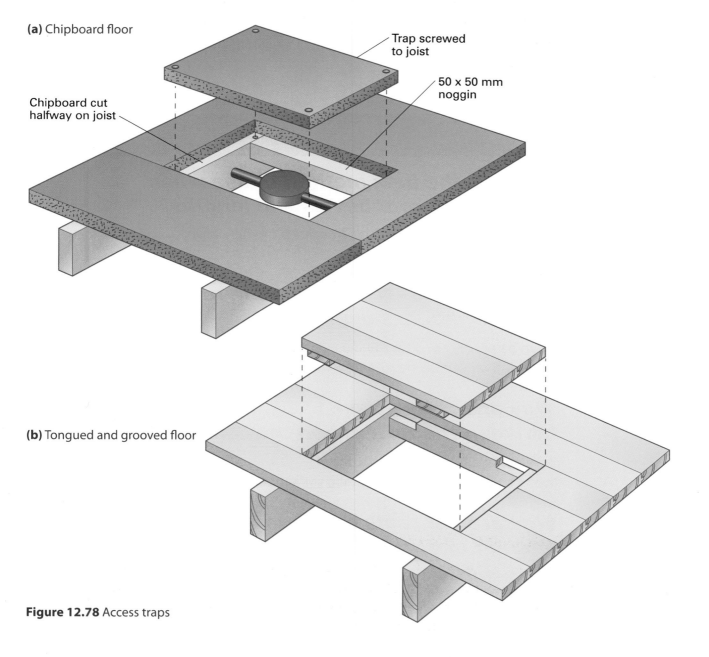

(a) Chipboard floor

Trap screwed to joist

50 x 50 mm noggin

Chipboard cut halfway on joist

(b) Tongued and grooved floor

Figure 12.78 Access traps

On the job: Cutting in a garage roof

Chloe and Tyrone have been tasked with cutting in a garage roof with gable ends. They work out the angles and lengths, cut the pattern rafter, which they use to mark out all the others, then fit them. They are almost finished when the foreman turns up and tells them to stop and check it for level. Chloe and Tyrone put a level on the ridge board and find that it is way out of level. How could this have happened? What could have been done to prevent it?

FAQ

Which is the best method to use when working out roof bevels and lengths?

There is no single best way: it all depends on the individual and what they find the easiest method to use.

When it comes to covering the roof, which method should I use?

The type of roof covering to be used depends on what the client and architect want, and on what is needed to meet planning and the *Building Regulations*.

Why does there have to be a 50 mm gap between the rafters and the chimney?

Building Regulations state that there must be a 50 mm gap to prevent the heat from the chimney combusting the timber.

Why are flat roofs only guaranteed for a certain amount of time?

Most things that you buy or have fitted have a guarantee for a certain amount of time and building work is no different. If the flat roof had a guarantee for 50 years, the builder would be responsible for any maintenance work on the roof free of charge for 50 years. Since the average life of a flat roof is 12–15 years, the builder will only offer a guarantee for 10 years.

FAQ

Which type of flat roof decking is the best?

There is no specific best or worst but some materials are better than others. All the materials stated serve a purpose, but only if they are finished correctly, for example a chipboard-covered roof that is poorly felted will leak, as will a metal roof if the screw or bolt holes are not sealed correctly.

How can I get the lead on a flat roof to fit into the brick/block work?

The lead is fitted into a channel or groove that is cut into the brick/block work by a bricklayer. The groove is cut using either a disc cutter or a hammer and cold chisel. Once the groove is cut the lead is fed into the groove, wedged and sealed with a suitable mastic or silicon.

I have laid chipboard flooring and the floor is squeaking. What causes this and how can I stop it?

The squeaking is caused by the floorboards rubbing against the nails – something that happens after a while as the nails eventually work themselves loose. The best way to prevent this is to put a few screws into the floor to prevent movement – but be cautious: there may be wires or pipes under the floor.

Knowledge check

1. Describe the difference between a single and double roof.

2. Explain the reason for having a mansard roof.

3. Explain the difference between a hip and a gable end.

4. What is a pitch line and where is it marked?

5. State a suitable way of fitting a wall plate.

6. State the distance the first and last common rafters must be away from the gable wall.

7. What is the dihedral angle (backing bevel) and what purpose does it serve?

8. What is the purpose of a roof ladder?

9. What type and size of nail should be used to fix bargeboard?

10. Why is it important that a carpenter learns Pythagoras' theorem?

11. Why do you not need to work out the plumb and seat cuts for the jack /cripple rafters?

12. State two different ways of forming a valley.

13. Why do you need to add extra rafters when trimming an opening?

14. State two different ways of providing a fall on a flat roof.

15. Name four different materials used when decking a flat roof?

16. What is bitumen?

17. What is the purpose of strutting a floor or flat roof?

18. What is the purpose of access hatches or traps in floors?

chapter 13

Maintaining components

OVERVIEW

In time, all components within a building will deteriorate. Timber will rot, brickwork crumble and even the ironmongery fail through wear and tear. These components need to be maintained or repaired to prevent the building falling into a state of disrepair. General building maintenance is now a recognised qualification, but traditionally this work is done by carpenters, as carpenters have the best links to other trades and the best understanding of what they do.

The maintenance of a building can range from periodically repainting the woodwork to replacing broken windows. This chapter will not cover simpler remedial tasks such as replacing a door handle, but will cover the following topics:

- Repairing and replacing mouldings
- Repairing a door that is binding
- Repairing damaged windowsills and door frames
- Replacing window sash cords
- Repairing structural timbers
- Minor repairs to plaster and brickwork
- Replacing gutters and downpipes.

These topics can be found in the following modules:

CC 1001K	CC 2011K
CC 1001S	CC 2011S

Repairing and replacing mouldings

Mouldings such as skirting or architrave rarely need repairing – usually only because of damp or through damage when moving furniture – and in most cases it is easier to replace them than repair them.

Architrave

To replace architrave, you simply remove the damaged piece and fit a new piece.

First check that the piece you are removing is not nailed to an existing piece. Then run a sharp utility knife down both sides of the architrave so that when it is removed it does not damage the surrounding decorations or remaining architrave.

Once the old piece is removed, clean the frame of old paint, give it a light sanding with sandpaper, then fit the new piece. Finally paint, stain or finish the new piece to match the rest.

Skirting board

Replacing skirting boards is slightly more difficult than replacing architrave. The way skirting boards are fitted could mean that the board you wish to replace is held in place by other skirting boards. Rather than remove the other skirting boards, you should cut or drill a series of holes in the middle of the board, splitting it in two so that you can remove it that way. Again, running a utility knife along the top of the skirting board will avoid unnecessary damage to surrounding decorations. Once the old skirting has been removed, the new piece can be fitted and finished to match the existing skirting.

You can replace other mouldings such as picture and dado rails in the same way as skirting boards, taking care not to damage the existing decorations.

Remember

When replacing mouldings, you may need to rub down the existing moulding work and re-apply a finish to both the old and the new to get a good match

Safety tip

Take care if gripper rods are fitted in front of the skirting, as this will make the replacing of the skirting more difficult and more dangerous

Repairing a door that is binding

One of the most common problems that occur with doors is that they **bind**. A door can bind at several different points, as you can see in the illustration below.

These binding problems are simple to fix.

Definition

bind – a door 'binds' when it will not close properly and the door springs open slightly when it is pushed closed

- If the door is binding at the hinges, it is usually caused by either a screw head sticking out too far or by a screw that has been put in squint. In these cases, screw the screw in fully or replace it straight. If the hinge is bent, fit a new one.

- If the door is binding on the hanging stile, it may not have been back-bevelled when hung. In this case, take the door off, remove the hinges, plane the door with a back bevel and then re-hang it. Alternatively, it may be expansion or swelling due to changes in temperature that is causing the problem. In this case, take off the door, plane it and re-hang it.

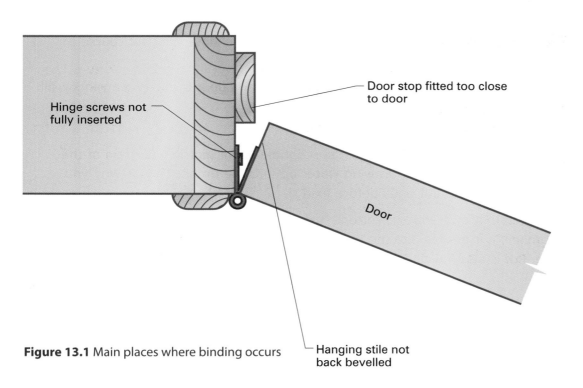

Figure 13.1 Main places where binding occurs

- If the door is binding at the stop on the hanging side, the fitter may not have left a 1–2 mm gap between the stop and the door to allow for paint, in which case the stop will have to be moved. If there is a door frame rather than a door lining, there is no stop to remove: simply plane the rebate using a rebate plane.

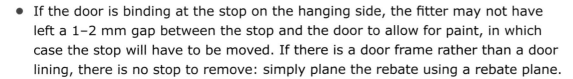

Repairing damaged windowsills and door frames

Provided they are painted or treated from time to time, exterior door frames and windows should last a long time, but certain areas of frames and windows are more susceptible to damage than others. The base of a door frame is where water can be absorbed into the frame; with windows, the sill is the most likely to suffer damage as water can sit on the sill and slowly penetrate it.

With damaged areas like these, the first thing to consider is whether to repair the damage or replace the component. With door frames, replacement may be the best option but you must take into account the extra work required to make good: repairing the frame may cost less and require less work. With a rotten windowsill, replacement involves taking out the entire window so that the new sill can be fitted, so repair may be the better option.

The choice usually comes down to a balance of cost and longevity: as the experienced tradesperson, you must help the client choose the best course of action.

In this section we will look at repairing the base of a door frame and a windowsill.

Repairing a door frame

The base of the door frame is most susceptible to rot as the end grain of the timber acts like a sponge, drawing water up into the timber (this is why you should always treat cut ends before fixing).

First check that the timber is rotten. Push a blunt instrument like a screwdriver into the timber; if it pushes into the timber, the rot is evident; if it does not, the frame is fine. Rot is also indicated by the paint or finish flaking off, or a musty smell.

If there is rot, repair it by using a splice, as follows.

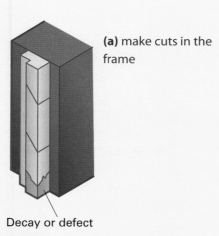

(a) make cuts in the frame

Decay or defect

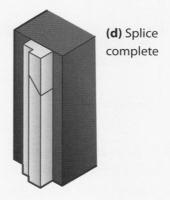

(b) remove the defect

Step 1 Make 45-degree cuts in the frame with the cuts sloping down outwards, to stop surface water running into the joints.

Step 2 Remove the defective piece and use a sharp chisel to chop away the waste, forming the scarf.

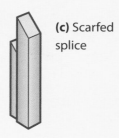

(c) Scarfed splice

(d) Splice complete

Step 3 Use either a piece of similar stock or, if none is available, a square piece planed to the same size and shape as the original, making sure the ends are thoroughly treated before fitting to prevent a recurrence.

Step 4 Fix the splice in place, dress the joint using sharp planes and make good any plaster or render.

Figure 13.2 Splicing a door frame

This method can also be used to repair door frames damaged by having large furniture moved through them, but the higher up the damage on the frame, the more likely the frame will be need to be replaced rather than repaired.

Repairing a windowsill

As with a door frame, this method uses a splice, as follows:

Step 1 Make a cut at 45 degrees, then carefully saw along the front of the windowsill to remove the rotten area.

Step 2 Clean the removed area using a sharp chisel and then, as before, use either a piece of similar stock or, if none is available, a square piece planed to the same size and shape as the removed area.

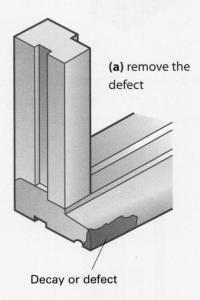

(a) remove the defect

Decay or defect

(b) Clean the area

Step 3 Give all cut areas a thorough treating with preservative, then attach the splice to match the existing windowsill shape and paint or finish to match the rest.

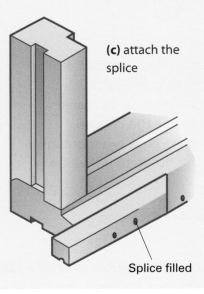

(c) attach the splice

Splice filled

Figure 13.3 Splicing a windowsill

Replacing window sash cords

As you saw in Chapter 10, box sash windows use weights attached to cords that run over a pulley system, to hold the sashes open and closed. The sash cord will eventually break through wear and tear, and will need to be replaced. This is not a large job so, if one cord needs attention, it is most cost-effective to replace them all at the same time.

There are various ways to replace sash cords. The following method is quick and easy and is done from the inside.

Step 1 Remove the staff bead, taking care not to damage the rest of the window. Cut the cords supporting the bottom sash, lowering the weight gently to avoid damaging the case. Take out the bottom sash, removing the old nails and bits of cord, and put to one side.

Step 2 Pull the top sash down and cut the cords carefully. Remove the parting bead and then the top sash, again removing the old nails and bits of cord, and put to one side.

Step 3 Remove the pockets and take out the weights, removing the cord from them and laying them at the appropriate side of the window.

Step 4 Slide the top sash into place at the bottom of the frame and, using chalk, mark the position of the sash cord onto the face of the pulley stiles (see distance A in Figure 13.4). Mark the bottom sash in the same way (distance B). Put the sashes to one side.

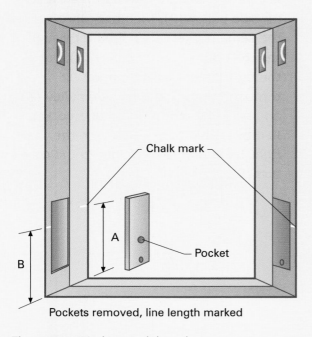

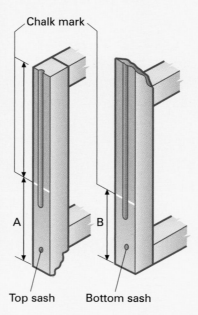

Figure 13.4 Marking sash lengths

Step 5 Next attach a **mouse** to the cord end, making sure it is long enough for the weighted end to reach the pocket before the sash cord reaches the pulley.

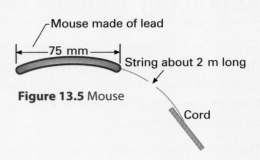

Figure 13.5 Mouse

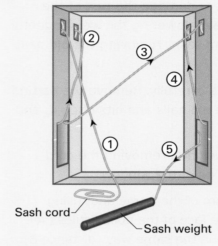

Figure 13.6 Cording a window

Step 6 Now cord the window. If only one cord is to be replaced, you can do this by feeding one pulley, then cutting the sash to length, re-attaching the end of the mouse to the sash cord and then feeding the next pulley. If you are replacing more than one cord, it is more efficient to feed the pulleys in succession and cut the cords after. In Figure 13.6 you can see one method for cording the window.

Feed the cord through the top left nearside pulley (1), out through the left-hand pocket, in through the top left far side pulley (2), then back through the same pocket. Then feed it through the top right nearside pulley (3), out through the right-hand pocket, in through the top right far side pulley (4) and back out of the right-hand pocket. Remove the mouse and attach the right-hand rear sash weight to the cord end (5).

Step 7 Working on the right-hand rear pulley, pull one weight through the pocket into the box and up until it is just short of the pulley. Lightly force a wedge into the pulley to prevent the weight from falling.

Pull the cord down the pulley stile and cut it to length, 50 mm above the chalk line (when the window is closed and the weight falls back down to the bottom of the box, the weight will stop 50 mm from the bottom of the frame as shown in Figure 13.7).

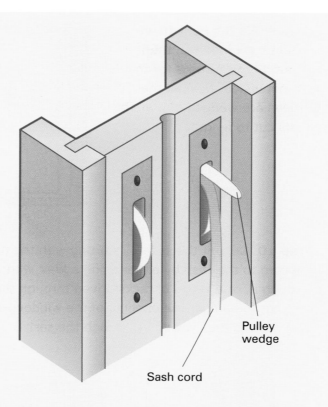

Pulley wedge

Sash cord

Figure 13.7 Wedged pulley

Step 8 Tie the other right-hand weight to the loose end of the cord. Still on the right-hand near-side, pull the cord again so that the weight is just short of the pulley and wedge it in place. Cut the cord to length, this time 30 mm longer than the chalk line (when the window is closed the bottom sash weight will be 30 mm away from the pulley, as shown in Figure 13.8).

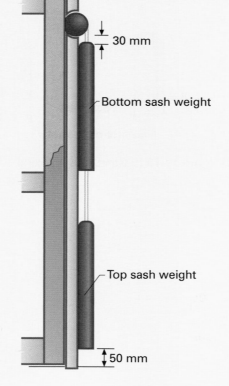

30 mm

Bottom sash weight

Top sash weight

50 mm

Figure 13.8 Sash weights at 50 mm and 30 mm

Step 9 Fix the right-hand pocket back in place, then repeat Steps 7 and 8 for the left-hand side. Now all the pulleys are wedged and all the cords cut to length.

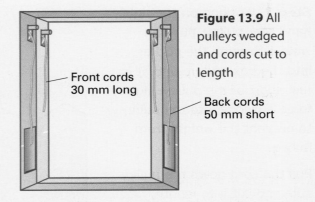

Figure 13.9 All pulleys wedged and cords cut to length

Front cords 30 mm long

Back cords 50 mm short

Step 10 Next fit the sashes, starting with the top sash. Fix the cord to the sash by either using a knot with a tack driven through it, or a series of tacks driven through the cord. Take care not to hamper the opening of the window, and do not use long nails as they will drive through the sash stile and damage the glass.

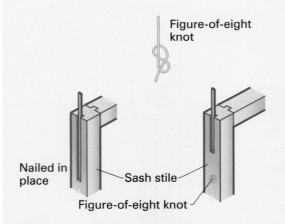

Figure-of-eight knot

Nailed in place

Sash stile

Figure-of-eight knot

Figure 13.10 Two methods for fixing cord to sashes

Step 11 Once the cord is fitted to both sides of the sash, slide the sash into place and carefully remove the wedges. Now test the top sash for movement and then re-fit the parting bead to keep the top sash in place. Now fit the bottom sash as in Step 10, test it and re-fit the staff beads, to secure the bottom sash in place.

Now test the whole window by sliding both sashes up and down. Finally, touch up any minor damage to the staff and parting beads.

Repairing structural timbers

The maintenance of structural timbers is vital, as they will almost certainly be carrying a load: joists carry the floors above, while rafters carry the weight of the roof. Because of this, structural repairs should be carried out by qualified specialists, so this section will give a brief understanding of the work that is involved.

Joists

Joist ends are susceptible to rot: they are close to the exterior walls and can be affected in areas like bathrooms if the floor gets soaked and does not dry out. Joists can also be attacked by wood-boring insects. For both rot and insect damage, the repair method is the same.

Shore the area, with the weight spread over the props, then lift the floorboards to see what the problem is (for the purposes of this example, we will use dry rot) and throw the old floorboards away.

Before making any repairs, you need to find and fix the cause of the problem – otherwise the same problems will keep on arising. Rot at the ends of joists is usually caused by poor ventilation: the cavity or the airbricks may be blocked, for example.

Next cut away the joists allowing 600 mm into sound timber, and treat the cut ends with a suitable preservative.

New pre-treated timbers should be laid and bolted onto the existing joists with at least 1 m overlap. New timbers should ideally be placed either side of the existing joist, and in place of the removed timber.

Safety tip

When repairing or replacing joists, you must first shore up the ceiling to carry the weight while you work. Follow the instructions found in Chapter 5, and do not alter or remove the shoring until the job is complete

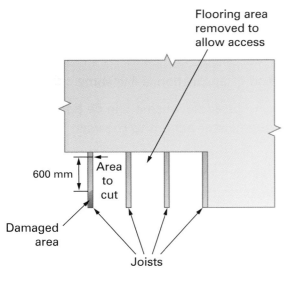

Figure 13.11 Cut away the rotten area of the joist

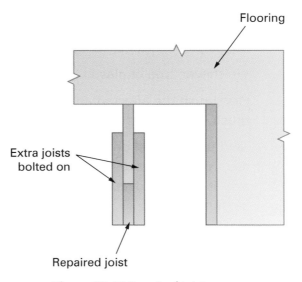

Figure 13.12 Repaired joist

Any new untreated timbers should be treated with a suitable preservative, before or after laying. Now slowly remove the props and make good as necessary.

Rafters

Rafters are susceptible to the same problems as joists, and especially to rot caused by a lack of ventilation in the roof space. Again the problem needs to be remedied before any repairs start.

Rafters are replaced or repaired in the same way as joists. They are easier to access, but shoring rafters can be difficult. In some cases it is best to strip the tiles and felt from the affected portion of the roof before starting.

For trussed roofs you should contact the manufacturer: trusses are stress graded to carry a certain weight, so attempting to modify them without expert advice could be disastrous.

Minor repairs to plaster and brickwork

During repair work there is always a chance that the interior plaster or brickwork may get damaged. Rather than call in a specialist most tradespeople will repair the damage themselves.

Repairing plasterwork

Plaster damage most often occurs when windows or door frames are removed – the plaster cracks or comes loose – and is simple to fix with either ready-mixed plaster (ideal for such tasks) or traditional bagged plaster, which is cheaper.

Bagged plaster needs to be mixed with water and stirred until it is the right consistency. Some people prefer a thinner mix as it can be worked for longer, while others prefer a consistency more like thick custard as it can be easier to use.

Whichever type of plaster you use, the method of application is the same.

First brush the affected area with a PVA mix, watered down so it can be applied by brush. This acts as a bonding agent to help the plaster adhere to the wall.

Next use a trowel or float to force the plaster into the damaged area. If it is quite deep you may have to part-fill the area and leave it to set, then put a finish skim over it. Filling a deep cavity in one go may result in the plaster running, leaving a bad finish.

When the plaster is almost dry, dampen the surface with water and use a wet trowel to skim over the plastered area, leaving a smooth finish. Once the plaster is dry, the area can be redecorated.

Repairs to exterior render are done in the same way, using cement instead of plaster.

Remember

Attempt only minor repairs unless you are fully trained. Otherwise you could end up doing more harm than good!

Plaster being applied and trowelled

Repairing brickwork

With repairs to brickwork, one of the first problems is finding bricks that match the existing ones, especially in older buildings. After that, the method is as follows.

Step 1 Make a hole in the centre of the area, to allow the removal of the bricks. Bricks rarely come out whole, but care must be taken not to damage the surrounding bricks.

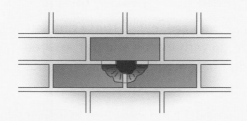

Step 2 Remove the bricks and carefully clean away the old mortar using a cold chisel.

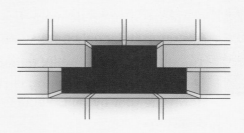

Step 3 Lay a mortar bed on the base of the opening.

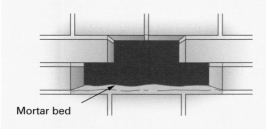

Mortar bed

Step 4 Place the first two bricks, making sure mortar is applied to all the joints.

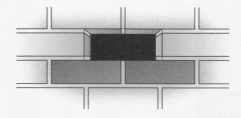

Step 5 Place the last brick, again making sure there is mortar between all the joints, then use a pointing trowel to point in the new bricks to match the others.

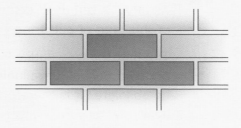

Figure 13.13 Stages in repairing brickwork

Replacing gutter and downpipes

Guttering is a vital part of a roof and must be maintained and replaced, if necessary, to ensure that the property remains waterproof. The purpose of guttering, which is attached to the fascia board, is to channel rainwater efficiently from the roof towards a downpipe, which in turn carries it down to ground level and then into the drainage system. This system helps to stop the walls of the house becoming soaked, causing problems with dampness.

Guttering used to be made from asbestos or cast iron, but is now made from pressed steel or, most commonly, uPVC.

The most common faults found in guttering are leaks. These are usually caused when the guttering is clogged up with leaves and dirt washed down the roof by rain. Regular clearing of the guttering will cure most problems.

In some cases, clearing the gutter may not be enough, and occasionally the guttering may need to be replaced. Possible causes are rusted steel or cracked pipes and, with uPVC guttering, damaged seals or brackets may mean that they have to be replaced.

Gutters must be properly maintained to prevent water damage to a building

Doing the work

When doing any work at height, you must ensure that a suitable scaffold is used – ladders are not ideal for replacing guttering or doing longer jobs.

The first thing to do is to assess the damage, as there is no need to replace the entire system if only one joint is leaking. Asbestos guttering must be removed by a specialist firm in line with COSHH regulations. With cast-iron guttering, which is usually connected via bolts, finding replacements may prove difficult, so in some instances it may be better to replace the entire system with uPVC.

As uPVC is the most common type, we will look at the replacement of this type of guttering.

The first thing to notice is that various guttering profiles and colours are commercially available – and some older systems may even be obsolete. It is therefore important to check that the parts are available for you to repair the system you have, otherwise it will all have to be replaced.

Gutters clogged with leaves and dirt can cause leaks

Once the damaged area has been identified, the next thing is to remove the old guttering. uPVC guttering is fixed by screwing clips to the fascia, which the guttering clips into. Unclipping the guttering allows it to be removed safely, and the replacement gutter can be clipped straight back into place.

If the whole gutter is to be replaced, first unclip all the guttering. Then either fix new brackets and clip in the new guttering, or clip the new guttering into the old clips.

Where one length of guttering meets another, a corner, a downpipe or a stop end, there are special brackets the gutter clips into.

The stop ends and corners simply clip onto the guttering, but the downpipe connector and straight connectors should be screwed to the fascia. If the whole system is to be replaced, you should screw the brackets and connectors to the fascia first, remembering to ensure that they are in line and running slightly downhill to where the water outlet is. Once the brackets are fitted, you can cut the guttering if needed and clip it into place, ideally running a bead of silicon along any joints.

Figure 13.14 Gutters are available in a variety of profiles and colours

The downpipe should be attached to the wall via brackets.

On the job: Repairing a binding door

Kelly has been called to an office block to repair a door that is opening by itself. She checks the hinges, hanging stile and door stop for binding but they are all fine.

1.   What could be causing the door to open?

2.   How could this be fixed?

FAQ

Can I do a scarf repair on skirting or architrave?

Yes, but it is often easier to replace mouldings than to repair them.

Should I go on a plastering course to help with repairs?

This is not necessary as the repairs you would be doing are not large, but if you feel that it would benefit you, yes.

Knowledge check

1. Give a reason why mouldings may need to be replaced.

2. What is the term for cutting in a repair piece at 45 degrees?

3. Name the component attached to the end of a sash cord and used to help feed the cord through the pulleys.

4. Why should trussed rafters not be repaired by a non-specialist?

5. What is the main cause of rot in rafters?

6. What is the first step to take before applying plaster?

7. What is the most common fault in guttering?

8. Why is it important to check the guttering profile and colour first?

9. When are special clips for guttering needed?

chapter 14

Marking and setting out joinery products

OVERVIEW

Many joinery items, such as doors, stairs and windows are now mass-produced under almost factory conditions using computer-programmed machinery. However, there will always be a place for bench joiners making small numbers of high-quality joinery items to a given specification. To do this, you must understand the main woodworking joints.

To turn the client's specification into a finished item of joinery, bench joiners must first produce detailed drawings. The information from the drawings can then be transferred to the timber in preparation for machining and assembly.

Production of the detailed drawings is known as 'setting out'. Transfer of this information to the timber is referred to as 'marking out'.

This chapter will cover the following topics:

- Woodworking joints
- Basic setting out
- Basic marking out
- Marking out basic components (windows, doors, stairs and basic units).

These topics can be found in the following modules:

CC 1001K	CC 2034K	CC 2035K
CC 1001S	CC 2034S	CC 2035S

Woodworking joints

At the end of this section you should be able to:

- understand simple jointing methods used on doors and windows
- identify the main joints used in the assembly of units and fitments
- state the correct jointing methods used on a common staircase.

Joints used on doors and windows

During the manufacture of doors and windows the mortise and tenon joint is extensively used. The type of mortise and tenon will depend on its location. Examples of this joint are described over the next few pages.

Through mortise and tenon

In a through mortise and tenon joint a single rectangular tenon is slotted into a mortise. See Figure 14.1.

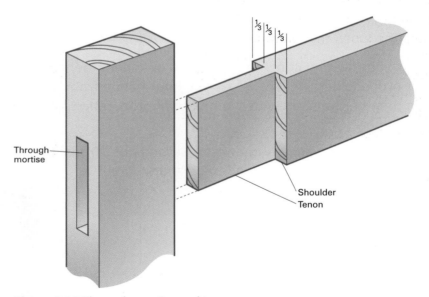

Figure 14.1 Through mortise and tenon

Stub mortise and tenon

In a stub mortise and tenon joint the tenon is stopped short to prevent it protruding through the member. See Figure 14.2.

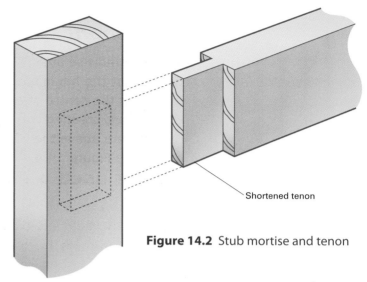

Figure 14.2 Stub mortise and tenon

Shortened tenon

Haunched mortise and tenon

In a haunched mortise and tenon joint the tenon is reduced in width, leaving a shortened portion of the tenon protruding which is referred to as a haunch. See Figure 14.3. The purpose of the haunch is to keep the tenon the full width of the timber at the top third of the joint. This will prevent twisting. A haunch at the end of the member will aid the wedging-up process and prevent the tenon becoming **bridled**. For a detailed description of the wedging-up process look at Chapter 15 Assembling joinery products.

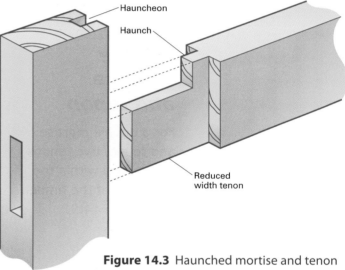

Hauncheon

Haunch

Reduced width tenon

Figure 14.3 Haunched mortise and tenon

Definition

Bridled – an open mortise and tenon joint. A tenon that has bridled is one that has no resistance and so is not secure

Twin mortise and tenon

In a twin mortise and tenon joint the haunch is formed in the centre of a wide tenon, creating two tenons, one above the other. See Figure 14.4.

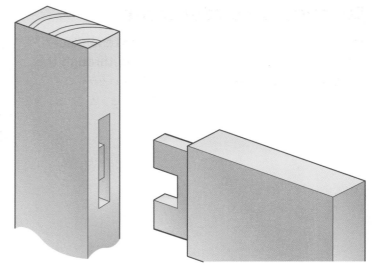

Figure 14.4 Twin mortise and tenon

Double mortise and tenon

For a double mortise and tenon, two tenons are formed within the thickness of the timber. See Figure 14.5.

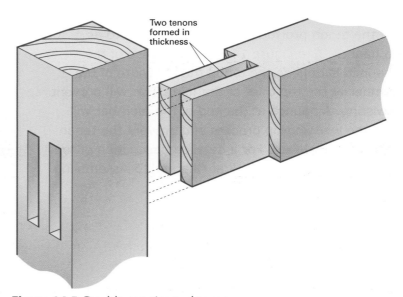

Two tenons formed in thickness

Figure 14.5 Double mortise and tenon

Stepped shoulder joint

Used on frames with rebates, a stepped shoulder joint has a shoulder stepped the depth of the rebate. This joint can also be combined with haunched, twin or stub tenons. See Figure 14.6.

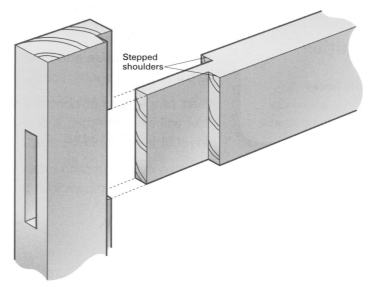

Figure 14.6 Stepped shoulder joint

Twin tenon with twin haunch

A twin tenon with twin haunch joint is used on the deep bottom rails of doors. See Figure 14.7.

Basic rules on mortise and tenon joints

The proportions of mortise and tenon joints are very important to their strength. Some basic rules are as follows:

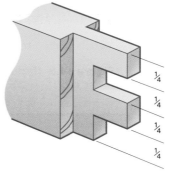

Figure 14.7 Twin tenon with twin haunch

- Tenon width should be no more than five times its thickness. This prevents shrinkage and movement in the joint. If more than five times then a haunch should be introduced.

- The tenon should be one-third of the thickness of the timber. If a chisel is not available to cut a mortise at one-third, the tenon should be adjusted to the nearest chisel size.

- When a haunch is being used to reduce the width of a tenon, then about one-third of the overall width should be removed. The depth of a haunch should be the same as its thickness.

- Although a tenon should be located in the middle third of a member, it can be moved either way slightly to stay in line with a rebate or groove.

Joints used in units and fitments

During the design and setting out of units and fitments the most common joint used is the mortise and tenon, but these are not the best if there are forces likely to try to pull the joint apart. These are called **tensile forces**.

Parts of a unit or fitment subject to such forces must incorporate a joint design that will allow for this. A drawer on a unit is often subject to tensile forces, so a dovetail joint would be used.

The two most common types of dovetail joint are through and lapped. A through dovetail joint is shown in Figure 14.8 and a lapped dovetail in Figure 14.9.

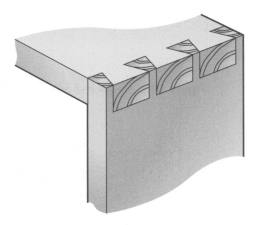

Figure 14.8 Through dovetail joint

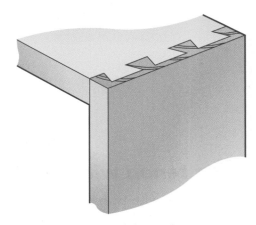

Figure 14.9 Lapped dovetail joint

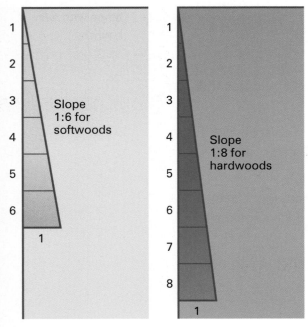

Dovetail joints should have a slope (sometimes called the pitch) of 1:6 for softwoods, or 1:8 for hardwoods. If the slope of the dovetail is excessive then the joint will be weak due to short grain. If the slope is insufficient the dovetail will have a tendency to pull apart. The slope (or pitch) is shown in Figure 14.10.

Figure 14.10 Slope of a dovetail joint

Correct jointing methods on staircases

The most common joint used in staircase construction is a **stopped housing joint**. This joint is used to locate or house the tread and riser of a step into the string. It will be stopped at the nosing of the tread. The minimum housing depth is 12 mm. See Figure 14.11.

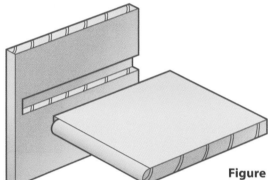

Figure 14.11 Stopped housing joint (staircase)

When the string of a stair meets a newel post, a stubbed and haunched mortise and tenon joint is used, as shown in Figure 14.12. More information on this type of joint can be found earlier in the chapter.

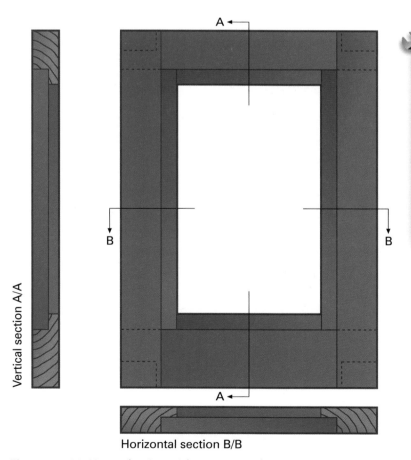

Vertical section A/A

Horizontal section B/B

Figure 14.12 Housed string with mortise and tenon

Basic setting out

At the end of this section you will understand:

- the principles of a setting out rod and its uses

- the purpose of a cutting list.

Setting out rod

A setting out rod will usually be a thin piece of plywood, hardboard or MDF, on which can be drawn the full size measurements of the item to be made. It is quite often painted white in order to aid the clarity of drawing.

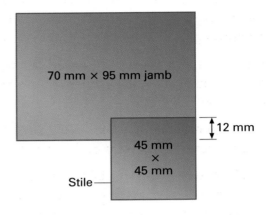

Figure 14.13 White setting out rod for small, four-pane sash

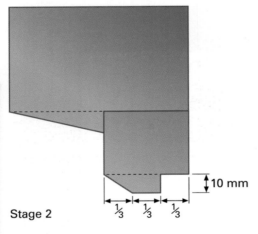

Stage 2

Figure 14.14 Height and width sections

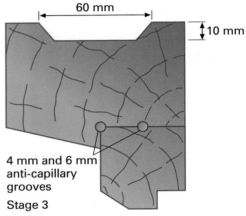

Stage 3

Figure 14.15 Rod with critical dimensions for a single-panel glazed door

Rods can be used time and time again, simply by repainting the surface upon completion of a task. If marked rods are to be kept for reuse they must be referenced and stored safely.

Upon receipt of scale drawings, specification and any on-site measurements the **setter out** will produce a full size, horizontal and vertical section through the item by drawing it on a setting out rod. See Figure 14.14.

Elevations may also be drawn on setting out rods. This is particularly valuable for shaped or curved work, as the setter out can get a 'true' visual image of a completed joinery item.

Although rods are marked up full size, certain critical dimensions can be added as a check against any errors or damage to the rod. These are usually:

- **sight size** – the size of the innermost edges of the component (usually the height and width of any glazed components and, therefore, sometimes referred to as 'daylight size')

- **shoulder size** – the length of any member between shoulders of tenons

- **overall size** – the extreme length and width of an item.

Definition

Setter out – an experienced bench joiner whose job is the setting out of joinery products

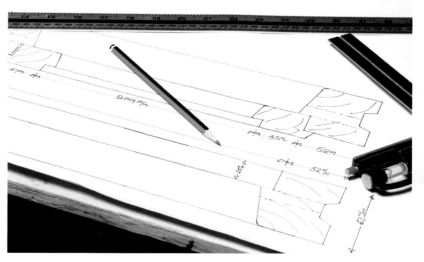

Rod marked up for a casement window

Developing drawn components

When producing workshop rods an inexperienced or apprentice joiner can sometimes have problems when building up a detailed section of timber. To overcome this, use the following step-by-step guidelines.

Step 1 Draw the components as a rectangular section

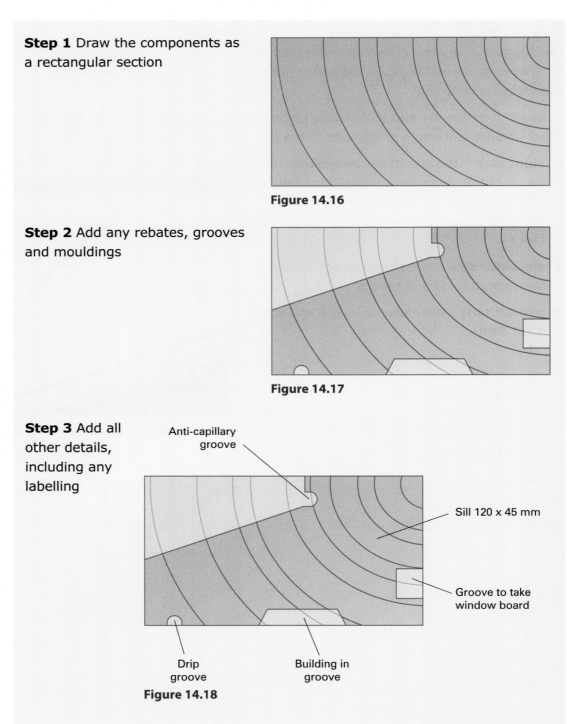

Figure 14.16

Step 2 Add any rebates, grooves and mouldings

Figure 14.17

Step 3 Add all other details, including any labelling

Anti-capillary groove

Sill 120 x 45 mm

Groove to take window board

Drip groove

Building in groove

Figure 14.18

Cutting lists

Once the setting out rod has been completed the cutting list can be compiled. The cutting list is an accurate, itemised list of all the timber required to complete the job shown on the rod.

The cutting list will need to be referred to throughout the manufacturing process. It is, therefore, good practice to include the cutting list on the actual rod wherever possible.

Although there is no set layout for a cutting list, certain information should be clearly given in all lists. It should include:

- reference for the setting out rod, i.e. rod number
- date the list was compiled
- brief job description
- quantity of items required
- component description (e.g. head, sill, stile etc.)
- component size, both sawn and finished (3 mm per face should be allowed for machining purposes)
- general remarks.

An example cutting list is shown in Figure 14.19.

Timber cutting list

Job description: Two panel door			Date: 8 Sept 2008			
Quantity	Description	Material	Length	Width	Thickness	Remarks
2	Stiles	S wood	1981	95	45	Mortise/groove for panel
1	Mid rail	"	760	195	45	Tenon/groove for panel
1	Btm rail	"	760	195	45	Tenon/groove for panel
1	Top rail	"	760	95	45	Tenon/groove for panel
1	Panel	Plywood	760	590	12	⟋
1	Panel	"	600	590	12	⟋

Figure 14.19 A cutting list

Basic marking out

At the end of this section you will be able to:

- select the correct sides of timber on which to mark out. These are known as the face and edge

- transfer information from setting out rod to timber.

Marking out is the transfer of the information on a setting out rod to the timber. It is a very important process and should be checked thoroughly. Wrong information transferred on to the timber at this stage will result in errors during assembly, which are likely to be time-consuming and costly.

Face and edge marks

The face and edge are the two most important of the four sides of a piece of timber. They are, therefore, usually selected as the two best adjacent sides.

The position and severity of any defects present in the timber should also be noted at this point. If these can be removed when cutting rebates or grooves then the sides containing them may still be the best sides to become the face and edge.

Face and edge marks are clearly applied to the relevant sides after careful inspection of the timber. They are used as a reference point from which all marking out is completed.

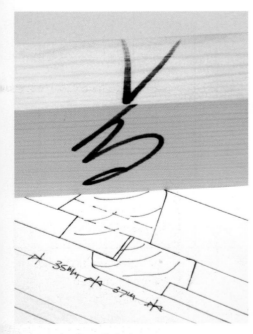

Face and edge of timber

Paired members

Items such as door stiles must be marked out in pairs, since they are not reversible. In this case the face and edge marks must always be opposite to each other.

Face side and edge are normally the front and inside edges of the framework. However, where one side of the frame is not flush with the other, such as a casement window and sill, then the flush side should be chosen as the face.

Transferring information from a setting out rod

Marking out on the timber should be as clear and simple as possible with no unnecessary lines, as these cause confusion and possible errors. Pencil lines should be clear, sharp and made with a hard pencil. Wherever possible all marking out should be completed in a single operation and any wrongly placed or double lines should be rectified immediately.

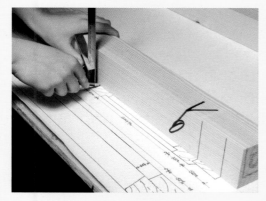

These two photos show a piece of timber being marked out from a door rod.

Note how the shoulder and mortise lines are directly transferred and that a completed section of timber is drawn across the member.

Marking out basic components

Windows

Windows are expected to stand up to the elements, so they usually need to be made from a solid construction.

In this section, our example will be a simple casement window, 900 mm high and 500 mm wide.

Did you know?

A 2H pencil is suitable for marking out, with the nib sharpened to a chisel point for better accuracy

Find out

What tools would be used in the marking out process?

Did you know?

The method of setting out and marking is virtually the same for every component. Remember this when you look at the following examples of manufactured components

Setting and marking out a window

First you need to make setting out rods drawn full size to show the sections of the height and width.

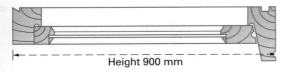

Height 900 mm

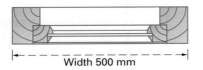

Width 500 mm

Figure14.20 Height and width rods

Use this drawing to produce the cutting list, then use the cutting list to prepare the materials for marking out.

When manufacturing a window, two different framing methods are used: one for the fixed parts and another for the moving parts. In our example, the fixed part is the window frame, and the moving part is the opening sash. For the frame, the horizontal members (head, sill) are mortised and the vertical members (jambs) are tenoned; for the moving sash, the opposite is true – the horizontal members are tenoned and the vertical members mortised.

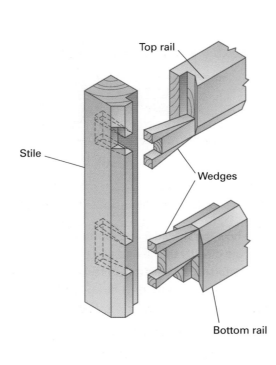

Top rail

Stile

Wedges

Bottom rail

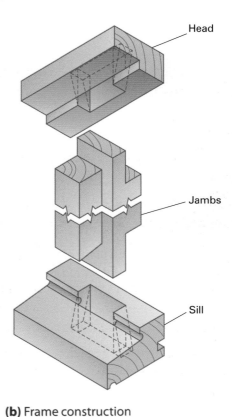

Head

Jambs

Sill

(a) Sash construction

(b) Frame construction

Figure14.21 Joints in frames and sashes

Remember

Don't forget that opposite members such as the head and sill or jambs should always be marked out in pairs

Once the materials are machined, you can mark them out by transferring the marks from the setting out rod to the various members.

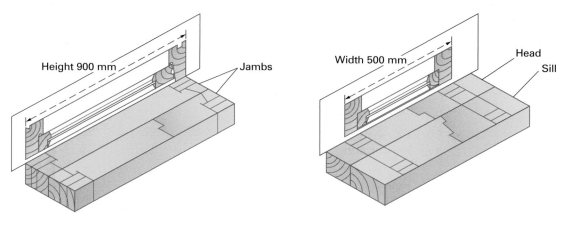

Figure14.22 Members marked out

Remember

It is good practice to remove any marks or pencil lines *prior* to assembly, as they can be difficult to remove once the component is fully assembled

Once all members are marked out, the joints, rebates, grooves, etc. can be machined ready for assembly.

Doors

There are several types of door, but all doors come under one of two main categories.

- **Flush/hollow core door** – as the name implies, this is a hollow door with a frame around the outside, clad usually with hardboard or plywood. The interior of the door is packed with cardboard for strength, and a lock block is fitted to one of the stiles to allow a lock or latch to be fitted.

- **Framed door** – this is made from hardwood or softwood and constructed using either mortise and tenon or dowel joints. The frame is normally rebated to incorporate solid wood panels or glass, onto which beads will be fitted.

Flush doors and dowel-construction framed doors are normally factory-produced, so here we will look at what is involved in the production of a mortise and tenon framed door.

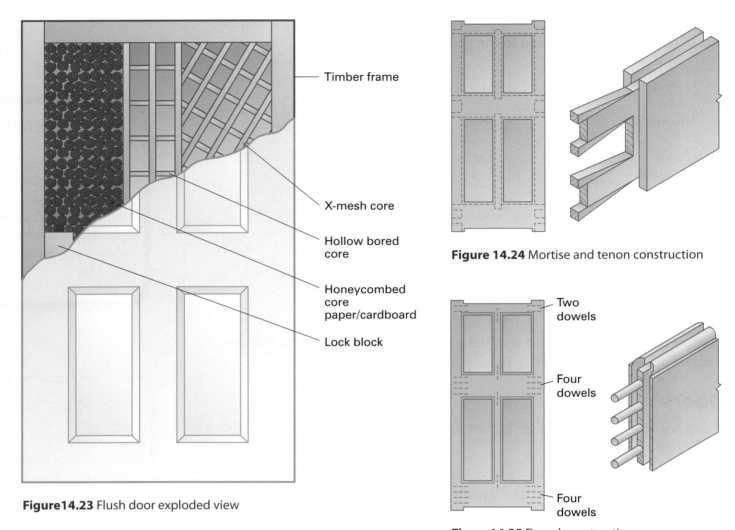

Figure14.23 Flush door exploded view

Timber frame

X-mesh core

Hollow bored core

Honeycombed core paper/cardboard

Lock block

Figure 14.24 Mortise and tenon construction

Two dowels

Four dowels

Four dowels

Figure14.25 Dowel construction

Setting out and marking a door

As always, you start with setting out rods showing a section through the height and width.

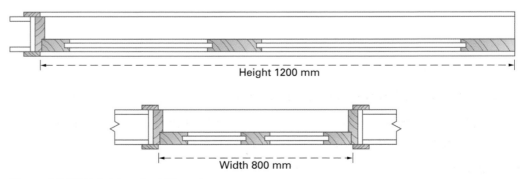

Height 1200 mm

Width 800 mm

Figure14.26 Height and width rods

You can use the rods to produce a cutting list, and use the cutting list to machine the members, remembering to apply the face and edge markings. Next transfer the marks from the rods to mark out all the members, remembering to mark out the horizontal members together and the vertical members together.

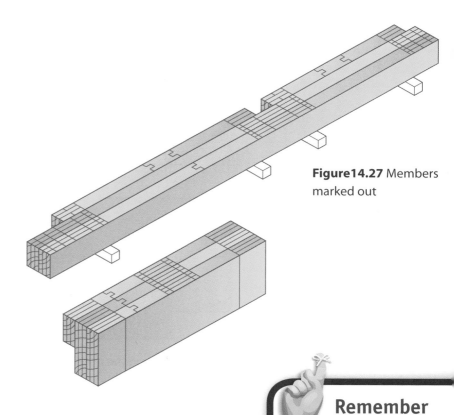

Figure14.27 Members marked out

Stairs

Stairs are set out and marked out differently from doors and windows. To set out stairs, you first need all the dimensions, including the rise, going, etc.

Two templates are needed to mark out the tread and riser positions on the strings: the pitch board and the tread and riser template.

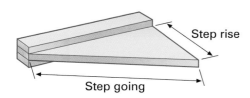

Step rise

Step going

Figure14.28 Pitch board

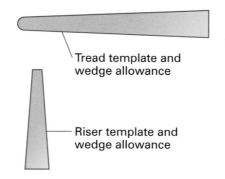

Tread template and wedge allowance

Riser template and wedge allowance

Figure14.29 Tread and riser template

Setting out and marking stairs

First machine the strings to the required sizes and set them out on the bench, remembering to mark the face and edge marks.

Mark the pitch line using a margin template for accuracy.

Remember

There are various regulations concerning stairs, and all dimensions must comply with these regulations. See Chapter 10 pages 252–253 for details

Remember

Take time and care when making templates so that they are as accurate as possible

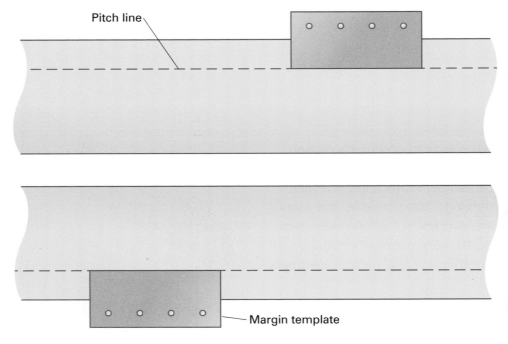

Figure 14.30 Marking the pitch line using a margin template

Set a pair of dividers to the hypotenuse of the pitch board, then mark this distance all the way along both the strings: the two points of intersection will establish the tread and riser points relevant to the pitch line.

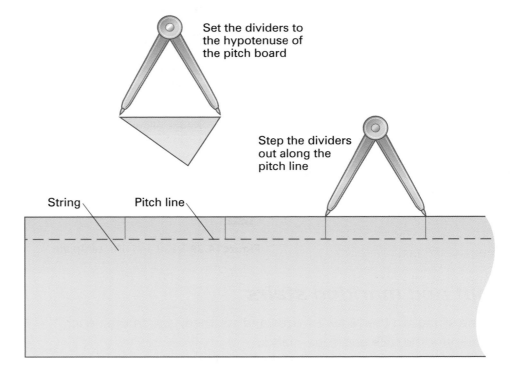

Figure 14.31 Marking the tread and riser points

The pitch board can now be used to mark the rise and going onto the strings and the riser and tread templates can be used to mark out the actual position of each step.

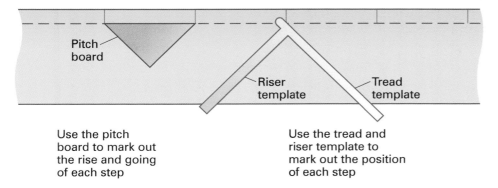

Pitch board

Riser template

Tread template

Use the pitch board to mark out the rise and going of each step

Use the tread and riser template to mark out the position of each step

Figure14.32 Marking the rise and going, and the position of the treads

Now router out the housings. It is best to use a stair-housing jig combined with a router for this job.

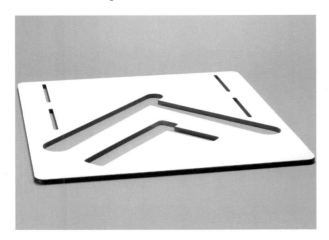

Router stair-housing jig

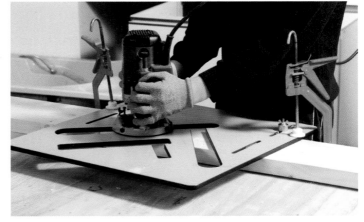

Stair-housing jig used with router

Next the treads and risers can be machined. The treads usually have a curved nosing and a groove on the underside, to allow the risers to be housed into the treads.

The stairs can now be assembled.

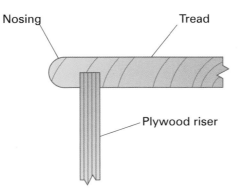

Nosing

Tread

Plywood riser

Figure14.33 Joint between tread and riser

On the job: Marking out mix-up

Molly, a second-year apprentice, has been given a task to manufacture a number of replacement sashes for a housing redevelopment project in the local area. All setting out rods had been drawn up by Phil, a recently-qualified bench joiner. Molly was given all the rods with a range of component drawings referenced to the specific rod. After marking out the first sash, Molly realised that the size written on the drawings did not match the size on the rod as drawn, but did however match another sash to be made.

1   What do you think has happened?

2   What action should Molly take to overcome this problem?

3   What should be done to prevent this happening again?

FAQ

Why use a double tenon joint? Would it not be easier to just put in a thicker single tenon?

Yes, it would be easier but it would not be as strong. Putting in a double tenon will increase the surface area of the joint. This will give the joint a larger area for adhesion (gluing), thus producing a stronger, well-proportioned joint.

When a stair string tenon is jointed into a newel post would you use a twin tenon or a twin tenon with a twin haunch?

Both. At the top of the stair the newel post will be deeper than the string, therefore a twin tenon should be used. At floor level the newel post will be cut flush, therefore a twin tenon with twin haunch would be the strongest joint.

Why would you need sawn sizes on a cutting list?

By putting on sawn sizes the machinist will quickly be able to determine the most cost-effective sections of stock to use from the timber rack.

When marking out for repetitive items of joinery, for example a large number of standard-size doors, would you need to mark each piece of stock from the rod separately?

No. If you transfer the information to one piece of stock you can then clamp all the pieces that will be the same size together. From there you can transfer joint lines to all pieces. You must remember to ensure that all pieces are in pairs using your face and edge marks.

Knowledge check

1. What is a haunch and why is it used?

2. When should a tenon with a stepped shoulder be used?

3. A door stile has a finished section of 95 mm x 44 mm. What size chisel would be used to cut the mortise?

4. What pitch should be used for a dovetail in softwood?

5. State the joints used in the following: stair tread to string; string to newel post; middle rail of a two-panel door; drawer on a kitchen unit.

6. What is the purpose of setting out?

7. Describe the purpose of cutting lists.

8. Explain the following: sight size; shoulder size; overall size.

9. What is the purpose of face and edge marks?

10. Explain what marking out entails.

11. State the best way to cut out the housings on a stair string.

12. What is the purpose of glue wedges?

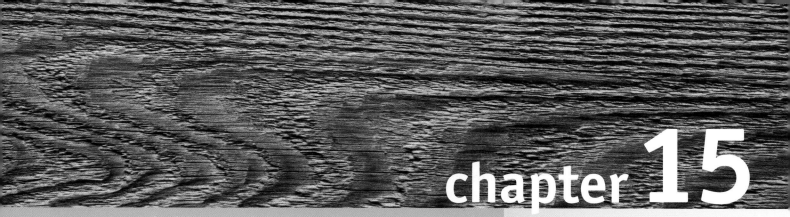

chapter 15

Assembling joinery products

OVERVIEW

Once a bench joiner has produced the setting out details for the specified task and transferred the information onto the timber stock to be used, a joiner can begin the task of cutting and assembling the required item of joinery.

This chapter will cover the following topics:

- Frames and linings
- Doors
- Stairs
- Units.

These topics can be found in the following modules:

CC 1001K	CC 2036K
CC 1001S	CC 2036S

Frames and linings

Window and door frames/linings are fitted into openings left in masonry and hold the window or door in place. It is very important to make sure the frame or lining fits the opening well and that it is level. If this is not done, you will end up with doors and windows that don't hang properly and don't open and close properly.

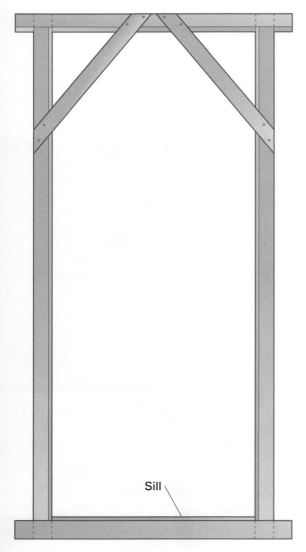

Sill

Figure 15.1 Door frame

Door frames and linings

Door frames are usually of a substantial and solid construction. They are mortised and tenoned and comprise heads, jambs and sills. They usually have a rebate cut from the solid timber. See Figure 15.1.

Door linings are of much lighter construction than frames and are used exclusively for internal doors. They don't usually have sills and are normally jointed between head and jambs with a form of housing joint, such as tongued housing (see Figure 15.2). Door linings usually have planted (nailed on) stops.

Door frames and linings are usually assembled in the joiner's shop using a tongued housing joint (see Chapter 14 Marking and setting out joinery products). To give additional strength and to pull the joint tight, a timber **dowel** is fixed through the face of the frame's head and into the tenon. The hole should be previously drilled in the face and then slightly off-centre in the tenon. When the dowel is knocked home the joint is pulled tight. This method is known as draw boring. The hole needs to be offset as shown in Figure 15.4.

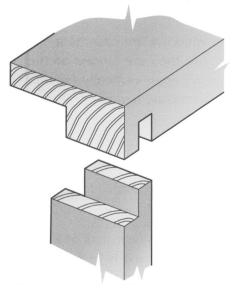

Figure 15.2 Tongued housing joint

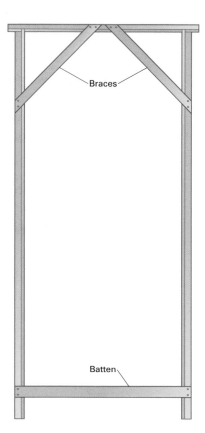

Figure 15.3 Door lining

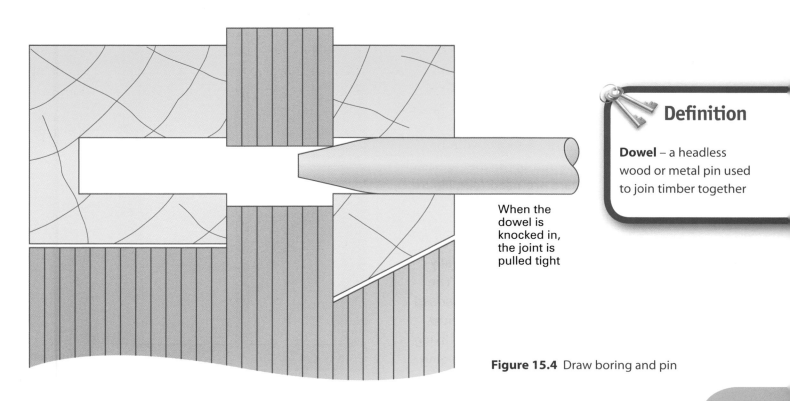

When the dowel is knocked in, the joint is pulled tight

Figure 15.4 Draw boring and pin

Definition

Dowel – a headless wood or metal pin used to join timber together

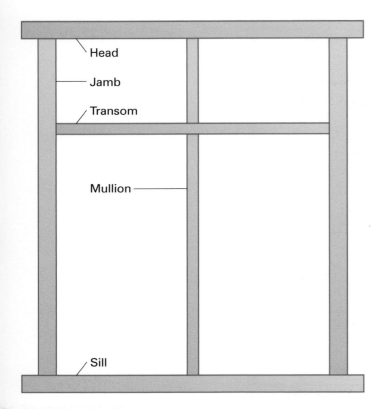

Figure 15.5 Window frame

Window frames

The majority of modern windows are casement windows, which means that they are hinged on the side allowing them to swing open vertically (similar to the way a door opens). A casement window is made up of two main components:

- the frame
- the opening casement.

The frame

Like a door frame, a window frame consists of a head, sill and jambs. When the frame is to be divided, members called **mullions** (vertical dividers) and **transoms** (horizontal dividers) are included (see Figure 15.5).

The opening casement

The opening part of the window, known as the sash, consists of a top rail, bottom rail and two **stiles**. When the opening of the casement sash is to be divided up further, glazing bars are used (see Figure 15.6). The procedure for hanging a casement sash is exactly the same as for a door, only usually on a smaller scale. The procedure is summarised below, but look back at Chapter 11 Second fixing for the full door hanging procedure (page 292).

Definition

Mullions– vertical dividers in a window frame

Transoms– horizontal dividers in a window frame

Stiles – the vertical members of the sash. The hinges are fitted on to one of them

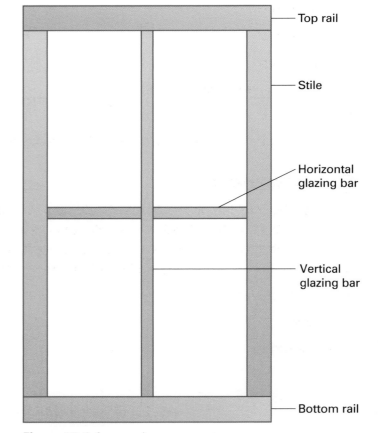

Figure 15.6 An opening casement

Hanging casement window sashes

- Mark the hanging side on both the frame and the sash.

- Cut off any horns (these are waste stock left overhanging on the stiles to aid cleaning up and prevent damage to sash corners prior to installation).

- Plane to fit the hanging stile.

- Plane the sash to the required width, running parallel with the side of the frame.

- Plane to fit the top and bottom of the frame.

- Mark out and cut the hinges.

- Screw one leaf of each hinge to the sash.

- Offer up the sash to the opening and screw the other leaves of the hinges to the frame.

- Make fine adjustments if needed and fit specified ironmongery.

General assembly procedure

There are some general points you should be aware of when assembling a door or window frame:

- dry assembly

- squaring up

- checking for winding

- wedging up.

We will look at each of these in turn over the next few pages.

Dry assembly

All the timber making up the frame should be knocked together dry prior to final assembly. 'Dry', in this situation, doesn't mean in dry conditions (i.e. out of the rain), but rather without the use of adhesive (i.e. a dry or practice run). This ensures that all joints are a good fit and that the frame is:

- the correct size

- square

- not **winding**.

Definition

Winding – twisted (wood), as when a wood frame is twisted or skewed

Remember

Once the frame is assembled dry, check the sizes one more time. As soon as the frame is glued together it will be too late to adjust anything

Squaring up

When the frame has been assembled, glued and cramped up (held together whilst drying with clamps – see Figure 15.7), it should be tested to make sure that it is square, that is that the corners are at right angles. The most accurate way of doing this is to compare the diagonals using a squaring rod, which is a piece of rectangular timber with a small nail or panel pin knocked into the end (see Figure 15.8). The protruding nail is placed in one corner of the frame and the corner of the opposite diagonal is marked on the rod with a pencil. This procedure is repeated for the two other corners of the frame. If both pencil lines match up, then the frame is square.

If the pencil lines do not match up, the frame needs to be adjusted by angling the cramps and pushing the frame. This should be done until the two pencil marks match up, meaning that the diagonals are the same length and the frame is square.

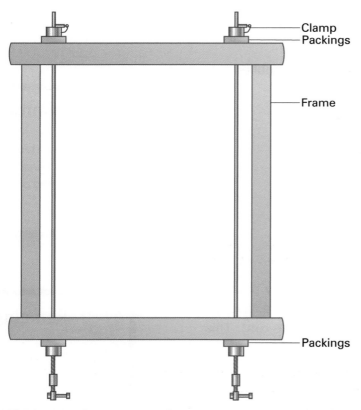

Figure 15.7 Cramping up a frame

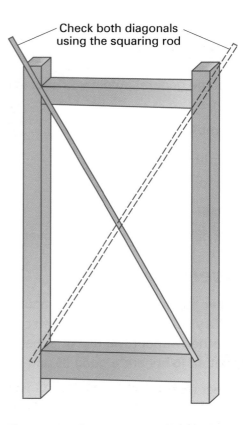

Figure 15.8 Squaring up with a squaring rod

Checking for winding

Winding is a term that describes a frame that is twisted. You will need to check that the frame you have assembled is not twisted by using winding rods. These are simply two pieces of timber which are laid parallel across the frame when it is lying flat on the workbench. Close one eye and look across the winding rods. They should be parallel. If they are, the frame has no twist. If the winding rods are not parallel, the frame is winding and adjustments will have to be made to the joints before further assembly is carried out.

Remember

When you have made any adjustments to remove winding, go back and make sure the frame is still square

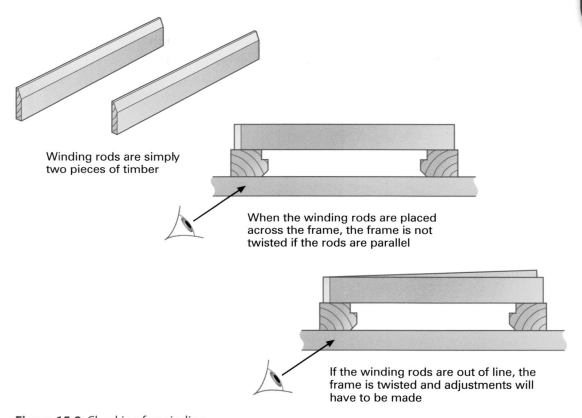

Winding rods are simply two pieces of timber

When the winding rods are placed across the frame, the frame is not twisted if the rods are parallel

If the winding rods are out of line, the frame is twisted and adjustments will have to be made

Figure 15.9 Checking for winding

Wedging up

The haunched mortise and tenon joints you have used in the assembly of your door or window frame are not only held in place by glue but also wedged. This should be done after the frame has been glued, squared and checked for winding. Wedging up involves placing a small wedge on either side of the tenon and carefully driving it in to ensure a good tight joint. To help keep the frame square and the joints pulled in tight, it is best to drive in the external wedge (the haunch side) first. Look back at Chapter 14 *Marking and setting out joinery products* to remind yourself about haunched mortise and tenon joints.

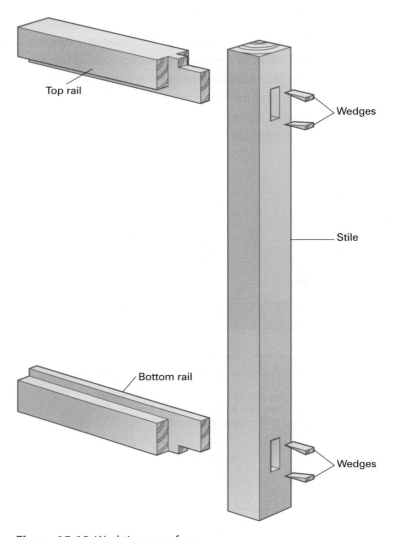

Figure 15.10 Wedging up a frame

Doors

Find out

How many different types of door are available?
Why do you think there are so many?

There are a number of different types of door assembled by carpenters and joiners. In this section we will only look at panelled and glazed doors, since they incorporate all the principles that you need to know about to be able to assemble any door.

Panelled doors

Panelled doors have a frame made from solid timber rails and stiles (see Figure 15.11). When made by the bench joiner in the workshop, they will almost certainly incorporate a mortise and tenon type of joint. The frame will either be grooved or rebated to receive a number of either plywood or timber panels.

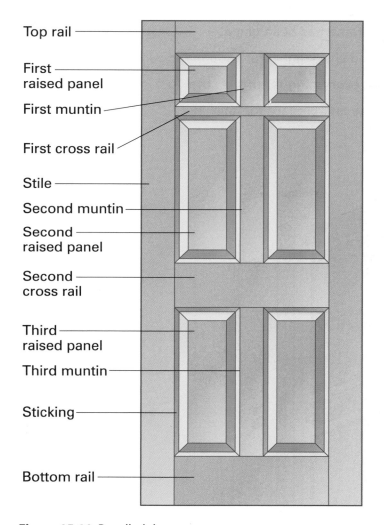

Labels on Figure 15.11 (Panelled door):
- Top rail
- First raised panel
- First muntin
- First cross rail
- Stile
- Second muntin
- Second raised panel
- Second cross rail
- Third raised panel
- Third muntin
- Sticking
- Bottom rail

Figure 15.11 Panelled door

Labels on Figure 15.12 (Glazed door):
- Top rail
- Glazing bars
- Stile
- Glass
- Bottom rail

Figure 15.12 Glazed door

Glazed doors

Glazed doors are made in a similar fashion to panelled doors, however one or more of the panels is replaced by glass.

General assembly procedure

The following assembly procedure can be carried out for all types of framed door.

- Assemble the frame dry to ensure that all joints are tight, the right size, square and not winding.

- Before final assembly, clean up the inside edges of all components as this will be extremely difficult once the frame has been glued.

- Glue, assemble, cramp up and check again for square and winding.

- When you are satisfied that everything is correct, the door can be wedged up.

- Clean up the rest of the frame and prepare for finishing.

Wedges

Figure 15.13 Door assembly

Stairs

We have already looked in some detail at stairs and their installation in Chapter 10 *First fixing*. Look back at pages 250–254 to remind yourself of the terminology and regulations governing the construction and installation of stairs. You may also want to look back at Chapter 14 *Marking and setting out joinery products* to refresh your understanding before reading on (see page 403).

The next section will cover the assembly of stairs.

Assembly

When all the work has been carried out on the strings and newels, the assembly can be carried out. Although the assembly procedure can be carried out using a proprietary cramping system, it is more often than not carried out on an adapted workbench. The bench must be sturdy and level as it is essential that the strings are totally straight and parallel.

The steps are held in position in the string housing with adhesive and timber wedges. First, the treads and risers should be slid into the string housing. Adhesive is then applied to the wedges, which are then pushed into both the tread and riser housing. Wedging must be done in a methodical order. It is easiest to start with the first riser and wedge both sides. Next the first tread can be fitted and wedged. This process continues until all risers and treads are fitted and wedged. During this process it is important that only the wedges are glued, not the treads and risers. Finally, the risers should be screwed to the back of the tread to prevent movement when the step is being used.

Remember

Strings are the main boards which the treads and risers are attached to. Treads are the flat, horizontal parts of the step; risers are the vertical parts of the step. Newel posts are the uprights at the bottom of the stairs and at each turn in the staircase

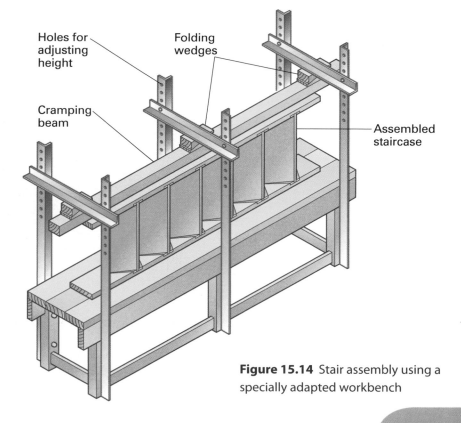

Holes for adjusting height

Folding wedges

Cramping beam

Assembled staircase

Figure 15.14 Stair assembly using a specially adapted workbench

Units

The most common type of unit you will probably work with is kitchen units, although if you understand the basic principles of assembling these, you will find that the assembly of other types of unit (e.g. bathroom, bedroom) can easily follow a similar procedure. Most types of unit are made from melamine faced chipboard, MDF, block board, plywood and, sometimes, solid timber. See Chapter 9 *Timber technology* for more information on these materials.

There are a number of different methods of unit construction but we will only look at probably the two most common: box and framed. We will also briefly look at different types of knock-down fittings, which are often used in unit assembly.

See Chapter 11 *Second fixing* for further information about kitchen units and their fitting.

Box construction

This is also known as 'slab construction'. The unit is composed of vertical standards, rails and shelves and the plinth and bottom shelf are often an integral part of the unit. The back panel holds the frame square.

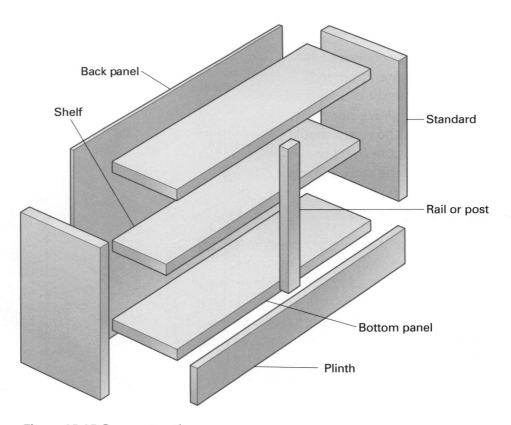

Figure 15.15 Box construction

Framed construction

This is also known as 'skeleton construction'. The units are composed of either a pair of frames (a front and a back frame) joined together with rails, or cross-frames joined together by rails at the front and back. See Figure 15.16. The plinths and drawers are usually built separately in this type of unit. The frames are mortise and tenoned and assembly follows the same procedure as doors or casement window sashes (see pages 420–426).

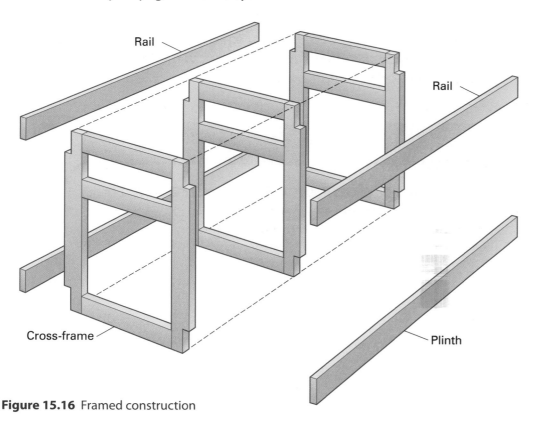

Figure 15.16 Framed construction

Knock-down fittings

Knock-down fittings were originally designed as temporary fittings, although they are often used as permanent fittings on units. The fittings are normally quite simple (a block with drilled holes) and often made from plastic. Because of their simplicity and the fact that they usually only need a few screws, they can be easily undone and the unit quickly dismantled. They are ideal for flat-pack self-assembly furniture and the joints they provide are relatively strong.

Example of knock-down fitting

On the job: Assembling a door

Thomas and Laura are both second-year apprentices and have been tasked with assembling a door that has been made by an experienced joiner. They glue and cramp up the door and leave it to go to lunch. When they return, their supervisor is not happy with them as the door is out of square and the joints are not tight-fitting.

How could this have happened and what could have been done to prevent it?

FAQ

What causes a door or window frame to wind and can it be fixed?

Winding can be caused by the frame not sitting flat when it is assembled and can also be caused by using timber with a high moisture content which warps when it dries out. Winding can only be prevented by ensuring the frame is flat when clamped up.

Why do frames always have to be dry assembled?

Every frame you make needs to be dry assembled to check that the joints are tight and that the frame is square. If a frame is not checked in this way and it is glued and clamped and then found to have gaps in the joints, it will be difficult to put right.

Knowledge check

1. What type of joint is used in door frames?

2. What is draw boring?

3. What are mullions and transoms?

4. Why is the dry assembly stage important with a window?

5. What is winding and how can you check for it?

6. Describe the general assembly procedure for doors.

7. Explain the difference between box and framed unit construction.

8. What are the advantages of using knock-down fittings?

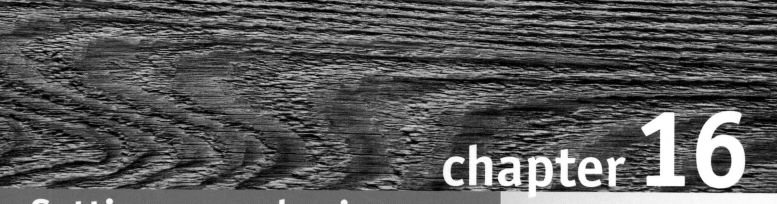

chapter 16

Setting up and using circular saws

OVERVIEW

This chapter has been designed to provide you with the knowledge and understanding to operate a hand-fed circular saw correctly and safely.

This chapter will cover the following topics:

- Safety regulations
- Function of the saw
- Component parts
- Saw blades
- Setting up for use
- Illegal methods of use.

These topics can be found in the following modules:

CC 1001K	CC 2012K
CC 1001S	CC 2012S

Safety regulations

At the end of this section you should be aware of your responsibilities in operating circular saws and the main regulations that govern their use.

There have been four main pieces of legislation which govern the use of circular saws:

- Woodworking Machines Regulations 1974
- Provision and Use of Work Equipment Regulations 1992
- Provision and Use of Work Equipment Regulations 1998
- Safe Use of Woodworking Machinery Approved Code of Practice and Guidance.

When the PUWER were introduced in 1992, most of the 1974 regulations were revoked. The PUWER were then reviewed in 1998, at which time the rest of the 1974 regulations were replaced by the Safe Use of Woodworking Machinery Approved Code of Practice. This code of practice gives employers further information on how to comply with the PUWER for woodworking machinery.

General safety requirements

As with all woodworking machines, circular saws can be dangerous unless handled properly. Paying close attention and taking care while you work will help you to avoid accidents. The following is a brief explanation of the main regulations governing the use of woodworking machines.

Compared to other industries, woodworking accounts for a large proportion of accidents. Woodworking machines often have high-speed cutters, and many cannot be fully enclosed owing to the nature of the work they do.

The use of woodworking machines was originally governed by the Woodworking Machine Regulations 1974. The introduction of the Provision and Use of Work Equipment Regulations (PUWER) 1992 superseded the 1974 Regulations, although regulations 13, 20 and 39 were still in use until the PUWER regulations were updated in 1998.

The PUWER regulations are explained in more detail in Chapter 2, but below are a few of the items relating to the safe use of all woodworking machines. Safety regulations relating specifically to a particular machine are noted in dedicated sections.

Did you know?

The PUWER regulations cover all working equipment, not just woodworking machines

Safety appliances

Safety appliances such as push sticks/blocks and jigs must be designed so as to keep the operator's hands safe. More modern machines use power feed systems, eliminating the need for an operator to go near the cutting action. Power feed systems should be used wherever possible; in the absence of a power feed system, the appropriate push sticks/blocks must be used.

Working area

An unobstructed area is vital for the safe use of woodworking machines. The positioning of any machine must be carefully thought through to allow the machine to be used as intended. In a workshop environment, where there are several machines, the layout should be arranged so that the materials follow a logical path. Adequate access routes between machines must be kept clear, and there should also be a suitable storage area next to each machine to store materials safely without impeding the operator or others.

Floors

The floors around machines must be kept flat and in a good condition, and must be kept free from debris such as chippings, waste wood and sawdust. Any electricity supply, dust collection ductwork, etc. must be run above head height or set into the floor in such a way that does not create a trip hazard. Polished surfaces must be avoided and any spills must be mopped up immediately. Non-slip matting around a machine is preferred, but the edges must not present a trip hazard.

Lighting

All areas must be adequately lit, whether by natural or artificial lighting, to ensure that all machine set-up gauges and dials are visible. Lighting must be strong enough to ensure a good view of the machine and its operations, and lights must be positioned to avoid glare and without shining into the operator's eyes.

Heating

The temperature in a workshop should be neither too warm nor too cold, and the area should be heated if needed. A temperature of 16°C is suitable for a workshop.

Controls

All machines should be fitted with a means of isolation from the electrical supply separate from the on/off buttons. The isolator should be positioned so that an operator can access it easily in an emergency. Ideally there should be a second cut-off switch accessible by others in case the operator is unable to reach the isolator. Machines must be fitted with an efficient starting/stopping mechanism, in easy reach of the operator. Machines must always be switched off when not in use and never left unattended until the cutter has come to a complete standstill.

Braking

All new machines must be fitted with an automatic braking system that ensures that the cutting tool stops within 10 seconds of the machine being switched off. Older machines are not required to have this, but PUWER states that all machines must be provided with controls that bring the machine to a controlled stop in a safe manner. The approved code of practice calls for employers to carry out a risk assessment to determine whether any machine requires a braking system to be fitted, and includes a list of machines for which braking will almost certainly be needed. If you are unsure, contact your local Health and Safety Executive.

Dust collection

Woodworking machines must be fitted with an efficient means of collecting the dust or chippings produced during the machining process.

Training

No person should use any woodworking machinery unless they have been suitably trained and deemed as competent in the use of that machine.

Maintenance

The machines should be maintained as per the manufacturer's instructions and should be checked over prior to use every day, with the inspection and any findings recorded in a log.

Regulations specific to circular saws

The code of practice gives the following guidance specific to the use of circular saws.

- While operating the saw, the operator's hands should never be in line with the saw blade.

- A riving knife should be used to reduce the risk of kickback of the work piece. It should be fixed securely below the table, and positioned directly behind and in line with the blade.

- The riving knife should be kept adjusted so that it is within 8 mm of the blade at table level.

- For saw blades 600 mm in diameter or less, the vertical distance between the top of the blade and the top of the riving knife should not exceed 25 mm.

- For saw blades more than 600 mm in diameter, the riving knife should be at least 225 mm above the top of the blade.

- When removing the cut piece from between the saw and the fence, you should always use a push stick (unless the cut piece is more than 150 mm wide).

- When using a push stick, move your left hand to a safe position along the plate of the saw, to avoid an accident if the piece moves unexpectedly.

- Every circular saw should have marked or displayed on it the diameter of the smallest saw blade that should be used.

- A circular saw should not be used for ripping unless the saw blade projects through the upper surface of the work piece at all times.

- If you use a circular saw for grooving, you should ensure that special guards are used to prevent access to the saw blade above the table.

Functions of the saw

The main function of a hand-fed circular saw is to **re-saw** timber. This involves ripping with the grain, and can be divided into the following four operations:

- flatting
- deeping
- bevel cutting
- angled cutting.

Flatting

Flatting is the term used to describe the cutting of timber to the required width on a hand fed circular saw. See Figure 16.1.

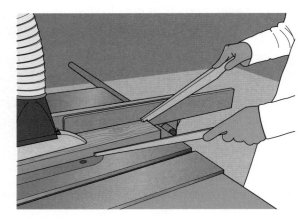

Figure 16.1 Flatting

Deeping

Deeping is the term used to describe the cutting of timber to the required thickness on a hand fed circular saw. See Figure 16.2.

Figure 16.2 Deeping

Bevel cutting

If a bevel is required to be cut on to a section of timber using a hand fed circular saw, then the saw operator will make a jig that will safely carry out the operation. These jigs are known as 'bed pieces' and 'saddles'.

Jigs for bevel cutting

Angled cutting

Angled cutting refers to cutting a piece of timber at an angle. The method is the same as cutting a bevel and, once again, the operator would make the appropriate jig for the task.

Did you know?

A jig is any device made to hold a specific piece of work and guide the tools being used on it

Safety tip

Remember to wear goggles and ear defenders when using a powered saw, as these will protect your eyes and ears from dust and noise

Component parts

To operate a hand fed circular saw you must be able to identify the main parts of the machine and understand how they function. This section looks at all the main parts in turn.

Figure 16.3 illustrates a typical hand fed circular saw.

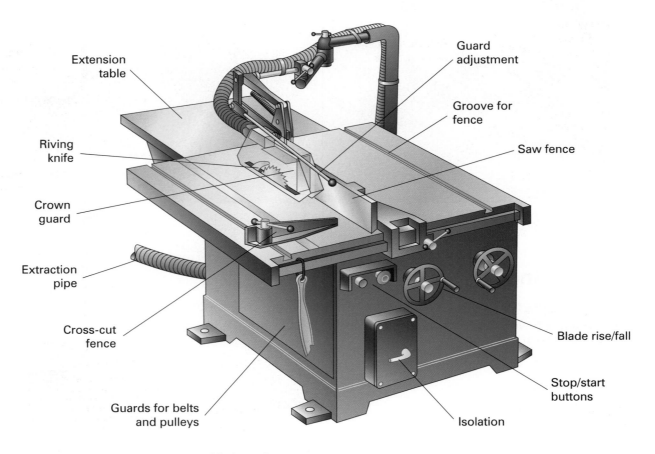

Figure 16.3 Hand fed circular saw

Riving knife

Riving knives are made from spring-tempered steel. The two main functions of the knife are:

- to prevent timber binding on the saw after cutting
- to act as a guard to the back of the saw teeth.

The photo shows a riving knife, indicating the maximum height and distances that the blade can move. Figure 16.4 illustrates the main parts.

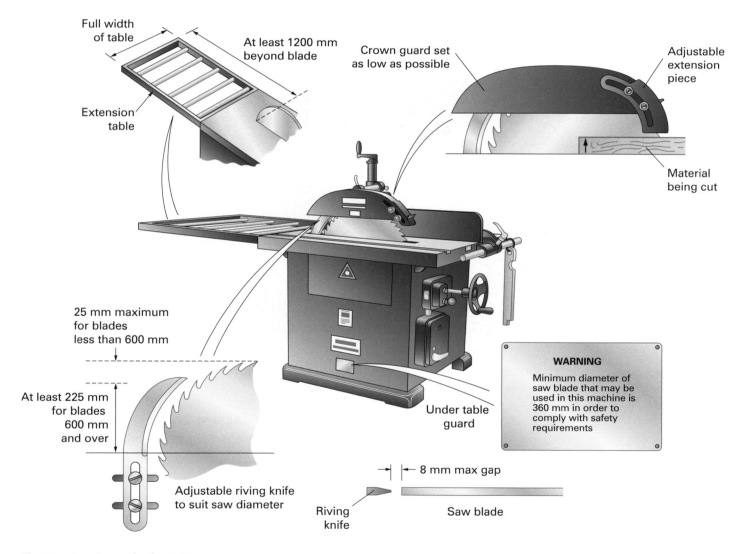

Figure 16.4 Riving knife parts

To avoid the timber binding on the saw blade, the riving knife must be thicker than the saw blade.

Crown guard and extension piece

To protect the operator from the top portion of the saw blade, a crown guard is fitted. This must cover the roots of the saw teeth. The crown guard is made of aluminium, steel or sometimes brass. On all hand fed circular saws, the blade should be guarded above and below the bench, and flanges should provide protection on both sides.

The function of the extension piece is to guard the part of the saw between the crown guard and the timber. It is made of brass or steel. The extension piece must be as close as possible to the timber in order to prevent the operator's fingers from contacting the saw blade. The flange should be on the side away from the fence. Both flanges must extend beyond the roots of the saw teeth. The guards and extension pieces should be kept adjusted to maintain this.

Circular saw showing crown guard, extension piece, mouthpiece, packings and finger plate

Definition

Packing box
– a recess in the finger plate that takes a suitable packing strip

Mouthpiece

The mouthpiece fits into the front portion of the **packing box**, between the packing and the front of the saw teeth. Its purpose is to prevent damage to the saw teeth and packing. It is made from a suitable hardwood or plywood. The mouthpiece is cut to an appropriate size to suit:

- the diameter of the saw being used
- the working height of the saw above the table.

Packings

The purpose of packings is to assist the saw to run true, and they can be made adjusted for tightness. They are made of gland felt or wood strips wound round with hemp.

Finger plate

When access to the saw blade is needed, the finger plate can be removed. The front section supports one side of the packing box. If the saw blade should wobble when running, damage to the teeth is prevented by a hardwood insert in the finger plate. Finger plates are made of cast iron.

Did you know?

Gland felt is a hard, absorbent material often used as packings. It can also be soaked with diesel or oil to assist the removal of resin build-up on the saw

Stop/start buttons

The stop/start buttons on the machine must be easily accessible and work efficiently. Usually, the stop button will be a 'mushroom head' type with a twist release. To prevent the machine being started accidentally, the start button is normally recessed.

Stop/start buttons

Extension table

An extension table is required to extend a minimum distance of 1200 mm behind the part of the saw blade that is running above the table, and the full width of the table. (This does not apply to movable machines with a maximum blade diameter of 450 mm.)

The purpose of this requirement is to protect an operator 'pulling off' timber behind the saw bench. When pulling off, you should stand behind the table.

Extension table

Saw blades

It is very important to select the correct type of saw blade for the job. At the end of this section you should be able to:

- identify types of blades
- understand terms used for parts of a blade
- understand how a blade fits on to a machine.

Types of blade

Although a circular saw blade may seem quite simple in construction, each part must be set up correctly for efficient performance in use. There are two main types used:

- tungsten carbide tipped (TCT) blade
- sprung steel blade.

TCT blades are more durable than the more traditional sprung steel blade, as the latter need to have each individual tooth set properly to work effectively.

The teeth on rip saws must have chisel edges, sloping towards the wood. This is called positive hook (see below). More hook is needed for ripping softwood than hardwood.

The cutting edge of sprung steel blades quickly become dull when ripping abrasive timbers, so TCT blades are recommended when cutting these, as they have greater resistance to wear and blunting.

Saw blade terminology

Key terms used in relation to blades are:

- **pitch** – the distance between the point of two teeth
- **hook** – the angle of the front of the tooth
- **positive hook** – when teeth incline towards the timber, as required for ripping (20°–25°, maximum 30°, for softwoods; 10°–15°, maximum 20°, for hardwoods)
- **clearance angle** – ensures the heel of the tooth clears the timber when cutting (15° is normal for softwood; 5°–10° for hardwoods)
- **top bevel** – angle across the top of the tooth (15° for softwoods; 5°–10° for hardwoods)
- **front bevel** – bevel across the face of the tooth (nil for a rip saw)
- **gullet** – space between each tooth to carry away sawdust

- **root** – the base of the tooth

- **kerf** – total width of the saw cut made by the blade (twice the set plus the thickness of the saw plate on a sprung steel blade; twice the overhang plus the thickness of the saw plate on a TCT blade)

- **set** – the amount on sprung steel blades that each tooth is bent, or 'sprung out', to give a clearance on the saw plate

- **overhang** – on TCT blades the amount that the cutting edge is wider than the saw plate.

Figure 16.5 shows the parts of a tooth.

Did you know?

Many sawmills now use TCT blades because they stay sharper for longer, especially on abrasive timbers, and do not require set as the tips overhang the saw plate

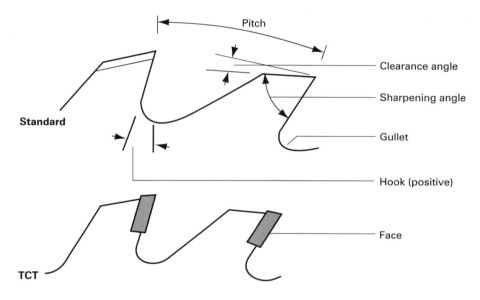

Figure 16.5 Parts of a saw tooth

Saw blade fitting

Saw blades are fitted on to a spindle between two flanges or collars. When a saw blade needs to be fitted or removed, the front collar can be taken off to allow access. The rear collar is fixed to the spindle. In some cases, circular saws will have a locating pin in the rear collar, which assists in the saw blade taking up the drive of the motor.

Saw blade fitting

Setting up for use

There are several preparations to be made, and factors to be considered, before using a hand fed circular saw to rip timber. These include:

- preliminary checks

- correct setting of the fence

- selecting additional safety work pieces required to feed timber through the machine

- identifying the safest position for the operator to stand while feeding timber into the machine.

Preliminary checks

Prior to starting the machine the operator must ensure that:

- the machine is isolated from the power supply

- the saw blade is suitable for the job, and is sharp and tensioned correctly

- the saw collars are clean and in good condition

- the saw blade is mounted correctly

- the riving knife is positioned correctly

- the finger plate is located correctly

- packings and mouthpiece are a good fit

- the height that the saw blade is set suits the depth of cut being made

- the saw blade runs free when slowly turned by hand

- the fence position is in line with the gullets of the saw at table level

- the extension piece and crown guard are correctly positioned

- the fine adjustment on the fence is working properly, that is not sticking, nor failing to operate on the thread

- dust extraction equipment is set up correctly

- additional safety work pieces are available

- the timber is free of nails, grit or other foreign bodies.

Setting the fence

The fence is a component part of a saw table, easily set to enable accurate cutting to the size of sheet materials or lengths of timber. The fence usually moves on a slide to the distance required between the blade and inside face of the fence. Some fences are fitted with a fine adjustment facility.

It is vital when setting the fence to ensure that the end of it does not exceed the position of the saw tooth gullets at table level. This is to eliminate the possibility of timber binding on the blade.

Determining the width of cut

The best way to set the width of the cut is to measure from the inside edge of the tooth to the fence. This will compensate for different amounts of set or overhang on the teeth of different saws. This is shown in Figure 16.6.

The rule attached to the table can be used for approximate setting, although the accuracy will vary according to the saw blade used. It is better to use a separate steel rule.

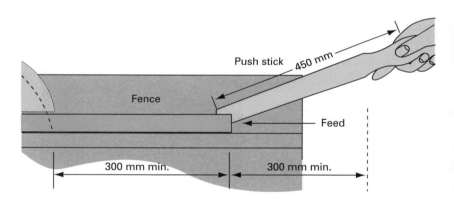

Figure 16.6 Method for setting the width of the cut

Using the rule to set the width of the cut

Safety work pieces

The greatest danger, when operating a hand fed circular saw, is getting hands or clothing too close to the rotating blade. In addition to ensuring that the machine is set up correctly, operators can use a variety of 'safety work pieces' to help in handling timber with minimal risk. These include:

- push sticks

- spikes

- wedges.

They need to be available for use before starting to cut timber with the saw.

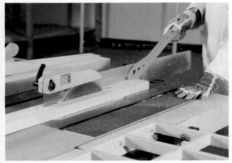

Push stick in use

Push sticks

A push stick can be made from any suitably sized scrap of timber. It should be a minimum of 300 mm in length.

A push stick should always be used if the material being fed into the machine is 300 mm or less in length or, for longer material, used for the final 300 mm of the cut. It should always be used to remove off-cuts of material remaining between the saw blade and the fence.

Spikes

A push spike is used to push timber together as it passes through the blade of the saw. The spike should be a minimum of 300 mm in length to ensure that hands are kept clear of the blade.

Spike in use

Wedge in use

Wedges

A chisel is used to drive wedges of timber into the wood being cut in order to split and part it, keeping the kerf (cut) open. This stops the wood from binding the blade. Without wedges, the wood can move back with considerable force towards the saw operator.

Operator position when operating the saw

Operators should always position themselves with one foot a comfortable distance in front of the other. This will ensure good balance and a strong stance. The operator should also make sure they are able to reach the emergency stop button easily.

Best position for the operator   Operator pulling timber through

If standing behind the machine to pull cut timber through, an operator should always make sure they are clear of the saw blade. The operator should support the timber being sawn and move backwards as it is fed through the saw. The two saw operators should always work together and listen carefully to each other's instructions.

Final checks

Having carried out the preliminary checks given above, the saw can now be started and a trial cut made, after which the size of the cut can be checked and adjustments made, if necessary, to the machine settings.

If there is any wobble on the blade the cut will be wider than expected. The trial cut should be measured with a steel rule, and adjustments made as necessary using the fine adjustment screw.

Illegal methods of use

It is dangerous to try to use a circular saw for operations for which it is not designed. In particular, the regulations state that they must not be used for cutting rebates, tenons, moulds or grooves. The use of a proper saw guard attachment is the only legal way of performing these tasks.

FAQ

What sign must be displayed on a circular saw?

The diameter of the smallest saw blade that can be safely used. A small diameter blade (less than 60 per cent of the diameter of the largest blade the saw can accommodate) will have a low peripheral blade speed and will cut inefficiently.

What type of saw blade would be most suitable for ripping abrasive timbers?

TCT (tungsten carbide tipped).

What suitable safety device should be used when ripping timber 300 mm or less?

A push stick.

What common tasks may be carried out on a circular saw bench?

Ripping, crosscutting, bevelling and angle cuts.

What action should be carried out when ripping hardwoods?

Use LEV (local exhaust ventilation system) and use suitable PPE (such as a dust mask).

While ripping timber the blade produces blue smoke, and the operator finds it difficult to push the timber through. What might be the cause?

Fence incorrectly set, blunt saw blade or no set on blade.

On the job: Cutting bevelled timber

Nathan has been asked to cut some bevelled timber using a circular saw. What checks should Nathan perform before operating the saw to make sure it is safe to use? How can Nathan produce bevelled timber with a circular saw? Do you think he needs any additional tools or equipment? Can Nathan do this job on his own?

Knowledge check

1. Name two pieces of legislation that cover the use of circular saws.

2. What are the four common cuts that can be made using a circular saw?

3. State the purpose of a riving knife.

4. What distance should the extension piece be from the timber, and why?

5. What is the function of a mouthpiece?

6. Why is the start button normally recessed?

7. What is the minimum distance of an extension table from the back of a blade?

8. What is the recommended length of a push stick?

9. What does TCT stand for?

10. What is 'kerf'?

11. Describe 'set' in relation to saw blades.

12. Name five parts of a saw tooth.

Technical Certificate: Module Codes

The technical certificate modules covered in each chapter are indicated on the chapter opener page by their code.

MODULE CODE	MODULE NAME
CC 1001K CC 1001S	Know how to carry out safe working practices Carry out safe working practices
CC 2002K CC 2002S	Knowledge of information, quantities and communicating with others 2 Information, quantities and communicating with others 2
CC 2003K CC 2003S	Knowledge of building methods and construction technology 2 Building methods and construction technology 2
CC 2008K CC 2008S	Know how to carry out first fixing operations Carry out first fixing operations
CC 2009K CC 2009S	Know how to carry out second fixing operations Carry out second fixing operations
CC 2010K CC 2010S	Know how to erect structural carcassing Erect structural carcassing
CC 2011K CC 2011S	Know how to carry out maintenance Carry out maintenance
CC 2012K CC 2012S	Know how to set up and operate a circular saw Set up and operate a circular saw
CC 2034K CC 2034S	Know how to produce setting out details for routine joinery products Produce setting out details for routine joinery products
CC 2035K CC 2035S	Know how to mark out efficiently from setting out details for routine joinery products Mark out efficiently from setting out details for routine joinery products
CC 2036K CC 2036S	Know how to manufacture routine joinery products Manufacture routine joinery products

Technical Certificate: Core Module Outcomes

This table outlines in detail the target outcomes for the core module training units and specifies in which chapter each outcome is met. The units that end in 'K' are particularly covered throughout the book, as they address the knowledge of certain criteria. This book will fully cover units that end in 'K' and will aid in the achievement of the units that end in 'S', which address the practical parts of the units.

For some outcomes, there are numerous instances throughout the book where sections or parts of the outcome are met. For example, health and safety is prevalent throughout the book in the form of safety tips, reminders and notes. These instances have not been indentified in this table.

MODULE	OUTCOME	CHAPTER
CC 1001K CC 1001S	1. Health and safety regulations	2 *Health and safety*
CC 1001K CC 1001S	2. Accident/first aid/emergency procedures and reporting	2 *Health and safety*
CC 1001K CC 1001S	3. Identify hazards	2 *Health and safety* Carried throughout book
CC 1001K CC 1001S	4. Health and hygiene	2 *Health and safety*
CC 1001K CC 1001S	5. Safe handling of materials and equipment	2 *Health and safety* 5 *Handling and storage of materials*
CC 1001K CC 1001S	6. Basic working platforms	3 *Working at height*
CC 1001K CC 1001S	7. Work with electricity	2 *Health and safety*
CC 1001K CC 1001S	8. PPE	2 *Health and safety* Carried throughout book
CC 1001K CC 1001S	9. Fire and emergency	2 *Health and safety*
CC 1001K CC 1001S	10. Signs and notices	2 *Health and safety*
CC 2002K CC 2002S	1. Interpret and produce building information	6 *Drawings*
CC 2002K CC 2002S	2. Estimate quantity of resources	7 *Numeracy*
CC 2002K CC 2002S	3. Communicate workplace requirements	1 *The construction industry*
CC 2003K CC 2003S	1. Principles behind walls, floors and roofs	4 *Principles of building*
CC 2003K CC 2003S	2. Principles behind internal work	4 *Principles of building* 5 *Handling and storage of materials*
CC 2003K CC 2003S	3. Materials storage	5 *Handling and storage of materials*

Technical Certificate: Bench and Site Module Outcomes

This table outlines the target outcomes for the certificate-specific module training units and specifies in which chapter each outcome is met. The units that end in 'K' are particularly covered throughout the book, as they address the knowledge of certain criteria. This book will fully cover units that end in 'K' and will aid in the achievement of the units that end in 'S', which address the practical parts of the units.

MODULE	OUTCOME	CHAPTER
CC2008K CC2008S	1. Fix frames and linings	10 *First fixing*
CC2008K CC2008S	2. Fit and fix floor coverings and flat roof decking	10 *First fixing* 12 *Structural carcassing*
CC2008K CC2008S	3. Erect timber stud partitions	10 *First fixing*
CC2008K CC2008S	4. Assemble, erect and fix straight flight of stairs	10 *First fixing* 15 *Assembling joinery products*
CC2009K CC2009S	1. Install side hung doors and ironmongery	5 *Handling and storage of materials* 11 *Second fixing*
CC2009K CC2009S	2. Install mouldings	11 *Second fixing*
CC2009K CC2009S	3. Install service encasements and cladding	11 *Second fixing*
CC2009K CC2009S	4. Install wall, floor units and fitments	11 *Second fixing*
CC2010K CC2010S	1. Erect truss rafter roofs	12 *Structural carcassing*
CC2010K CC2010S	2. Construct gables, verge and eaves	12 *Structural carcassing*
CC2010K CC2010S	3. Install floor joists	12 *Structural carcassing*
CC2011K CC2011S	1. Repair mouldings	13 *Maintenance of components*
CC2011K CC2011S	2. Repair doors and windows	13 *Maintenance of components*
CC2011K CC2011S	3. Replace gutters and down pipes	13 *Maintenance of components*

CC2011K CC2011S	4. Replace sash cords	13 *Maintenance of components*
CC2011K CC2011S	5. Make good plaster, brick and paintwork	13 *Maintenance of components*
CC2012K CC2012S	1. Set up fixed and transportable circular saws	16 *Setting up and using circular saws*
CC2012K CC2012S	2. Change saw blades	16 *Setting up and using circular saws*
CC2012K CC2012S	3. Cut timber and sheet material	16 *Setting up and using circular saws*
CC2034K CC2034S	1. Interpret information for setting out	7 *Drawings* 14 *Marking and setting out*
CC2034K CC2034S	2. Select resources for setting out	5 *Handling and storage of materials* 9 *Timber technology* 14 *Marking and setting out*
CC2034K CC2034S	3. Set out for bench joinery and site carpentry	14 *Marking and setting out*
CC2035K CC2035S	1. Produce marking out efficiently	14 *Marking and setting out*
CC2035K CC2035S	2. Produce accurate marking out	14 *Marking and setting out*
CC2036K CC2036S	1. Select correct materials	5 *Handling and storage of materials* 9 *Timber technology* 15 *Assembling joinery products*
CC2036K CC2036S	2. Manufacture joinery	15 *Assembling joinery products*

Glossary

access	entrance, a way in
adhesive	glue
aggregates	sand, gravel and crushed rock
architrave	a decorative moulding, usually made from timber, that is fitted around door and window frames to hide the gap between the frame and the wall
arris	the sharp edge formed when two flat or curved surfaces meet
balusters	vertical members forming the infill between the string and the handrails in stairs
balustrade	unit comprising handrail, newels and the infill between it and the string, which provides a barrier for the open side of the stair
bargeboards	fascia boards inclined like a pair of rafters and fixed to the face of the verge on a gable-ended roof; sometimes with soffit boards added if the bargeboards project from the wall
barrier cream	a cream used to protect the skin from damage or infection
beads	the shaped pieces of timber added to the end of other pieces of timber to give a desired finish
bind	a door 'binds' when it will not close properly and the door springs open slightly when it is pushed closed
binders	timbers fixed on edge in the roof space, at right angles to ceiling joists, to give support if the span is greater than 2.5 m
birdsmouth	notch cut out of the rafter to form a seating on the outside edge of the wall plate
bitumen	also known as pitch or tar, bitumen is a black sticky substance that turns into a liquid when heated, and is used on flat roofs to provide a waterproof seal
botanical	the classification of trees based upon scientific study
brick-on-edge	a way of laying bricks to form an attractive, projecting window or door sill
bridled	an open mortise and tenon joint. A tenon that has bridled is one that has no resistance and so is not secure
bull nose step	quarter-rounded step at the bottom of a stair
butt joint	the simplest joint between two pieces of wood, with the end grain of one meeting the long grain of the other and glued, screwed or nailed together
camber (crown)	the bow or round that occurs along some timbers depth
cap	shaped top of a newel post, which can be fixed on or turned on the solid newel
capping	fixed on the top edge of strings to take the fixing of a balustrade
carded scaffolder	someone who holds a recognised certificate showing competence in scaffold erection
cement bonded chipboard	a type of chipboard bonded with cement to make it stronger and more durable

chipboard	material made in rigid sheets from compressed wood chips and resin
clear span	the distance between joist supports
collars	sawn timber ties, sometimes used to give extra strength to prevent a roof spreading out at purlin level
common rafters	load-bearing ribs that pitch up opposite each other from the wall plate on each side of the roof span, and fixed to the ridge board
conservation	preservation of the environment and wildlife
contamination	when harmful chemicals or substances pollute something (e.g. water)
corrosive	a substance that can damage things it comes into contact with (e.g. material, skin)
coving	a decorative moulding that is fitted at the top of a wall where it meets the ceiling
cradling	L-shaped brackets used to provide fixing for the soffit
crawling board	a board or platform placed on roof joists which spread the weight of the worker allowing the work to be carried out safely
cripple rafters	pairs of rafters similar to jacks, spanning from the ridge to the valley rafter
crown rafter	the rafter that sits in the middle of a hip end, between the two hips
cube units	the product of a number multiplied by its square, represented by a 3 after the unit (see **square units**)
damp proof course	a substance that is used to prevent damp from penetrating a building

deciduous	the name given to a type of tree that sheds its leaves every year
denominator	the bottom number in a fraction
dermatitis	a skin condition where the affected area is red, itchy and sore
detached	a building that stands alone and is not connected to any other building
dovetailed nails	pairs of nails angled in towards each other
dowel	a headless wood or metal pin used to join timber together
dry rot safety line	when the moisture content of damp timber reaches 20 per cent. Dry rot is likely if the moisture content exceeds this
dust mask	a form of respiratory protection that covers the mouth and nose and is used when working in a dusty environment
eaves	lowest part of the roof surface where it meets the outside walls
edge cut	the angled cut on top of the jack rafter where it meets the hip
effective span	the distance between centres of joist bearings
egress	an exit or way out
employee	the worker
employer	the person or company you work for
enforced	making sure a law is obeyed
evergreen	a type of tree that keeps its leaves all year round
fascia board	board of about 175 mm × 20 mm p.a.r. (plane all round timber) fixed to the plumb (i.e. vertical) cuts of the rafters at the eaves to

	provide a finish and a fixing board for guttering
fibreglass	a material made from glass fibres and a resin that starts in liquid form then hardens to be very strong
firring/firring pieces	a long wedge, tapered where joists are parallel to the fall, or of variable depth for joists at right angles
frame	a main frame consists of a head sill and two vertical jambs
gable	the triangular part of end walls
gable-ends	the triangular ends of a roof
galvanised steel frame cramps	a fixing component which is screwed to a frame
gang nailed	galvanised plate with spikes used to secure butt joints
geometry	a form of mathematics using formulas, arithmetic and angles
going	the depth of a step (the measurement from a step's riser to the edge of the step)
gudgeons	tubes at the end of hinges to take the pin around which the hinge rotates
halving joint	the same amount is removed from each piece of timber so that when fixed together the joint is the same thickness as the uncut timber
hazard	a danger or risk
hazardous	dangerous or unsafe
header	a piece of timber bolted to the wall to carry the weight of the joists
Health and Safety Executive	the government organisation that enforces health and safety law in the UK

helical	in a spiral shape
helical hinge	a hinge with three leaves, with allows a door to be opened through 180 degrees
hip	part of the roof where two external sloping surfaces meet
hip rafters	pitched up from the wall plate corners of a hipped end to the saddle board at the ridge; acting as a spine for the heads of the jack rafters
horns	these are extensions to the frame that protect it during storage or transport, cut off before installation
hypotenuse	the longest side of a right-angled triangle
induction	a formal introduction you will receive when you start any new job, where you will be shown around, shown where the toilets and canteen etc. are, and told what to do if there is a fire
inverted	tipped and turned upside down
jack rafters	fixed in pairs on each side of the hip rafter, diminishing in size from ridge to wall plate
jig	any device made to hold a specific piece of work and help guide the tools being used
joint	a point at which parts are joined
key	the end result of a process that prepares a surface, usually by making it rough or grooved, so that paint or some other finish will stick to it

landings	used between floor levels to break up the overall length of a flight and can be used to change the direction of a flight of stairs	**newel**	heavy vertical member at each end of the stair to which the handrail is fixed
lay board	a piece of timber fitted to the common rafters of an existing roof to allow the cripple rafters to be fixed	**nogging**	a short length of timber, most often found fixed in a timber frame as a brace
lincrusta	a wall-hanging with a relief pattern used to imitate wood panelling	**nosing**	front edge of a tread, or loose narrow top tread which sits on the trimmer joist at the top of the stairs
LPG	liquefied petroleum gas	**noxious**	harmful or poisonous
making a risk assessment	measuring the dangers of an activity against the likelihood of accidents taking place	**numeracy skills**	abilities that demonstrate a good basic knowledge of arithmetic
manual handling	using the body to lift, carry, push or pull a load	**numerator**	the top number in a fraction
MDF	medium density fibreboard, used to form skirting boards and mouldings	**obligation**	something you have a duty or a responsibility to do
metal decking	a decking material using aluminium or galvanised steel	**omission**	something that has not been done or has been missed out
mid-pitch	in the middle of the pitch rather than at the apex or eaves	**opening casement**	consists of top and bottom rails with two stiles
mildew	a fungus that grows in damp conditions	**oriented strand board (OSB)**	oriented strand board is made by coating wood chips (known as strands) with MDI and/or phenol formaldehyde, then arranging consecutive strand layers roughly perpendicular to each other, and finally pressing the strands under high temperature and high pressure to form boards
moulding	decorative finishes around door openings and at floor and wall junctions		
mouse	a piece of strong, thin wire or rope with a weight such as a piece of lead attached to one end, used to feed the sash cord over the pulleys in a box sash window	**overall size**	the extreme length and width of an item
mullions	vertical dividers in a window frame	**packing**	any material, generally waste wood, used to fill a gap
muntin	vertical divider on a panelled door	**packing box**	a recess in the finger plate that takes a suitable packing strip
muster points	fire assembly points	**p.a.r.**	a term used for timber that has been 'planed all round'
		pattern rafter	a precisely cut rafter used as a template to mark out other rafters

perpendicular   at right angles to

pitch or pitch angle   the rise and run gives the pitch angle or, knowing the pitch angle and the run, we can calculate the rise and basic rafter length

pitch line   plumb line marked at the base of the setting out rafter (pattern rafter), marked down two-thirds of its depth to the top of the birdsmouth cut, which acts as a reference or datum point for the rafter's length

plant   industrial machinery

plaster skim   a thin layer of plaster that is put on to walls to give them a smooth and even finish

plumb cut   the vertical cut at the top of the rafter where it meets the ridge

plywood   a material made from thin layers of timber called veneers

porous   full of pores or holes: if a board is porous, it may soak up moisture or water and swell

PPE   personal protective equipment, which might include a helmet, safety glasses and gloves

proactive   taking action before something happens (e.g. an accident)

profile   the shape of moulding when you cut through it

proportionately   in relation to the size of something else

prosecute   to accuse someone of committing a crime, which usually results in being taken to court and, if found guilty, being punished

purlins   horizontal beams that support the rafters mid-way between the ridge and the wall plate when the rafters are longer than 2.5 m

quirk   the name used for the hollow in a moulding

ratio   something that describes a relationship between quantities

RCD   residual current device, a device that will shut the power down on a piece of electrical equipment if it detects a change in the current, thus preventing electrocution

reactive   taking action after something happens

regularised joists   joists that are all the same depth

re-saw   to cut large sections of timber into smaller sections

ridge   acts as a spine at the apex of the roof structure, running horizontally and against which the uppermost ends of the rafters are fixed

rise   distance measured from the outside of the wall plates at wall plate level to the apex of the pitch lines, which run at two-thirds of the depth of the rafters

riser   vertical part of a step

rising   the height of a step (the measurement from the top of one step to the top of another)

roofing ready reckoner   a set of mathematical tables giving a quick way to work out rafter lengths

run   equal to half of the span, it is used to reduce the roof shape to a right-angled triangle

saddle board	a timber board that is fixed to the common rafter and used as a bearing for the crown and hip rafters to fix to
sand course	where the bricklayer beds in certain bricks with sand instead of cement so that they can be easily removed to accommodate things like purlins
scale drawing	a drawing of a building or component in the right proportions, but scaled down to fit on a piece of paper. On a drawing at a scale of 1:50, a line 10 mm long would represent 500 mm on the actual object
seat cut	the 90° cut in a rafter where it meets the horizontal wall plate
semi-detached	a building that is joined to one other building and shares a dividing wall called a party wall
setter out	an experienced bench joiner whose job is the setting out of joinery products
setting out, or pattern, rafter	rafter selected as the basic rafter from which all setting out is done
shelf life	how long something will remain fit for its purpose whilst being stored
sherardised holdfast	a fixing component that is fixed quickly and easily driven in by hammer
sherardising	a process of covering metal with zinc, a non-ferrous metal, to reduce rusting
shoulder size	the length of any member between shoulders of tenons

side cut	the angled cut at each end of the purlin where it meets the hip
sight size	the size of the innermost edges of the component (usually the height and width of any glazed components and, therefore, sometimes referred to as 'daylight size')
skew-nailing	knocking nails in at angles so that the wood cannot be pulled away easily
skinning	the formation of a skin which occurs when the top layer dries out
skirting	a decorative moulding that is fitted at the bottom of a wall to hide the gap between the wall and the floor
sliding bevel	a tool that can be set so that the user can mark out any angle
soffit board	board, similar to a fascia, fixed to the underside of the cradling in closed-eaves design
span	the distance measured in the direction of ceiling joists, from the outside of one wall plate to the other; known as the overall (O/A) span
spandrel framing	where the triangular area is formed under stairs, which can be framed to form a cupboard
sprocket	a piece of timber bolted to the side of the rafter to reduce the pitch at the eaves

sprocket pieces long, wedge-shaped pieces of rafter material fixed on top of each rafter so that the eaves create an upward tilt, thus reducing the slope on a steep roof to ease flow of rainwater into guttering

square units a unit of measurement equal to the area of a square whose side is of the unit given

squaring rod any piece of straight timber long enough to lie across the diagonal

stair well opening formed in a floor layout to accommodate a staircase

step combination of one tread and one riser

stiles the side pieces of a stepladder into which the steps are set. Also, the longest vertical timbers in the frame of a door, window, etc.

still green mortar that is unset and not at full strength

stopped housing joint where a cut does not go completely through the timber

straining pieces sole plates fixed to the ceiling joists between the base of struts

stress a body that has a constant force or system of forces exerted upon it resulting in strain or deformation

string main board to which treads and risers are fixed

struts timbers that support purlins at every fourth or fifth pair of rafters, used to transfer roof load to ceiling joists and onto a load-bearing wall or partition

swing an arc set a compass to the required radius and lightly draw a circle or arc

symptom a sign of illness or disease (e.g. difficulty breathing, a sore hand or a lump under the skin)

tensile force a force that is trying to pull something apart

terraced a row of three or more buildings that are joined together, of which the inner buildings share two party walls

tie-rods metal rods underneath the rungs of a ladder that give extra support to the rungs

tile or slate battens sawn battens fixed at regular intervals on top of the lapped roofing felt; usually fixed by a slater and tiler, not the carpenter

tilt fillet triangular-shaped timber fixed behind a raised fascia to give it support

tongued and grooved board a decking material, rarely used these days

tongues the projecting pieces of timber that fit into a groove in a tongue and groove joint

toxic poisonous

translucent allowing some light through, but not clear like glass

translucent sheeting corrugated or flat decking material

transoms horizontal dividers in a window frame

tread flat, horizontal part of a step

trimmer/ trimming/ trimmed joists joists affected by an opening in the floor such as a stairwell

trimming an opening	removing structural timbers to allow a component such as a chimney or staircase to be fitted, and adding extra load-bearing timbers to spread the additional load
truss	a prefabricated component of a roof which spreads the load of a roof over the outer walls and forms its shape
tusk tenon joint	a kind of mortise and tenon joint that uses a wedge-shaped key to hold the joint together
valley	part of a roof where two sloping surfaces meet
valley boards	used to form a gutter in the valley
valley rafters	like hip rafters but forming an internal angle, acting as a spine for fixing cripple rafters
veneer	a very thin layer of wood used as a ply (usually the finishing layer)
verge	where a roof overhangs at the gable
vibration white finger	a condition that can be caused by using vibrating machinery (usually for very long periods of time). The blood supply to the fingers is reduced which causes pain, tingling and sometimes spasms (shaking)
vitrified	a material that has been converted into a glass-like substance via exposure to high temperatures
volatile	quick to evaporate (turn to a gas)
volume	the amount of space taken up by a 3-D or solid shape
wall plates	timber plates laid flat and bedded on mortar, running along a wall to carry the feet of all rafters and ceiling joists; anchored down with restraint straps to prevent movement
WBP	weather/water boil proof
weather proofing	to make something resistant to the effects of bad weather, especially rain
winding	twisted (wood), as when a wood frame is twisted or skewed
zinc-plated screw tie	a fixing component that is screwed to a frame; use avoids vibration from hammering

Index